XIX International School on Nuclear Physics, Neutron Physics and Applications

(VARNA 2011)

Journal of Physics: Conference Series Volume 366

Varna, Bulgaria
19-25 September 2011

ISBN: 978-1-62276-783-0
ISSN: 1742-6588

Printed from e-media with permission by:

Curran Associates, Inc.
57 Morehouse Lane
Red Hook, NY 12571

Some format issues inherent in the e-media version may also appear in this print version.

Printed by Curran Associates, Inc. (2013)

For permission requests, please contact the Institute of Physics
at the address below.

Institute of Physics
Dirac House, Temple Back
Bristol BS1 6BE UK

Phone: 44 1 17 929 7481
Fax: 44 1 17 920 0979

techtracking@iop.org

Additional copies of this publication are available from:

Curran Associates, Inc.
57 Morehouse Lane
Red Hook, NY 12571 USA
Phone: 845-758-0400
Fax: 845-758-2634
Email: curran@proceedings.com
Web: www.proceedings.com

TABLE OF CONTENTS

PAPERS

Journal of Physics: Conference Series **366** (2012) 011001 doi:10.1088/1742-6596/366/1/011001

Preface

This volume contains the lectures and short talks given at the XIX International School on Nuclear Physics, Neutron Physics and Applications. The School was held from 19–25 September 2011 in 'Club Hotel Bolero' located in the 'Golden Sands' (Zlatni Pyasaci) Resort Complex on the Black Sea coast, near Varna, Bulgaria. The School was organized by the Institute for Nuclear Research and Nuclear Energy of the Bulgarian Academy of Sciences. The co-organizer of the School was the Bulgarian Nuclear Regulatory Agency and the Bogoliubov Laboratory of Theoretical Physics of the Joint Institute for Nuclear Research – Dubna.

According to long-standing tradition the School has been held every second year since 1973. The School's program has been restructured according to our enlarged new international links and today it is more similar to an international conference than to a classical nuclear physics school. This new image attracts many young scientists and students from around the world. This year – 2011, we had the pleasure of welcoming more than 50 distinguished scientists as lecturers. Additionally, 14 young colleagues received the opportunity to each present a short contribution. The program ranged from recent achievements in areas such as nuclear structure and reactions to the hot topics of the application of nuclear methods, reactor physics and nuclear safety. The 94 participants enjoyed the scientific presentations and discussions as well as the relaxing atmosphere at the beach and during the pleasant evenings.

The main topics were as follows:

- Nuclear excitations at various energies
- Nuclei at high angular moments and temperature
- Structure and reactions far from stability
- Symmetries and collective phenomena
- Methods for lifetime measurements
- Astrophysical aspects of nuclear structure
- Neutron nuclear physics
- Nuclear data
- Advanced methods in nuclear waste treatment
- Nuclear methods for applications

Several colleagues helped with the organization of the School. We would like to give our thanks to them, especially to the Scientific Secretary of the School, Dr Elena Stefanova, and to Dr Dimitar Tarpanov and the members of the Organizing Committee for their cordiality and high level assistance.

Sofia, 12 March 2012

Co-chairpersons of the Organizing Committee
Professor Dr Sc Ch Stoyanov
Professor Dr Sc S Dimitrova

Journal of Physics: Conference Series **366** (2012) 011001 doi:10.1088/1742-6596/366/1/011001

Organization and Committees

Organizers
Institute for Nuclear Research and Nuclear Energy
Bulgarian Academy of Sciences

Co-organizers
Bulgarian Nuclear Regulatory Agency
Joint Institute for Nuclear Research, Dubna, Russia

International Advisory Committee

A Bracco	University of Milan
D Tonev	INRNE, Bulgaria
J Dobaczewski	Warsaw University
J J Benlliure	U. Santiago de Compostela
M Itkis	JINR, Dubna
M Ploszajczak	GANIL
N Pietralla	TU, Darmstadt
R Casten	Yale, USA
S Tyutyunnikov	Dubna
V Zelevinsky	MSU
A Covello	U. Napoli, Italy
G De Angelis	LNL
J Draayer	SURA
L Kostov	NRA, Bulgaria
M Harakeh	KVI Groningen, Netherlands
N Janeva	INRNE, Bulgaria
N V Zamfir	NIPNE-Bucharest
R Julin	JYI
V Voronov	BLTP, Dubna
W Westmeier	U. Marburg

Local Organizing Committee

Ch Stoyanov (Chairperson)
stoyanov@inrne.bas.bg

S Dimitrova (Co-chairperson)
sevdim@inrne.bas.bg

E Stefanova (Scientific Secretary)
elena1stef@yahoo.com

D Tarpanov
tarpanov@inrne.bas.bg

Journal of Physics: Conference Series **366** (2012) 011002 doi:10.1088/1742-6596/366/1/011002

List of Participants

1. Anton Antonov
INRNE-BAS, 72 Tzarigradsko chaussee, Blvd.
BG – 1784 , Sofia, Bulgaria
aantonov@inrne.bas.bg

2. Naftali Auerbach
School of Physics and Astronomy, Tel Aviv University
69978 TEL AVIV Israel
auerbach@post.tau.ac.il

3. Evgeny Balbutsev
BLTP, JINR, Dubna, Joliot-Curie 6, 141980 Dubna, Moscow region, Russia
balbuts@theor.jinr.ru

4. Jacob Beller
IKP, TU Darmstadt
Schlossgartenstr. 9, Darmstadt, 64289 Hessen, Germany
beller@ikp.tu-darmstadt.de

5. Davide Bianco
U. Napoli Federico II, Corso Umberto I, 80138 Napoli, Italy
bianco@na.infn.it

6. Jonathan Billowes
School of Physics & Astronomy, Unuversity of Manchester
M13 9PL Manchester, UK
j.billowes@manchester.ac.uk

7. Dennis Bonatsos
N.C.S.R. Demokritos, 153 10 - Agia Paraskevi Athens P.O. 60228, Greece
bonat@inp.demokritos.gr

8. Plamen Boutachkov
GSI Helmholtzzentrum fur Schwerionenforschung GmbH
Planckstr. 1, Hessen, D-64291,Germany
P.Boutachkov@gsi.de

9. Juan Caballero
University of Seville, Avda.Reina Mercedes s/n
41012 Sevilla, Spain
jac@us.es

Journal of Physics: Conference Series **366** (2012) 011002 doi:10.1088/1742-6596/366/1/011002

10. Rick Casten
Yale University
272 Whitney Avenue, PO Box 208120, New Haven 06520-8120, USA
richard.casten@yale.edu

11. Jozsef Cseh
MTA ATOMKI, H-4001 Debrecen, PO Box 51, Hungary
cseh@atomki.hu

12. Miroslav Danchev
U. Sofia, 5 James Bourchier Blvd., 1164 Sofia, Bulgaria
miroslav@phys.uni-sofia.bg

13. Judit Darai
U. Debrecen, Pf. 105, 4010 Debrecen, Hungary
darai@atomki.hu

14. Stijn de Baerdemacker
Ghent University, Department of Physics and Astronomy
Proeftuinstraat 86, Ghent, 9000 Oost-Vlaanderen, Belgium
stijn.debaerdemacker@ugent.be

15. Elvira Moya de Guerra
Universidad Complutense de Madrid (UCM)
Dpto. FAMN / Facultad de CC. Fisicas / Avda. Complutense s/n
28040 Madrid, Spain
emoyaval@fis.ucm.es

16. Isabelle Deloncle
CSNSM, Orsay Bâtiments 104 et 108, 91405 Orsay CAMPUS, France
isabelle.deloncle@csnsm.in2p3.fr

17. Vera Derya
Institut fuer Kernphysik, Universitat zu Koeln
Zuelpicher Str. 77, 50937 Cologne, Germany
derya@ikp.uni-koeln.de

18. Alessia Di Pietro
INFN, via S. Sofia 62, I-95125, Catania, Italy
Dipietro@lns.infn.it

19. Sevdalina Dimitrova
INRNE-BAS, 72 Tzarigradsko chaussee, Blvd
BG – 1784, Sofia, Bulgaria
sevdim@inrne.bas.bg

20. Martin Djongolov
Sofia University St. Kliment Ohridski, Physics Department
5 James Bourchier Blvd., 1164 Sofia, Bulgaria
martin.djongolov.4d@gmail.com

21. Jerry Draayer
DPA-LSU, Baton Rouge, LA 70803, USA
draayer@sura.org

22. Cristian Dragolici
30 Reactorului Street, Magurele, P.O. Box MG-6
077125 Ilfov, Romania
adrag@nipne.ro

23. Kalin Drumev
INRNE-BAS, 72, Tzarigradsko chaussee, Blvd
BG – 1784 , Sofia, Bulgaria
Kdrumev2000@yahoo.com

24. Enrico Fioretto
INFN- LNL, Viale dell'Università 2, 35020 Legnaro (PADOVA), Italy
fioretto@lnl.infn.it

25. Mitko Gaidarov
INRNE-BAS, 72 Tzarigradsko chaussee, Blvd.
BG – 1784 , Sofia, Bulgaria
gaidarov@inrne.bas.bg

26. Huben Ganev
INRNE-BAS, 72 Tzarigradsko chaussee, Blvd.
BG – 1784 , Sofia, Bulgaria
huben@inrne.bas.bg

27. Georgi Gavrilov
INRNE-BAS, 72 Tzarigradsko chaussee, Blvd.
BG – 1784 , Sofia, Bulgaria
georgi.gavrilov@gmail.com

28. Panagiotis Georgoudis
N.C.S.R. Demokritos, 153 10 - Agia Paraskevi Athens P.O. 60228, Greece
potoulis@gmail.com

29. Agnese Giaz
INFN Sezione di Milano - Università degli studi di Milano
Via Celoria, 16, Milano 20133, Italy
agnese.giaz@mi.infn.it

30. Carlotta Giusti
Universita di Pavia, Dipartimento di Fisica Nucleare e Teorica
Via Bassi 6 Pavia, I 27100 Italy
Carlotta.Giusti@pv.infn.it

31. Kalin Gladnishki
Sofia University St. Kliment Ohridski, Physics Department
5 James Bourchier Blvd., 1164 Sofia, Bulgaria
kag@phys.uni-sofia.bg

32. Ernest Grodner
University of Warsaw, Faculty of Physics
ul. Hoza 69, 00-681 Warszawa, Poland
ernest.grodner@fuw.edu.pl

Journal of Physics: Conference Series **366** (2012) 011002 doi:10.1088/1742-6596/366/1/011002

33. Nikolaj Gutev
INRNE-BAS, 72 Tzarigradsko chaussee, Blvd.
BG – 1784 , Sofia, Bulgaria
nikolay.v.goutev@gmail.com

34. Jan Jolie
University of Cologne, Institut für Kernphysik
Zülpicher Str 77, D-50937 Cologne, Germany
jolie@ikp.uni-koeln.de

35. Katalin Juhasz
Department of Information Technology, University of Debrecen
1, Egyetem tér, H-4032 DEBRECEN, Hungary
juhasz.katalin@inf.unideb.hu

36. Mikhail Kadykov
Joint institute for Nuclear Research, Dubna
Moscow region Joliot-Curie 6, 141980 Russia
kadykov@jinr.ru

37. Shuhrat Kalandrov
JINR, Joliot-Curie 6, Dubna, Moscow region, 141980 Russia
shuhrat@theor.jinr.ru

38. Anu Kankainen
U. Jyväskylä, P.O. Box 35, 40014 Jyväskylän yliopisto, Finland
anu.k.kankainen@jyu.fi

39. Sofia Karampagia
N.C.S.R. Demokritos, 153 10 - Agia Paraskevi Athens P.O. 60228, Greece
skaram@central.ntua.gr

40. Segei Khabarov
JINR, Joliot-Curie 6, Dubna, Moscow region, 141980 Russia
Sergei.Khabarov@mail.ru

41. Jurgen Kiener
CSNSM, Orsay Bâtiments 104 et 108, 91405 Orsay CAMPUS, France
jurgen.kiener@csnsm.in2p3.fr

42. Stanimir Kisyov
U. Sofia, 5 James Bourchier Blvd., 1164 Sofia, Bulgaria
Stanimir.kisyov@gmail.com

43. Amel Korichi
CSNSM, Orsay Bâtiments 104 et 108, 91405 Orsay CAMPUS, France
amel.korichi@csnsm.in2p3.fr

44. Pavlos Koseoglou
Vironos 8, Stavros Verias, 59100 Macedonia, Greece
pkoseogl@physics.auth.gr

45. Lachezar Kostov
INRNE-BAS, 72 Tzarigradsko chaussee, Blvd.
BG – 1784 , Sofia, Bulgaria
lkostov@inrne.bas.bg

46. Gerfried Kumbartzki
Rutgers, The State University of New Jersey
136 Frelingshuysen Rd., Piscataway, NJ 08854, USA
kum@physics.rutgers.edu

47. Hristo Laftchiev
INRNE-BAS, 72 Tzarigradsko chaussee, Blvd.
BG – 1784 , Sofia, Bulgaria
laftchiev@gmail.com

48. Stefan Lalkovski
INRNE-BAS, 72 Tzarigradsko chaussee, Blvd.
BG – 1784 , Sofia, Bulgaria
s.p.lalkovski@googlemail.com

49. Silvia Leoni
University of Milano and INFN, Via Celoria 16, 20133 Milano, Italy
silvia.leoni@mi.infn.it

50. Jean Libert
IPN - Orsay, bat. 100, 91406 Orsay, France
libert@ipno.in2p3.fr

51. Nicola Lo Iudice
U. Napoli Federico II, Corso Umberto I, 80138 Napoli, Italy
nicola.loiudice@na.infn.it

52. Konstantin Lukyanov
LoIT, JINR, Joliot-Curie 6, Dubna, 141980 Moscow region, Russia
luky@jinr.ru

53. Metaxia Manolopoulou
Nuclear Physics Department, School of Physics, Faculty of Sciences
Aristotle University of Thessaloniki, Thessaloniki, 541 24 Greece
metaxia@auth.gr

54. Konstantin Mihai
IFIN-HH, Reactorului 30, Magurele
077125 Ilfov, Romania
cmihai@tandem.nipne.ro

55. Thomas Moeller
Darmstadt Technical University
Schlossgartenstr. 9, 64289 Darmstadt, Germany
tmoeller@ikp.tu-darmstadt.de

56. Iain David Moore
Department of Physics, University of Jyväskylä
Survontie 9, Jyväskylä, FI-40014, Finland
iain.d.moore@jyu.fi

57. Nguyen Dinh Dang
Theoretical Nuclear Physics Laboratory
Nishina Center for Accelerator-Based Science
Main Research Bldg. R. 159, RIKEN
2-1 Hirosawa, Wako city, 351-0198 Saitama, Japan
dang@riken.jp

58. Barna Nyako
ATOMKI (Inst. of Nucl. Research)
18/c BEM ter, H-4026, Hungary
nyako@atomki.hu

59. Chavdar Palev
INRNE-BAS, 72 Tzarigradsko chaussee, Blvd.
BG – 1784 , Sofia, Bulgaria
tpalev@inrne.bas.bg

60. Pavel Petkov
INRNE-BAS, 72 Tzarigradsko chaussee, Blvd.
BG – 1784 , Sofia, Bulgaria
pp@inrne.bas.bg

61. Norbert Pietralla
TU Darmstadt, D-64289 Darmstadt, Germany
pietralla@ikp.tu-darmstadt.de

62. Georgi Rainovski
U. Sofia, 5 James Bourchier Blvd., 1164 Sofia, Bulgaria
rig@phys.uni-sofia.bg

63. Patrick Regan
Department of Physics, University of Surrey, Guildford, GU2 5XH Surrey, UK
p.regan@surrey.ac.uk

64. Brigitte Roussiere
Institut de Physique Nucleare
15 rue Georges Clemenceau, 91406 Orsay Sedex, France
roussier@ipno.in2p3.fr

65. Pedro Sarriguren
IEM-CSIC, Serrano 123, E-28006 Madrid, Spain
sarriguren@iem.cfmac.csic.es

66. Markus Scheck
IKP, TU Darmstadt
Schlossgartenstr. 9, Darmstadt, 64289 Hessen, Germany
mscheck@ikp.tu-darmstadt.de

67. Konstantin Shegunov
INRNE-BAS, 72 Tzarigradsko chaussee, Blvd
BG – 1784 , Sofia, Bulgaria

68. Fedor Simkovic
JINR Dubna and Comenius University
BLTP, JINR, Dubna, 141980 Moscow region, Russia
simkovic@theor.jinr.ru

69. Ralitsa Stanoeva
South-West University 'Neofit Rilski', 66 Ivan Michailov st., 2700 Blagoevgrad, Bulgaria
rali.stanoeva@gmail.com

70. Elena Stefanova
INRNE-BAS, 72 Tzarigradsko chaussee, Blvd.
BG – 1784 , Sofia, Bulgaria
elenas@inrne.bas.bg

71. Chavdar Stoyanov
INRNE-BAS, 72 Tzarigradsko chaussee, Blvd.
BG – 1784 , Sofia, Bulgaria
stoyanov@inrne.bas.bg

72. Elena Strekalovskaia
LHEP JINR, Joliot-Curie 6, Dubna, 141980 Moscow region, Russia
strekale@jinr.ru; strekale@sunse.jinr.ru

73. Andrew Stuchbery
The Australian National University, Department of Nuclear Physics, Building 57
Canberra ACT 0200, Australia
andrew.stuchbery@anu.edu.au

74. Dimitar Tarpanov
INRNE-BAS, 72 Tzarigradsko chaussee, Blvd.
BG – 1784, Sofia, Bulgaria
dimitert@yahoo.co.uk

75. Dimitar Tonev
INRNE-BAS, 72 Tzarigradsko chaussee, Blvd.
BG – 1784, Sofia, Bulgaria
dimitar.tonev@lnl.infn.it

76. Nadia Tsoneva
Institut für Theoretische Physik
Heinrich-Buff-Ring 16, 35390 Giessen, Germany
tsoneva@theo.physik.uni-giessen.de

77. Sergey Tyutunnikov
Joint institute for Nuclear Research, Dubna
141980 Russia Moscow region Joliot-curie 6, Russia
tsu@sunse.jinr.ru

78. Calin Ur
INFN and U. Padova, Italy
calin.ur@pd.infn.it

79. Eleni Vagena
Aristotle University Thessaloniki
Aristotle University Campus, 54124 Thessaloniki, Greece
evagena@auth.gr

80. Valeria Vandone
Universita degli Studi e INFN – Milano
Via Celoria 16, 20133 Milano, Italy
Valeria.vandone@mi.infn.it

81. Alberto Ventura
ENEA, Via Martiri di Monte Sole 4, 40129 Bologna, Italy
alberto.ventura@enea.it

82. Vladimir Wagner
Nuclear Physics Institute of ASCR
250 68 Rež, Czech Republic
wagner@ujf.cas.cz

83. Volker Werner
U. Yale P.O. Box 208120, New Haven, CT 06520, USA
volker.werner@yale.edu

84. Wolfram Westmeier
Dr. Westmeier GmbH
Gesellschaft fur Kernspektrometrie
Mollner Weg 32
D-35085 Ebsdorfergrund-Molln, Germany
webmaster@westmeier.com

85. Mariya Yavahchova
INRNE-BAS, 72 Tzarigradsko chaussee, Blvd.
BG – 1784 , Sofia, Bulgaria
m.yavahchova@gmail.com

86. Orlin Yordanov
INRNE-BAS, 72 Tzarigradsko chaussee, Blvd.
BG – 1784 , Sofia, Bulgaria
orlin_yordanov@abv.bg

87. Nicolae Victor Zamfir
National Institute of Physics and Nuclear Engineering
30 Reactorului str., 077125 Bucharest-Magurele, Romania
zamfir@tandem.nipne.ro

88. Petar Zhivkov
INRNE-BAS, 72 Tzarigradsko chaussee, Blvd.
BG – 1784 , Sofia, Bulgaria
Petar.zhivkov@gmail.com

89. Markus Zweidinger
IKP, TU Darmstadt
Schlossgartenstr. 9, Darmstadt, 64289 Hessen, Germany
zweidinger@ikp.tu-darmstadt.de

Peer review statement

All papers published in this volume of *Journal of Physics: Conference Series* have been peer reviewed through processes administered by the proceedings Editors. Reviews were conducted by expert referees to the professional and scientific standards expected of a proceedings journal published by IOP Publishing.

Journal of Physics: Conference Series **366** (2012) 012001 doi:10.1088/1742-6596/366/1/012001

Super-radiance and the widths of neutron resonances in the compound nucleus

N. Auerbach

School of Physics and Astronomy,

Tel Aviv University, Tel Aviv 69978, Israel

E-mail: auerbach@post.tau.ac.il

Abstract. In the 1950s the possibility of forming a ''super-radiant'' (SR) state in a gas of atoms confined to a volume of a size smaller than the wave length of radiation was suggested by Dicke. During the years this mechanism was applied to many phenomena in many different fields. Here it is used in the discussion of the statistics of resonance widths in a many-body system with open decay channels. Depending on the strength of the coupling to the continuum such systems show deviations from the Porter-Thomas distribution. In the limit of very strong coupling this leads to super-radiance. The results presented are important for the understanding of recent experimental data concerning the widths distribution of neutron resonances in nuclei.

1. Introduction

In 1953 the possibility of forming a ''super-radiant'' (SR) state in a gas of atoms confined to a volume of a size smaller than the wave length of radiation was proposed by Dicke [1]. In the absence of a direct interaction, the atoms are coupled through their common radiation field. This indirect interaction through the continuum leads to a redistribution of lifetimes among unstable intrinsic states. A rapidly decaying SR state is created at the expense of the rest of the states of the system that are ''robbed'' of their decay width and become narrow. This mechanism has a general origin and analogous phenomena should appear in many quantum systems when quasi-bound states are strongly coupled through common decay channels. The SR approach has been used in many different fields. For example, it was applied in chemistry [2], atomic physics [3], condensed matter physics [4, 5], intermediate energy nuclear physics [6, 7], in particle physics [8], in the theory of nuclear reactions [5, 9, 10]. In the next section we will develop the SR formalism for some simple cases in order to arrive in a heuristic fashion to the idea of a super-radiant state and its connection to the concept of a doorway.

2. The effective Hamiltonian.

Following the projection formalism [6,11], we divide the Hilbert space of nuclear states into two parts, the $\{Q\}$-subspace involving complicated many-body states $|q\rangle$, and the subspace $\{P\}$ of open channels $|c\rangle$. We use the notations Q and P for the corresponding projection operators onto the above subspaces. The wave function of the total system,

$$|\Psi\rangle = Q|\Psi\rangle + P|\Psi\rangle \qquad (1)$$

obeys the Schrödinger equation

$$H|\Psi\rangle = E|\Psi\rangle \tag{2}$$

that can be written as a set of coupled equations,

$$(E - H_{QQ})Q|\Psi\rangle = H_{QP}\,P|\Psi\rangle \tag{3}$$

and

$$(E - H_{PP})P|\Psi\rangle = H_{PQ}Q|\Psi\rangle \tag{4}$$

where we use the notations $H_{AB} = AHB$. Eliminating the part $P|\Psi\rangle$, we obtain an equation in the Q-space,

$$(E - \widetilde{H}_{QQ})Q|\Psi\rangle = 0 \tag{5}$$

with the effective Hamiltonian

$$\widetilde{H}_{QQ} = H_{QQ} + H_{QP}\frac{1}{E^{(+)} - H_{PP}}H_{PQ} \tag{6}$$

Here $E^{(+)} \equiv E + i0$ contains the infinitesimal imaginary term $+i0$ ensuring correct asymptotic conditions for the continuum wave functions. The second term of the effective Hamiltonian (6) contains a real and imaginary part of the propagator

$$G^{(+)}(E) = \frac{1}{E^{(+)} - H_{PP}} \tag{7}$$

arising from the principal value and the delta-function $\delta(E - H_{PP})$ (on-shell contributions from channels c open at energy E), respectively. The imaginary part of the effective Hamiltonian is $-(i/2)W$, with

$$W = 2\pi\sum_c H_{QP}|c\rangle\langle c|H_{PQ} \tag{8}$$

Thus, the effective Hamiltonian (7) in Q-space is anti-Hermitian,

$$\widetilde{H}_{QQ} = \overline{H}_{QQ} - \frac{i}{2}W \tag{9}$$

where $\overline{H}_{QQ}$ is a symmetric and real matrix that includes, apart from the original Hamiltonian in the Q-space-H_{QQ}, the principal value contribution of the PQ-coupling. The second part is anti-Hermitian. The eigenvalues of $\widetilde{H}$, $\widetilde{E} = E - (i/2)\Gamma$ are complex poles of the scattering matrix corresponding to the resonances in the cross sections.

To demonstrate in a simple way the role of the anti-Hermitian term we assume that only one channel is open. Then the matrix W, Eq. (8), has a completely separable form,

$$\langle q|W|q'\rangle = 2\pi A_q^c A_{q'}^{c*} \tag{10}$$

where:

$$A_q^c = \langle q|H_{QP}|c\rangle \tag{11}$$

The rank of the matrix W is 1, so that all the eigenvalues of this matrix are zero except one that has the value equal to the trace of the matrix:

$$\Gamma_s = \sum_q \langle q|W|q\rangle = 2\pi\sum_q |A_q^c|^2 \equiv \sum_q \Gamma_q^{\uparrow} \tag{12}$$

with $\Gamma_q^{\uparrow}$ denoting the escape width of the individual levels before the W-matrix is diagonalized. The special unstable state with width Γ_s is often referred to as the *super-radiant* (SR), in analogy to the Dicke coherent state [1] of a set of two-level atoms coupled

Journal of Physics: Conference Series **366** (2012) 012001 doi:10.1088/1742-6596/366/1/012001

through the common radiation field. Here the coherence is generated by the *common decay channel.* The stable states are trapped and decoupled from the continuum.

In the more general case of N intrinsic states and N_c open channels with $N_c << N$ the super-radiant mechanism survives if the mean level spacing D of internal states and their decay widths $\Gamma_q^{\uparrow}$ obey:

$$\kappa^c = \frac{\Gamma_q^{\uparrow}}{D} > 1 \tag{13}$$

In this case we have N_c broad states, while the rest of states $N - N_c$ become very narrow.

3. Doorway States

Often only a subset of intrinsic states $\{Q\}$ connects directly to the $\{P\}\}$ space of channels. The rest of states in $\{Q\}$ will connect to $\{P\}$ only through admixtures of these selected states. The special states directly coupled to continuum are the *doorways*, $|d\rangle$. They form the doorway subspace $\{D\}$ within $\{Q\}$, and the corresponding projection operator will be denoted here as *D*. The remaining states in {Q} will be denoted as $|\tilde{q}\rangle$ and the subspace as $\{\tilde{Q}\}$.

The full Hamiltonian can be decomposed the following way:

$$H = \left(H_{\tilde{Q}\tilde{Q}} + H_{DD} + H_{\tilde{Q}D} + H_{D\tilde{Q}}\right) + \left(H_{PP} + H_{DP} + H_{PD}\right) \tag{14}$$

Note that the terms $H_{P\tilde{Q}}$ and $H_{\tilde{Q}P}$ are missing because in accordance with the doorway hypothesis they are zero. Also note that diagonalizing the operator in the upper line of (14) would give back the states $|q\rangle$ with the components of $|d\rangle$ mixed with $|\tilde{q}\rangle$ states. The two last items in the above equation couple the doorway states, and therefore all the $|q\rangle$ states to the open channels.

In the case when there is only one important doorway state $|d\rangle$ the matrix elements of the effective operator W in the intrinsic space are given by:

$$\langle q|W|q'\rangle = 2\pi \sum_{c=1}^{N_c} \langle q|H_{DP}|c\rangle\langle c|H_{PD}|q'\rangle \tag{15}$$

with the doorway assumption:

$$\langle q|H_{DP}|c\rangle = \langle q|d\rangle\langle d|H_{DP}|c\rangle \tag{16}$$

where $\langle q|d\rangle$ is the amplitude of the admixture of the doorway into the $|q\rangle$ state. Eq. (16) becomes:

$$\langle q|W|q'\rangle = 2\pi\langle q|d\rangle\langle d|q'\rangle \sum_{c} \left|\langle d|H_{DP}|c\rangle\right|^2 \tag{17}$$

again we have separable matrix elements of the matrix W, this time irrespective of the number of open channels $- N_c$. Diagonalizing this matrix one finds again a single state that has the widths:

$$\Gamma_s = 2\pi \sum_q \left|\langle q|d\rangle\right|^2 \sum_c \left|\langle d|H_{DP}|c\rangle\right|^2 \tag{18}$$

This width is the decay width of the doorway $\Gamma_d^{\uparrow}$.

As discussed above, the criterion of validity of the SR mechanism is that the average spacing between the levels in $\{Q\}$ is smaller than the decay width of such a state "before" the SR mechanism takes effect. This can be expressed, in the case of doorways, the following way. Consider the *spreading width*, $\Gamma_d^{\downarrow}$ of the doorway state representing the fragmentation of $|d\rangle$ into compound states $|\tilde{q}\rangle$. If N_q is the number of compound states in the interval covered by the spreading width, their average energy spacing is:

$$\overline{D} \approx \frac{\Gamma_d^{\downarrow}}{N_q} \tag{19}$$

Before the SR mechanism is turned on, the average decay width of a typical $|q\rangle$ state is:

$$\Gamma_q^{\uparrow} = 2\pi \sum_c \left|\langle q|H_{QP}|c\rangle\right|^2 \tag{20}$$

that can be estimated as:

$$\Gamma_q^{\uparrow} = \frac{\Gamma_s}{N_q} \tag{21}$$

Thus:

$$\frac{\Gamma_q^{\uparrow}}{D_q} \approx \frac{\Gamma_s}{\Gamma_d^{\downarrow}} \approx \frac{\Gamma_d^{\uparrow}}{\Gamma_d^{\downarrow}} \tag{22}$$

We conclude that the requirement for the SR doorway mechanism to be valid can be formulated as:

$$\frac{\Gamma_d^{\uparrow}}{\Gamma_d^{\uparrow}} > 1 \tag{23}$$

4. Porter-Thomas Distribution and the Super-Radiant Mechanism.

We will now discuss an example that illustrates the use of the super-radiant approach to a complicated problem in nuclear physics. One the best examples in which the interplay between intrinsic dynamics and decay channels is manifested are the low-energy neutron resonances. These resonances are interpreted as quasi-stationary levels of the compound nucleus formed after the neuron is captured. These resonances were described, with certain degree of success, by the theory of random matrices [12]. With exceedingly complicated wave functions of the compound states their components are Gaussian distributed. This applies also to the specific component related to the channel, thus the neutron in the continuum and the residual nucleus in the ground state. The decay width of the neutron resonance is proportional to the amplitude squared of this component and the width distribution is therefore given by the χ_ν^2 function with $\nu = 1$, that is the Porter-Thomas distribution (PTD). Recently experiments found significant departures from the PTD [13]. Attempts to still use the χ_ν^2 distribution in order to fit the data require $\nu < 1$. This result has been interpreted as a breakdown of the random matrix theory.

In a paper published recently [14] it was shown that a proper description of unstable quantum states using the super-radiant mechanism can in fact produce deviations from the PTD of the same type as observed in experiments. The coupling to open channels is essential in the redistribution of the decay widths of the compound states. The details of the calculations are described in reference [14]. When the strength of the coupling between compound states and the open channel(s) (denoted by κ) is very small ($\kappa << 1$) one finds that the distribution of the calculated widths follows the PTD. With increasing coupling the deviations become more pronounced. At $\kappa \sim 1$ the super-radiant stage is reached and the distribution of widths becomes singular. In Figure 1 are shown some the results from reference [14].

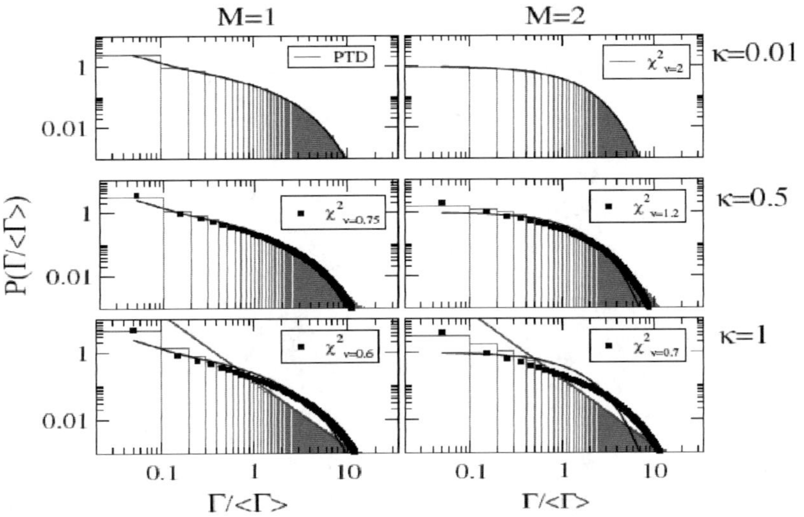

Figure 1. Distribution of widths for one open channel (M=1) and for two channels (M=2) for an intrinsic Gaussian Orthogonal Ensemble Hamiltonian and for different continuum coupling strength, κ. The numerical results are given by histograms; the PTD for M=1 and the $\chi^2_{\nu=2}$ distribution for M=2 are shown as a smooth curve. The full squares stand for the best fit to a χ^2_ν distribution.

In summary, the interpretation of the widths of compound resonances as a strength of the pure neutron component in the wave function fails due to the coupling to the continuum that has to be taken into account in a proper statistical description. This phenomenon is of general nature and might influence many other quantum systems.

I wish to thank L.Celardo, F. Izrailev and V. Zelevinsky for their collaboration.

References

[1] Dicke R H 1954 *Phys. Rev.* **93** 99.
[2] Pavlov-Verevkin V B 1988 *Phys. Lett.* A **129** 168.
[3] Flambaum V V, Gribakina A A, Gribakin G F 1996 *Phys. Rev.* A **54** 2066.
[4] Pichugin K, Schanz H, Šeba P, 2001 *Phys. Rev.* E **64** 056227.
[5] Volya A, Zelevinsky V, in: Zelevinsky V (Ed.), *Nuclei and Mesoscopic Physics,* in: *AIP Conference Proceedings*, 2005 vol. 777, p. 229.
[6] Auerbach N, 1994 *Phys. Rev. C* **50** 1606.
[7] Auerbach N, Zelevinsky V, 2002 *Phys. Rev.* C **65** 034601.
[8] Auerbach N, Zelevinsky V, Volya A, 2004 *Phys. Lett.* B **590** 45.
[9]. Sokolov V V, Zelevinsky V, 1989 *Nucl. Phys.* A **504** 562.
[10] Rotter I, 1991 *Rep. Prog. Phys.* **54** 635.
[11] Feshbach H, 1958 *Ann. Phys.* **5** 357; Feshbach H, 1962 *Ann. Phys.* **19** 287.
[12] Porter C E, 1965 *Statistical Theories.of Spectra: Fluctuations*, Academic, New York
[13] Koehler P E, Becvar F, Krticka M, Harvey J and Gruber K H, 2010 *Phys. Rev. Lett.* **105** 072502
[14] Celardo G L, Auerbach N, Izrailev F, and Zelevinsky V G 2011. *Phys. Rev. Lett.* **106** 042501

Journal of Physics: Conference Series **366** (2012) 012002 doi:10.1088/1742-6596/366/1/012002

Fine structure of the nuclear scissors mode (new type of collective motion)

E B Balbutsev, I V Molodtsova

Joint Institute for Nuclear Research, 141980 Dubna, Moscow Region, Russia

E-mail: balbuts@theor.jinr.ru

Abstract. The coupled dynamics of low-lying modes (including scissors) and various giant quadrupole resonances are studied with the help of Wigner Function Moments method generalized to take into account spin degrees of freedom. Equations of motion for collective variables are derived on the basis of Time Dependent Hartree-Fock equations in the model of harmonic oscillator with spin orbital interaction plus quadrupole-quadrupole residual interaction. Enlisting the spin allows one to introduce into consideration new types of nuclear collective motion.

1. Introduction

The nuclear scissors mode was predicted [1, 2, 3] as the collective motion of two types of nucleons – all protons accomplish rotational vibrations with respect of all neutrons. Its theoretical and experimental aspects were discussed in many papers from various points of view (see the review [4] and references therein.) Here we will consider the Wigner Function Moments (WFM) method. On the one hand it is the pure microscopic method, because it is based on the Time Dependent Hartree-Fock (TDHF) equation. On the other hand the method works with average values (moments) of operators which have the direct relation to the considered phenomenon, that makes it the ideal instrument to describe the basic characteristics (energies and excitation probabilities) of collective excitations [5] and, in particular, the scissors mode [6]. Our investigations have shown that already the minimal set of collective variables (the nucleus angular momentum and nucleus quadrupole moments in the coordinate and momentum spaces) is sufficient to reproduce the most important properties of the scissors mode: its inevitable coexistence with the IsoVector Giant Quadrupole Resonance (IVGQR) and with the Fermi surface deformation.

Further improvement of WFM method, namely, the switch-over from TDHF to TDHFB equation, i.e. taking into account pair correlations, allowed us to improve considerably the quantitative description of the scissors mode: for rare earth nuclei the energies are reproduced with $\sim$ 10% accuracy and B(M1) factors decreased practically two times (though exceed nevertheless experimental values practically two times). The reason of the last discrepancy is hidden probably in spin degrees of freedom, which were ignored by WFM method until now. One can not exclude, that due to spin dependent interactions some part of the force of M1 transitions is shifted to the energy region of 5-10 MeV, where 1^+ resonance of the spin nature is observed. The generalization of WFM method to take into account spin degrees of freedom is the goal of this paper. At a first step we include in the consideration only the spin orbital interaction, as the most important one among all possible spin dependent interactions because it enters into the mean field. It became clear already on the stage of formulation of the equations of motion for new collective variables, that we have met the new (at least not discussed until now) type of collective motion, or maybe several ones.

2. TDHF equation and model Hamiltonian

One starts with the time dependent HF equation in matrix form [7]

$$i\hbar < \mathbf{r}, s|\dot{\hat{\rho}}|\mathbf{r}'', s'' >= \sum_{s'} \int d^3r' \left(< \mathbf{r}, s|\hat{h}|\mathbf{r}', s' >< \mathbf{r}', s'|\hat{\rho}|\mathbf{r}'', s'' > - < \mathbf{r}, s|\hat{\rho}|\mathbf{r}', s' >< \mathbf{r}', s'|\hat{h}|\mathbf{r}'', s'' >\right). \quad (1)$$

By means of the Wigner transformation the following set of equations for the Wigner functions with specified spin indeces is derived

$$\begin{aligned} i\hbar\dot{f}^{\uparrow\uparrow} &= i\hbar\{h^{\uparrow\uparrow}, f^{\uparrow\uparrow}\} + h^{\uparrow\downarrow}f^{\downarrow\uparrow} - f^{\uparrow\downarrow}h^{\downarrow\uparrow} + \frac{i\hbar}{2}\{h^{\uparrow\downarrow}, f^{\downarrow\uparrow}\} - \frac{i\hbar}{2}\{f^{\uparrow\downarrow}, h^{\downarrow\uparrow}\} \\ &\quad -\frac{\hbar^2}{8}\{\{h^{\uparrow\downarrow}, f^{\downarrow\uparrow}\}\} + \frac{\hbar^2}{8}\{\{f^{\uparrow\downarrow}, h^{\downarrow\uparrow}\}\} + ..., \\ i\hbar\dot{f}^{\uparrow\downarrow} &= f^{\uparrow\downarrow}(h^{\uparrow\uparrow} - h^{\downarrow\downarrow}) + \frac{i\hbar}{2}\{(h^{\uparrow\uparrow} + h^{\downarrow\downarrow}), f^{\uparrow\downarrow}\} - \frac{\hbar^2}{8}\{\{(h^{\uparrow\uparrow} - h^{\downarrow\downarrow}), f^{\uparrow\downarrow}\}\} \\ &\quad -h^{\uparrow\downarrow}(f^{\uparrow\uparrow} - f^{\downarrow\downarrow}) + \frac{i\hbar}{2}\{h^{\uparrow\downarrow}, (f^{\uparrow\uparrow} + f^{\downarrow\downarrow})\} + \frac{\hbar^2}{8}\{\{h^{\uparrow\downarrow}, (f^{\uparrow\uparrow} - f^{\downarrow\downarrow})\}\} + ..., \end{aligned} \quad (2)$$

where $\uparrow$ means $s = 1/2$ and $\downarrow$ means $s = -1/2$, the functions $h(\mathbf{r}, \mathbf{p})$, $f(\mathbf{r}, \mathbf{p})$ are the Wigner transforms of $\hat{h}$, $\hat{\rho}$ respectively, $\{f, h\}$ is the Poisson bracket of the functions f and h and $\{\{f, h\}\}$ is their double Poisson bracket; the dots stand for terms proportional to higher powers of $\hbar$. Hereinafter we omit $(\mathbf{r}, \mathbf{p})$-dependence and isotopic indices of variables. Two more equations are obtained by the obvious change of arrows $\uparrow\leftrightarrow\downarrow$.

We define the functions $f^+ = f^{\uparrow\uparrow} + f^{\downarrow\downarrow}$ and $f^- = f^{\uparrow\uparrow} - f^{\downarrow\downarrow}$ as spin-scalar and spin-vector ones respectively and introduce the notation $f^u = f^{\uparrow\downarrow}$, $f^d = f^{\downarrow\uparrow}$, $h^\pm = h^{\uparrow\uparrow} \pm h^{\downarrow\downarrow}$. We get:

$$\begin{aligned} i\hbar\dot{f}^+ &= \frac{i\hbar}{2}\{h^+, f^+\} + \frac{i\hbar}{2}\{h^-, f^-\} + i\hbar\{h^{\uparrow\downarrow}, f^d\} + i\hbar\{h^{\downarrow\uparrow}, f^u\} + ..., \\ i\hbar\dot{f}^- &= \frac{i\hbar}{2}\{h^+, f^-\} + \frac{i\hbar}{2}\{h^-, f^+\} - 2h^{\downarrow\uparrow}f^u + 2h^{\uparrow\downarrow}f^d + \frac{\hbar^2}{4}\{\{h^{\downarrow\uparrow}, f^u\}\} - \frac{\hbar^2}{4}\{\{h^{\uparrow\downarrow}, f^d\}\} + ..., \\ i\hbar\dot{f}^u &= -h^{\uparrow\downarrow}f^- + h^-f^u + \frac{i\hbar}{2}\{h^{\uparrow\downarrow}, f^+\} + \frac{i\hbar}{2}\{h^+, f^u\} + \frac{\hbar^2}{8}\{\{h^{\uparrow\downarrow}, f^-\}\} - \frac{\hbar^2}{8}\{\{h^-, f^u\}\} + ..., \\ i\hbar\dot{f}^d &= h^{\downarrow\uparrow}f^- - h^-f^d + \frac{i\hbar}{2}\{h^{\downarrow\uparrow}, f^+\} + \frac{i\hbar}{2}\{h^+, f^d\} - \frac{\hbar^2}{8}\{\{h^{\downarrow\uparrow}, f^-\}\} + \frac{\hbar^2}{8}\{\{h^-, f^d\}\} + ... \end{aligned} \quad (3)$$

The microscopic Hamiltonian of the model is the harmonic oscillator with spin-orbital interaction plus separable quadrupole-quadrupole residual interaction

$$\begin{aligned} H &= \sum_{i=1}^{A}\left[\frac{\hat{\mathbf{p}}_i^2}{2m} + \frac{1}{2}m\omega^2\mathbf{r}_i^2 - \eta(\mathbf{r}_i)\hat{\mathbf{l}}_i\hat{\mathbf{S}}_i\right] + \bar{\kappa}\sum_{\mu=-2}^{2}(-1)^\mu\sum_i^Z\sum_j^N q_{2-\mu}(\mathbf{r}_i)q_{2\mu}(\mathbf{r}_j) \\ &\quad +\frac{1}{2}\kappa\sum_{\mu=-2}^{2}(-1)^\mu\left\{\sum_{i\neq j}^{Z} q_{2-\mu}(\mathbf{r}_i)q_{2\mu}(\mathbf{r}_j) + \sum_{i\neq j}^{N} q_{2-\mu}(\mathbf{r}_i)q_{2\mu}(\mathbf{r}_j)\right\}, \end{aligned} \quad (4)$$

where N and Z are the numbers of neutrons and protons, respectively. The quadrupole operator $q_{2\mu}$ can be written as the tensor product: $q_{2\mu}(\mathbf{r}) = \sqrt{16\pi/5}\,r^2Y_{2\mu}(\theta, \phi) = \sqrt{6}\{r \otimes r\}_{2\mu}$, where

$\{r\otimes r\}_{\lambda\mu}=\sum_{\sigma,\nu}C^{\lambda\mu}_{1\sigma,1\nu}r_\sigma r_\nu$, $C^{\lambda\mu}_{1\sigma,1\nu}$ is the Clebsch-Gordan coefficient and r_1, r_0, r_{-1} are cyclic coordinates [8]. The mean field potential for protons (or neutrons) is

$$V^\tau=\frac{1}{2}m\,\omega^2r^2+\sum_{\mu=-2}^{2}(-1)^\mu Z^{\tau+}_{2-\mu}\{r\otimes r\}_{2\mu}-\eta(\mathbf{r})\sum_{\mu=-1}^{1}(-)^\mu\hat{l}_\mu\hat{S}_{-\mu},\qquad \tau=n,p. \tag{5}$$

Here $\hat{l}_\mu$ are angular momentum components, $\hat{S}_\mu$ are spin matrices [8] and

$$Z^{\rm n+}_{2\mu}=\chi R^{\rm n+}_{2\mu}+\bar{\chi}R^{\rm p+}_{2\mu}\,,\ Z^{\rm p+}_{2\mu}=\chi R^{\rm p+}_{2\mu}+\bar{\chi}R^{\rm n+}_{2\mu}\,,\quad \chi=6\kappa,\ \bar{\chi}=6\bar{\kappa},$$
$$R^{\tau+}_{\lambda\mu}(t)=(2\pi\hbar)^{-3}\int d^3p\int d^3r\{r\otimes r\}_{\lambda\mu}f^{\tau+}(\mathbf{r},\mathbf{p},t). \tag{6}$$

3. Equations of motion

Integrating the set of equations (3) over the phase space with the weights $W=\{r\otimes p\}_{\lambda\mu}$, $\{r\otimes r\}_{\lambda\mu}$, $\{p\otimes p\}_{\lambda\mu}$ and 1 one gets the dynamical equations for the following moments of second and zero orders – collective variables of the method:

$$L^\sigma_{\lambda\mu}(t)=\int d(\mathbf{p},\mathbf{r})\{r\otimes p\}_{\lambda\mu}f^\sigma(\mathbf{r},\mathbf{p},t),\quad R^\sigma_{\lambda\mu}(t)=\int d(\mathbf{p},\mathbf{r})\{r\otimes r\}_{\lambda\mu}f^\sigma(\mathbf{r},\mathbf{p},t),$$
$$P^\sigma_{\lambda\mu}(t)=\int d(\mathbf{p},\mathbf{r})\{p\otimes p\}_{\lambda\mu}f^\sigma(\mathbf{r},\mathbf{p},t),\quad F^\sigma(t)=\int d(\mathbf{p},\mathbf{r})f^\sigma(\mathbf{r},\mathbf{p},t), \tag{7}$$

where $\int d(\mathbf{p},\mathbf{r})\equiv(2\pi\hbar)^{-3}\int d^3p\int d^3r$ and $\sigma=+,-,u,d$. The required expressions for $h^\pm$, $h^{\uparrow\downarrow}$ and $h^{\downarrow\uparrow}$ are

$$h^+=\frac{p^2}{m}+m\,\omega^2r^2+2\sum_\mu(-1)^\mu Z^{\tau+}_{2\mu}\{r\otimes r\}_{2-\mu},\quad h^-=-\hbar\eta l_0,\ h^{\uparrow\downarrow}=-\frac{\hbar}{\sqrt{2}}\eta l_{-1},\ h^{\downarrow\uparrow}=\frac{\hbar}{\sqrt{2}}\eta l_1.$$

We are interested in the scissors mode, the excitation with $K^\pi=1^+$, therefore we need only the set of equations with $\mu=1$. These equations are nonlinear and are solved in a small amplitude approximation.

Let us remind about isospin indices $\tau=$ n, p. One implies that all variables and equilibrium characteristics $R^+_{20}(eq)$ and $Z^+_{20}(eq)$ in equations have such indices. All difference between neutron and proton systems sits in the mean field characteristic $Z^{\tau+}_{20}(eq)$, which is different for neutrons and protons (see (6)). It is convenient to rewrite the equations in terms of the isovector $\bar{X}_{\lambda\mu}=X^{\rm n}_{\lambda\mu}-X^{\rm p}_{\lambda\mu}$ and isoscalar $X_{\lambda\mu}=X^{\rm n}_{\lambda\mu}+X^{\rm p}_{\lambda\mu}$ variables. Isovector and isoscalar strength constants $\chi_1=(\chi-\bar{\chi})/2$ and $\chi_0=(\chi+\bar{\chi})/2$ are connected by the relation $\chi_1=\alpha\chi_0$ [6]. Neglecting by the fourth order moments, generated by the term $h^{\uparrow\downarrow}f^{\downarrow\uparrow}-f^{\uparrow\downarrow}h^{\downarrow\uparrow}$ of (2) we get the closed set of 19 equations. As an example we demonstrate here only the small part of them, which in the case of $\eta=0$ becomes closed:

$$\begin{aligned}
\dot{\bar{\mathcal{L}}}^+_{21} &= \frac{1}{m}\bar{\mathcal{P}}^+_{21}-\left[m\,\omega^2-\frac{2}{\sqrt{3}}\alpha\chi_0R^+_{00}(eq)+\frac{1}{\sqrt{6}}(1+\alpha)\chi_0R^+_{20}(eq)\right]\bar{\mathcal{R}}^+_{21}\\
&\quad -i\hbar\frac{\eta}{2}[\bar{\mathcal{L}}^-_{21}+2\bar{\mathcal{L}}^u_{22}+2\sqrt{\frac{3}{2}}\bar{\mathcal{L}}^d_{20}],\\
\dot{\bar{\mathcal{L}}}^+_{11} &= -\sqrt{\frac{3}{2}}(1-\alpha)\chi_0R_{20}(eq)\bar{\mathcal{R}}^+_{21}-i\hbar\frac{\eta}{2}[\bar{\mathcal{L}}^-_{11}+\sqrt{2}\bar{\mathcal{L}}^d_{10}],\\
\dot{\bar{\mathcal{R}}}^+_{21} &= \frac{2}{m}\bar{\mathcal{L}}^+_{21}-i\hbar\frac{\eta}{2}[\bar{\mathcal{R}}^-_{21}+2\bar{\mathcal{R}}^u_{22}+2\sqrt{\frac{3}{2}}\bar{\mathcal{R}}^d_{20}],\\
\dot{\bar{\mathcal{P}}}^+_{21} &= -[2m\,\omega^2+\sqrt{\frac{2}{3}}\chi_0R_{20}(eq)]\bar{\mathcal{L}}^+_{21}+\sqrt{6}\chi_0R_{20}(eq)\bar{\mathcal{L}}^+_{11}-i\hbar\frac{\eta}{2}[\bar{\mathcal{P}}^-_{21}+2\bar{\mathcal{P}}^u_{22}+2\sqrt{\frac{3}{2}}\bar{\mathcal{P}}^d_{20}].
\end{aligned} \tag{8}$$

Journal of Physics: Conference Series **366** (2012) 012002 doi:10.1088/1742-6596/366/1/012002

The isoscalar set of equations is easily obtained from (8) by taking $\alpha = 1$.

4. Energies and excitation probabilities

Imposing the time evolution via $e^{i\Omega t}$ for all variables one transforms dynamical equations into a set of algebraic equations. Eigenfrequencies are found as the zeros of its determinant. Excitation probabilities are calculated with the help of the linear response theory.

Table 1. Energies and excitation probabilities of the isovector system, calculated for ^{164}Er with two values of the spin-orbital interaction constant η. The quantum numbers (including indices $+$, $-$, u, d) of variables responsible for the generation of the present level are shown in a first column of Table 1b.

Table 1a. $\eta = 0.361$ MeV			**Table 1b.** $\eta = 0$ MeV			
$E_{\rm iv}$, MeV	$B(M1)$, μ_N^2	$B(E2)$, B_W	$(\lambda,\mu)^\sigma$	$E_{\rm iv}$, MeV	$B(M1)$, μ_N^2	$B(E2)$, B_W
1.61	3.54	0.12	$(1,1)^-$	1.70	0	0
2.18	5.33	1.02	$(1,1)^+$	2.07	9.03	1.17
12.80	0.01	0.04	$(0,0)^d$	12.81	0	0
14.50	0.01	0.03	$(2,1)^-$	14.50	0	0
16.18	0.02	0.18	$(2,2)^u$	16.20	0	0
16.20	0	0	$(2,0)^d$	16.20	0	0
20.59	2.78	35.45	$(2,1)^+$	20.55	2.81	35.70

The results of calculations for the **isovector** system are demonstrated in the table 1a. They will be discussed in comparison with the results for $\eta = 0$ (table 1b). In this case the isovector set of equations splits into four independent subsets.

The first one (8) operates with spin-scalar variables $X^+_{\lambda\mu}$ and describes the joint dynamics of the "standard" nuclear scissors mode (E=2.07 MeV) and IVGQR (E=20.55MeV) [6].

The second subset operates with spin-vector variables $X^-_{\lambda\mu}$ and describes the relative motion of spin up nucleons with respect of spin down nucleons. The respective eigenfrequencies are

$$\Omega^2_\pm = 2\,\bar{\omega}^2\left[1 + \delta/3 \pm \sqrt{(1+\delta/3)^2 - \delta^2}\,\right]. \tag{9}$$

The low-lying level has $E^-_{11}(low) = \hbar\Omega_- = 1.70$ MeV and the high-lying level has $E^-_{21}(high) = \hbar\Omega_+ = 14.50$ MeV. So, in addition to the well known ("standard") scissors mode with energy $E^+_{11} = 2.07$ MeV we get the new type of the scissors mode – spin scissors. The inclusion of the spin-orbital interaction only slightly repulses these levels. However, the main role of the spin-orbital interaction consists in the excitation of the new scissors mode. This mode is excited only via the coupling with the "standard" scissors mode by means of the spin-orbital interaction. The new scissors mode is accompanied by the high lying excitation, which can be called the isovector spin-vector giant quadrupole resonance.

Third and fourth subsets operate with variables $X^u_{\lambda\mu}$ and $X^d_{\lambda\mu}$. They describe the spin-flip modes: giant quadrupole and monopole spin-flip resonances with energies $E_{20} = E_{22} = 16.20$ MeV and $E_{00} = 12.81$ MeV respectively.

The results of calculations for the **isoscalar** system are demonstrated in table 2.

The most interesting fact here is the appearance of the new low-lying mode with E=0.39 MeV which describes the relative motion of the orbital angular momentum and spin of the nucleus. Another new excitation with E=1.73 MeV corresponds to the rotational oscillations of all spin up nucleons out of phase with all spin down nucleons.

All high lying modes are just the isoscalar analogues of the respective isovector modes.

Table 2. The same as in table 1 for the isoscalar system.

Table 2a. $\eta = 0.361$ MeV			**Table 2b.** $\eta = 0$ MeV			
$E_{\rm is}$, MeV	$B(M1)$, μ_N^2	$B(E2)$, B_W	$(\lambda,\mu)^\sigma$	$E_{\rm is}$, MeV	$B(M1)$, μ_N^2	$B(E2)$, B_W
1.73	−0.07	1.12	$(1,1)^-$	1.70	0	0
0.39	0.24	117.19	$(1,1)^+$	0	–	–
12.83	0	0.66	$(0,0)^d$	12.81	0	0
14.51	0	0.12	$(2,1)^-$	14.51	0	0
16.20	0	0	$(2,2)^u$	16.20	0	0
16.22	0	0.20	$(2,0)^d$	16.20	0	0
10.28	0	66.50	$(2,1)^+$	10.33	0	67.47

5. Conclusion

The WFM method is applied for a first time to solve the HF equation with spin. The model Hamiltonian is a harmonic oscillator with the spin-orbital interaction plus quadrupole-quadrupole residual interaction. Spin dependent collective variables are defined and the closed set of nonlinear dynamical equations for them is derived. Equations are solved in a small amplitude approximation. Two isovector and two isoscalar low-lying eigenfrequencies and five isovector and five isoscalar high-lying eigenfrequencies are found. Three low-lying levels correspond to the excitations of the new nature, which was not known earlier. For example the isovector level with energy E=1.61 MeV describes the rotational oscillations of all nucleons with the spin projection "up" with respect of all nucleons with the spin projection "down", i.e. one more nuclear scissors. It is necessary to note the curious feature of the new mode. Being isovector one it describes also the motion of protons with respect of neutrons and, therefore, the resulting motion can be characterized as the rotational oscillations of proton "spin" scissors with respect of neutron "spin" scissors. In the analogous isoscalar mode with the energy E=1.73 MeV the proton and neutron "spin" scissors move in phase.

One more new low lying mode with energy E=0.39 MeV is generated by relative motion of the orbital angular momentum and spin of the nucleus (they can change their absolute values and directions keeping the total angular momentum unchanged).

There are ten high-lying excitation, the two of them being really new: isovector and isoscalar "spin-vector" resonances with energies E=14.5 MeV. Another six high-lying modes, according to the physical sense of variables which generate them, are spin-flip ones.

The study of all mentioned excitations with pair correlations and proper residual interactions included will be the natural continuation of this work.

References

[1] Hilton R R 1976 *Int. Conf. on Nuclear Structure (Dubna), A possible vibrational mode in heavy nuclei,* unpublished
[2] Suzuki T, Rowe D J 1977 *Nucl. Phys.* A **289** 461
[3] Lo Iudice N, Palumbo F 1978 *Phys. Rev. Lett.* **41** 1532
[4] Heyde K, von Neuman-Cosel P and Richter A 2010 *Rev. Mod. Phys.* **82** 2365
[5] Balbutsev E B 1991 *Sov. J. Part. Nucl.* **22** 159
[6] Balbutsev E B, Schuck P 2003 *Nucl. Phys.* A **720** 293; 2003 *Nucl. Phys.* A **728** 471
[7] Ring P and Schuck P 1980 *The Nuclear Many-Body Problem* (Berlin: Springer)
[8] Varshalovich D A, Moskalev A N and Khersonskii V K 1975 *Quantum Theory of Angular Momentum* (Leningrad: Nauka), 1988 (Singapore: World Scientific)

Energy separation of the $1^+/1^-$ parity doublet in ^{20}Ne

J. Beller[1], J. Wagner[1], M. Ahmed[2], D. Deleanu[3], D.M. Filipescu[3], T. Glodariu[3], J. Isaak[1], J. H. Kelley[4], E. Kwan[2], N. Pietralla[1], R. Raut[2], C. Romig[1], G. Rusev[2], M. Scheck[1], S.C. Stave[2], A. Tonchev[2], W. Tornow[2], H.R. Weller[2], N.-V. Zamfir[3] and M. Zweidinger[1]

[1] Institut für Kernphysik, TU Darmstadt, Schlossgartenstraße 9, D-64289 Darmstadt, Germany

[2] Department of Physics, Duke University, Durham, NC 27708-0308, USA

[3] National Institute for Physics and Nuclear Engineering, Str. Atomistilor no. 407, 077125 Bucharest - Magurele, Romania

[4] Department of Physics, North Carolina State University, Raleigh, NC 27695-8202, USA

E-mail: `beller@ikp.tu-darmstadt.de`

Abstract. The parity doublet of $1^+/1^-$ states of Ne-20 at 11.26 MeV excitation energy is one of the best known test cases to study the weak part of the nuclear Hamiltonian. The feasibility of parity violation experiments depend on the effective nuclear enhancement factor ($R_N/|E(1^+)-E(1^-)|$) which amplifies the impact of the matrix element of the weak interaction on observables indicating parity mixing. An extreme large value of $R_N/|E(1^+)-E(1^-)| = (670 \pm 7000)$ MeV^{-1} was reported for the doublet in ^{20}Ne. The large uncertainty depends amongst others on the large uncertainty of $|E(1^+)-E(1^-)| = 7.7\pm5.5$ keV of the parity doublet. Nuclear resonance fluorescence (NRF) experiments with linearly and circularly polarized photon beams were performed at the High Intensity Gamma-Ray Source at Duke University, Durham, NC, USA, to determine the energy difference of the parity doublet with higher precision. The different angular distributions for $0^+ \to 1^- \to 0^+$ and $0^+ \to 1^+ \to 0^+$ NRF cascades in polarized γ-ray beams were used to determine the energy difference of the parity doublet to 2.9(13) keV.

1. Introduction

Since 1956, when Lee and Yang postulated a mirror symmetry violation in β-decays [1] and 1957 when Wu experimentally verified the symmetry violation effect [2] parity non-conservation is well known. It has fundamental importance for studies of the impact of the weak interaction on nuclear structure. Hence, various theoretical and experimental approaches have been done to investigate parity violation (details are reviewed in [3, 4]). However, the weak meson-nucleon coupling constants deduced from various experiments are not consistent. Consequently a further investigation of parity violation is desirable.

The parity doublet of $1^+/1^-$ levels of ^{20}Ne at 11.26 MeV excitation energy has been suggested as one of the best cases to study parity violation of isolated nuclear eigenstates. The parity doublet in ^{20}Ne offers one of the very rare cases, where the effects of weak interaction can be studied for nuclear states with comparatively simple wave functions.The effect of parity violation in elementary quantum systems is known to result from contributions of the weak interaction [2]. Usually, contributions of the weak part to the nuclear effective Hamiltonian are not well

Published under licence by IOP Publishing Ltd

known, small, and neglected.
The $T = 1$, $1^+/1^-$ parity doublet represents the isospin analogue structure of the lowest-lying 1^+ and 1^- states near 1 MeV excitation energy of the $T = 1$ isobar ^{20}F. The structure of ^{20}F can comparatively easily be calculated in microscopic theoretical approaches such as the nuclear shell model. Therefore, this doublet allows to test the effect of weak interaction on nuclear two-particle states whose structure can be calculated quite reliable in the nuclear shell model.
Recently, scattering off circularly polarized photons on strongly excited parity doublets has been proposed [5] as a promising tool for such studies. Despite that the degree of the expected parity violation is small, in the order of 10^{-7}, dynamical and kinematical nuclear enhancements are expected to make its observation possible. However, as the large error demonstrates, the effective nuclear enhancement factor $|R_N/\Delta E| = (670 \pm 7000)$ as given by [5] is so far not known to a sufficient precision for making quantitative proposals for such investigations. The major contribution to the large uncertainty of the nuclear enhancement factor is due to the excitation energy difference of the parity doublet, which is only measured to an accuracy of (7.7 ± 5.5) keV. In order to improve this value, a combination of circular and linear polarized photons was used to increase the accuracy of the excitation energy difference of the parity doublet.

2. Experiment

The linearly and circularly polarized photons were provided by the High Intensity γ-Ray Source (HIγS). The deexiting transitions of ^{20}Ne were studied with the NRF setup consisting of four HPGe detectors surrounding the target. Each of the detectors has an efficiency of 60 % relative to a 3 inch by 3 inch NaI detector.

The angular distribution function for a $0^+ \xrightarrow{\vec{\gamma}} 1^\pm \xrightarrow{\gamma} 0^+$ NRF cascade as it occurs when ^{20}Ne is excited from its $J_0^\pi = 0^+$ ground state to a $J^\pi = 1^\pm$ level by linearly polarized photons and the subsequent decay back into the ground state is given by [6]

$$W(\vartheta, \varphi) = 1 + \frac{1}{2}\left[P_2(\cos\vartheta) + \frac{1}{2}\pi \cdot \cos(2\varphi) \cdot P_2^{(2)}(\cos\vartheta)\right], \quad (1)$$

with π being the parity of the excited state and ϑ and φ being the polar and azimuthal angle, respectively. The terms P_2 and $P_2^{(2)}$ denote the second order ordinary and unnormalized associated Legendre polynomial. Figure 1a illustrates the relevant angles. The angular distribution function for E1 ($\pi = -$) and M1 ($\pi = +$) transitions as given in Eq. (1) together with a schematic plot of the detector setup is shown in Fig. 1b.

Photons emitted in the $0^+ \xrightarrow{\vec{\gamma}} 1^+ \xrightarrow{\gamma} 0^+$ cascade are preferentially detected within the polarization plane. While the decay photons of a $0^+ \xrightarrow{\vec{\gamma}} 1^- \xrightarrow{\gamma} 0^+$ cascade are preferentially emitted in the plane perpendicular to the polarization plane. Thus, by measuring the NRF intensities in both planes it is possible to separate $J^\pi = 1^\pm$ states with respect to their parity. Since the states of the parity doublet have different parities they can be separated by this method. Due to the finite target and detector sizes, the separation will, of course, not be 100%. The analysing power of the setup has been reported to be 0.9 in previous experiments (see e.g. [7]). Thus, the peak of the $0^+ \to 1^+ \to 0^+$ ($0^+ \to 1^- \to 0^+$) transition will be supressed by a factor of ten in the vertical (horizontal) spectrum. Therefore, even though the $0^+ \to 1^+ \to 0^+$ transition is expected to be about a factor 30 stronger [8], a clear identification of the peak corresponding to the $0^+ \to 1^- \to 0^+$ transition is possible. The transition strength of the two states to the ground state were measured with ^{16}O(α, γ) and ^{16}O(α, α) reactions before.
During the experiment two measurements were performed on ^{20}Ne at a γ-beam energy of 11.26 MeV and a spectral width of 3%. A ^{28}Si target was placed together with the ^{20}Ne gas target in the beam. ^{28}Si has a 1^+ state at 11.45 MeV [9] which is used to extract the analysing power of the setup. In addition the ^{28}Si state together with its single escape peak

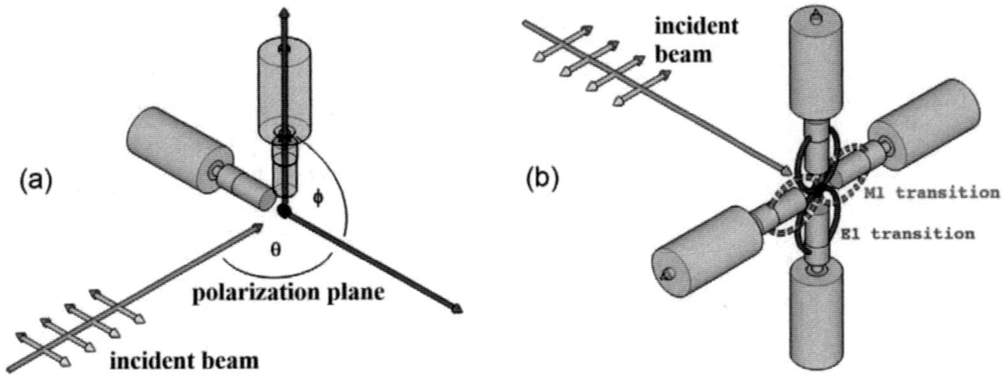

Figure 1. (a) Definition of the polar angle ϑ measured with respect to the incident horizontally polarized γ-ray-beam and of the azimuthal angle φ. (b) Schematic view of the NRF setup. The incoming beam is horizontally polarized and hits the target in the middle of the detector setup. The dashed curve shows the emission in case of a M1 transition and the solid curve illustrates the emission in case of an E1 transition.

deliver a consistent energy calibration for each detector. In one measurement a linearly polarized photon beam was used to separate the parity doublet. To get a consistent energy calibration for all detectors we performed a second measurement with circularly polarized photon beams. Here, the angular distributions of the scattered γ-rays have no azimuthal dependence and the deexciting transitions of the parity doublet and of the excited states in the calibration target have the same intensities in every detector. The obtained spectra are shown in Fig. 2. The HPGe detectors have an energy resolution of 7.5 keV at 11.4 MeV.

3. Analysis

The parity doublet states are too close in energy to resolve them individually within the detector resolution of 7.5 keV. However, the positions of the centroids of the parity doublet shift when switching from circularly to linearly polarised photon beams because depending on the azimuthal observation angles with respect to the polarization plane of the incident γ-ray beam, the intensities of the $1^{\pi} \to 0_1^+$ E1 or M1 transition will be different. Their energy centroids $E_{vertical/horizontal}$ can be expressed with the help of known values and the energies of the doublet states

$$E_{vertical} = \frac{E(1^-)\frac{I(1^-)}{I(1^+)} + E(1^+)\frac{W_{vertical}}{W_{horizontal}}}{\frac{I(1^-)}{I(1^+)} + \frac{W_{vertical}}{W_{horizontal}}} \quad (2)$$

$$E_{horizontal} = \frac{E(1^-)\frac{I(1^-)}{I(1^+)} \cdot \frac{W_{vertical}}{W_{horizontal}} + E(1^+)}{\frac{I(1^-)}{I(1^+)} \cdot \frac{W_{vertical}}{W_{horizontal}} + 1} \quad (3)$$

with $I(1^{+/-})$ being the integrated cross sections of the $1^{+/-}$ levels and $W_{vertical/horizontal}$ representing the angular distributions function for NRF M1 $1^+ \to 0_1^+$ transitions in vertical and horizontal direction, respectively. We could derive the polarization sensitivity of our setup from the ^{28}Si $1^+ \to 0_1^+$ transition. The ratio of the integrated cross sections of the parity doublet could be determined to $\frac{I(1^-)}{I(1^+)} = 0.034(14)$ with the help of the peak areas of the parity doublet and the analysing power of the setup which was measured with the calibration target ^{28}Si. Combining this value with equations (2) and (3) we calculate the energy position of the

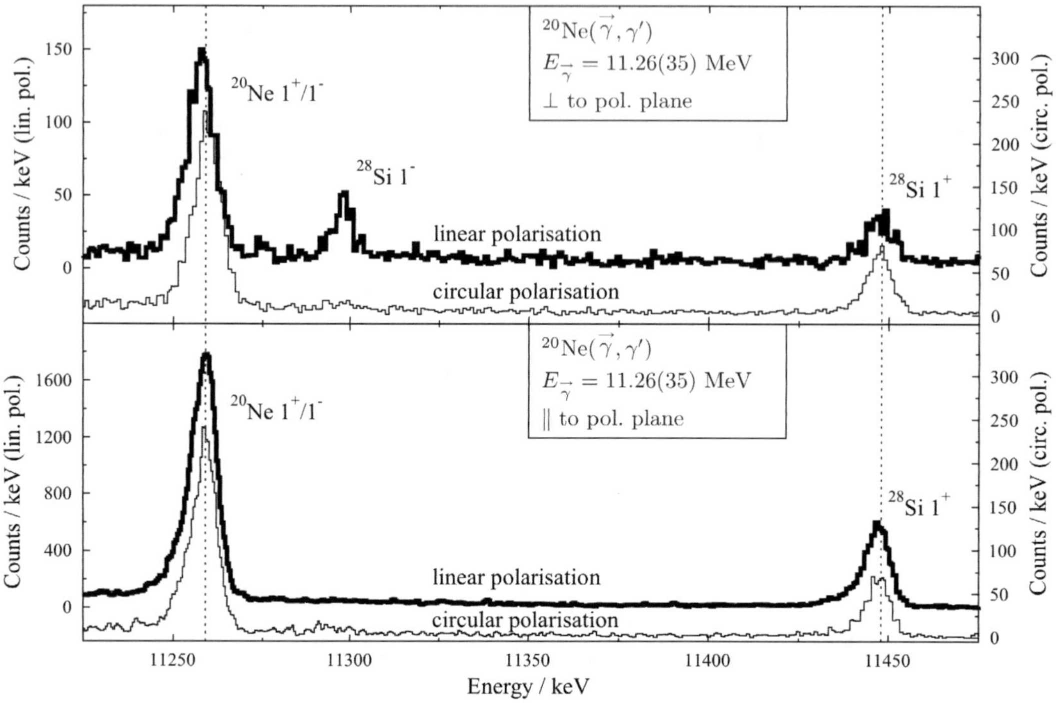

Figure 2. The summed spectra for the vertical (top) and horizontal (bottom) detectors. For a better comparison between spectra obtained either with linear polarization or with circular polarization two different scales are used to account for varying count rates data acquisition time periods. The spectra obtained with an incoming linearly polarized beam (bold) use the left axis, while the spectra obtained with an incoming circularly polarized beam (thin) use the right axis. We determine the energy difference between the ^{20}Ne doublet states from the small energy shift between linear and circular polarisation.

doublet states to:

$$\begin{aligned} E(1^+) &= \quad 11257.8(1)\ \text{keV} \\ E(1^-) &= \quad 11254.9(12)\ \text{keV} \\ \Delta E &= \quad 2.9(13)\ \text{keV}. \end{aligned} \tag{4}$$

Despite the fact, that we were able to measure the energy difference of the parity doublet with a higher precision, we also state that the energetic ordering of the two states was wrong. The 1^- state is lower in energy than the 1^+. This was only possible because we excited both states simultaneously and, therefore, were able to perform a high-precision relative measurement.

4. Summary

We reported on the status of our analysis of the energy separation of the $1^+/1^-$ parity doublet states in ^{20}Ne. Our preliminary value amounts to $|E(1^+) - E(1^-)| = 2.9(13)$ keV. This leads to a preliminary value for the effective nuclear enhancement factor of $R_N/|E(1^+) - E(1^-)| \approx$ 1800 MeV^{-1}. In addition, the energetic ordering of the two levels could be corrected. These achievements were only possible with the combination of linearly and circularly polarized photon beams and the high resolution of HPGe detectors.

This work was supported by the Deutsche Forschungsgemeinschaft (Contract No. SFB 634).

[1] Lee T D and Yang C N 1956 *Phys. Rev.* **104** 254.
[2] Wu C S *et al.* 1957 *Phys. Rev.* **105** 1413.
[3] Adelberger E G and Haxton W C 1985 *An. Rev. Nucl. Sci.* **35** 501.
[4] Desplanques B 1998 *Phys. Rep.* **297** 1.
[5] Titov A I *et al.* 2006 *J. Phys. G: Nucl. Part. Phys.* **32** 1097.
[6] Pietralla N, Weller H R, Litvinenko V N, Ahmed M W, and Tonchev A P 2002 *Nucl. Instr. and Meth. A* **483** 556.
[7] Isaak J *et al.* 2011 *Phys. Rev. C* **83** 034304.
[8] Tilley D R *et al.* 1998 *Phys. Rev.* **A 636** 249.
[9] Endt P M 1998 *Nucl. Phys. A* **633** 1.

XIX International School on Nuclear Physics, Neutron Physics and Applications IOP Publishing
Journal of Physics: Conference Series **366** (2012) 012004 doi:10.1088/1742-6596/366/1/012004

Complex nuclear spectra in a large scale shell model approach

Bianco D, Andreozzi F, Lo Iudice N, Porrino A, Knapp F §

Dipartimento di Scienze Fisiche, Universitá di Napoli Federico II, and INFN Sezione di Napoli, Napoli, Italy

E-mail: loiudice@na.infn.it

Abstract. We report on a shell model implementation of an iterative matrix diagonalization algorithm in the spin uncoupled scheme. A new importance sampling is adopted which brings the eigenvalues to convergence with about 10% of the basis states. The method is shown to be able to provide an exhaustive description of the low-energy spectroscopic properties of $^{132-134}$Xe isotopes and of the spectrum of ^{130}Xe.

1. Introduction

Large scale shell model (SM) calculations require the diagonalization of the nuclear Hamiltonian in spaces whose dimensions may exceed the capability of the most powerful computers. Hence, the necessity of developing more and more efficient matrix diagonalization codes.

Practically all of them are implementations of the Lanczos [1] algorithm. Some codes, like Nathan [2, 3] and Nushell [4], exploit a spin-coupled SM basis, others, like Antoine [3], adopt an uncoupled m-scheme. The SM Hamiltonian matrices obtained in the coupled spin have relatively small dimensions. They, however, are highly dense and composed of quite involved matrix elements. The matrices constructed in the m-scheme, though of much larger dimensions, are sparse and easy to compute.

The performance of these codes may be enhanced by resorting to sampling techniques which effectively reduce the dimensions of the Hamiltonian matrices. A successful importance sampling is the quantum Monte Carlo diagonalization (QMCD) method [5] which generates stochastically a truncated basis by exploiting the Monte Carlo technique, first adopted for studying the nuclear ground state properties [6]. The wavefunctions so generated do not have well defined spin, isospin and parity. Specific projection techniques are to be adopted in order to extract the good quantum numbers.

A new iterative algorithm, endowed with an importance sampling, was proposed few years ago [7, 8]. We have now implemented the algorithm in the m-scheme and proposed a new importance sampling [9, 10]. The upgraded method yields a large number of levels

§ Present address: Faculty of Mathematics and Physics Charles University, Prague, Czech Republic for each angular momentum and, therefore, offers a complete description of the low-lying nuclear spectroscopic properties. An illustrative example is provided by the study of the spectra of some heavy Xe isotopes, discussed in [10]. After review briefly the algorithm, we will discuss its convergence properties and present the most meaningful results of the study performed on $^{130-134}$Xe.

2. The algorithm

Let us consider a symmetric matrix $A = \{a_{ij}\}$ representing a self-adjoint operator $\hat{A}$ in an orthonormal basis $\{\mid 1\rangle, \mid 2\rangle, \ldots, |i>, \ldots, \mid N\rangle\}$. The diagonalization algorithm prescribes a sequence of iteration loops. The first loop consists of:
1a) Consider the lowest n_0 basis states $|i>$ $(n_0 << N)$, construct and diagonalize the $n_0 \times n_0$ submatrix $A_0 = (a_{ij})$,
1b) select the lowest v eigenvalues $\lambda_1^{(0)}, \ldots, \lambda_k^{(0)}, \ldots, \lambda_v^{(0)}$ and the corresponding eigenvectors $\mid \varphi_k^{(0)}\rangle = \sum_{i=1}^{n_0} c_i^{(0)} \mid i\rangle$,
1c) construct and diagonalize the matrix

$$A_k^{(1)} = \begin{pmatrix} \Lambda_{k-1}^{(1)} & B_k^{(1)} \\ B_k^{(1)^T} & A_k^{(1)^c} \end{pmatrix}, \tag{1}$$

where $\Lambda_k^{(1)} \equiv \{\lambda_1, \ldots, \lambda_i, \ldots, \lambda_v\}$ is a v-dimensional diagonal matrix and $A_k^{(1)^c} = \{a_{ij}\}$ $(i,j = (k-1)p+1, \ldots, kp)$ is a p-dimensional submatrix. The off diagonal block $B_k^{(1)}$ is composed of the matrix elements $b_{ij}^{(k)} = \langle\lambda_{(k-1)}^{(1)}(i) \mid \hat{A} \mid j\rangle$ $(i = 1, \ldots, v;\ j = (k-1)p+1, \ldots, kp)$, $B_k^{(1)^T}$ is its transpose.
1d) diagonalize $A^{(1)}$ and extract the new lowest v eigenvalues $\lambda_k^{(1)}$ $(k = 1, v)$ and the corresponding eigenvectors $\mid \varphi_1^{(1)}\rangle, \ldots \mid \varphi_v^{(1)}\rangle$,
1e) consider now the new subspace spanned by $\mid \varphi_1^{(1)}\rangle, \ldots \mid \varphi_v^{(1)}\rangle$ plus the basis states $|j>=|n_1+1>, \ldots |n_2>$, construct the new submatrix $A^{(2)}$ just as done in point 1c) for $A^{(1)}$ and, after its diagonalization, extract the new lowest v eigenvalues $\lambda_1^{(2)}, \ldots \lambda_v^{(2)}$ and the corresponding eigenvectors $\mid \varphi_1^{(2)}\rangle, \ldots \mid \varphi_v^{(2)}\rangle$.

The just outlined procedure is iterated until the full basis is exhausted. This zero approximation loop yields the approximate v eigenvalues and eigenvectors

$$E_k^{(1)} \equiv \lambda_k^{(N)}, \qquad \mid \psi_k^{(1)}\rangle \equiv \mid \varphi_k^{(N)}\rangle = \sum_{i=1}^{N} c_k^{(N)}(i) \mid i\rangle. \tag{2}$$

The basis composed of the eigenvectors $\psi_k^{(1)}$ plus the original basis states $\{\mid j\rangle\}$ are the new entries for a new iteration loop. The loops are iterated until a sequence of v vectors $\psi_1^{(i)}, \ldots, \psi_v^{(i)}$ converges to the exact eigenstates [7].

3. Implementation of the algorithm in the m-scheme

We add to the nuclear Hamiltonian H a spin term multiplied by a positive constant c

$$H_J = H + c[\mathbf{J}^2 - J(J+1)]^2. \tag{3}$$

Journal of Physics: Conference Series **366** (2012) 012004 doi:10.1088/1742-6596/366/1/012004

The Hamiltonian acts in a model space spanned by the states $|i>=|\alpha_1, \alpha_2, \ldots, \alpha_p>$, where p denotes the number of valence nucleons, $\alpha_i = \{a_i m_i\}$ the single particle (s.p.) quantum numbers and $a_i = \{n_i l_i j_i\}$ the s.p. shells. The states $|i>$ have a good magnetic quantum number $M = m_1 + \ldots + m_i + \ldots + m_p$.

Let us decompose the full space as a direct sum of subspaces $\sum_k \oplus \mathcal{H}_k$, each composed of a set of partitions $\{n_i\}_k = \{a_1^{n_1}, \ldots a_i^{n_i}, \ldots\}_k$, where $\sum_i n_i = p$. The partitions in $\mathcal{H}_k$ differ from those in $\mathcal{H}_{k-1}$ by at most two single particle shells a_i. Let us also fix a sequence of positive small numbers of decreasing values $\epsilon_1 > \ldots \epsilon_k > \ldots > \epsilon_F$.

We first diagonalize the Hamiltonian H_J in $\mathcal{H}_0$ obtaining v lowest eigenvalues $E_1^{(0)}, \ldots, E_v^{(0)}$ and eigenvectors $\psi_1^{(0)}, \ldots, \psi_v^{(0)}$ spanning a subspace $\mathcal{E}_0$. These eigenstates are exact in this subspace and have all the same spin J if the constant c is chosen so as to push the states with $J' \neq J$ up in energy. Because of its two-body nature, the Hamiltonian couples the subspace $\mathcal{E}_0$ to $\mathcal{H}_1$ only. We pick up only the basis states $|j>$ of $\mathcal{H}_1$ that fulfill the condition

$$\frac{|<j|H_J|\varphi_i^{(k-1)}>|^2}{a_{jj} - E_i^{(k-1)}} > \epsilon_k \tag{4}$$

with $k = 1$. The eigenvalue problem is thus solved in $\mathcal{E}_0 \oplus \mathcal{H}_1(\epsilon_1)$ yielding new v eigensolutions $E_k^{(1)}(\epsilon_1)$, $\psi_k^{(1)}(\epsilon_1)$ defining the subspace $\mathcal{E}_1^{(\epsilon_1)}$. We now explore the full subspace complementary to $\mathcal{H}_0 \oplus \mathcal{H}_1(\epsilon_1)$ and select all the states $|j>$ that fulfill the condition (4) with $k = 2$. The states so selected span a subspace $\mathcal{H}_2^{(\epsilon_2)} \in \mathcal{H}_0 \oplus \mathcal{H}_1 \oplus \mathcal{H}_2$. The above procedure is iterated with updated eigensolutions and decreasing sampling values ϵ_k until the full space is covered. The states obtained by such a sampling process are orthonormal and have a good **J** for a sufficiently small ϵ.

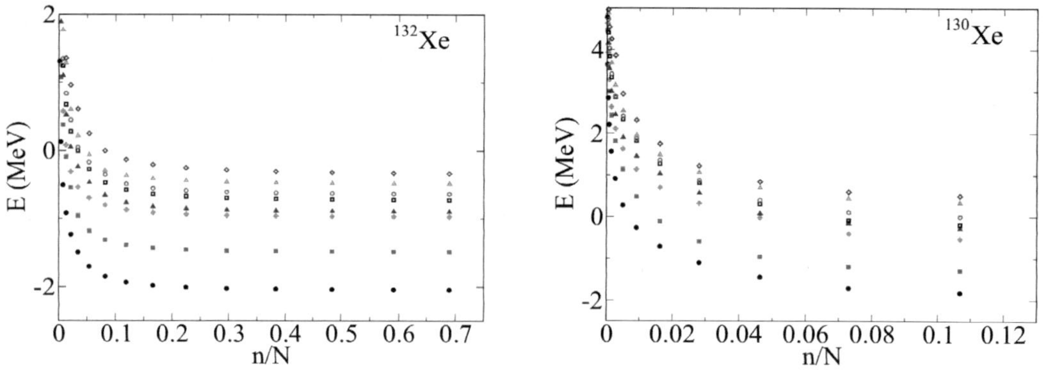

Figure 1. Convergence rate of the eigenvalues in ^{130}Xe and ^{132}Xe

4. Application to Xe isotopes

The method was applied to $^{134-130}$Xe. A series of experiments [11, 12, 13] have produced a large variety of data for these isotopes, which allowed to perform a thorough investigation of their low-energy states, and, especially, of those having a

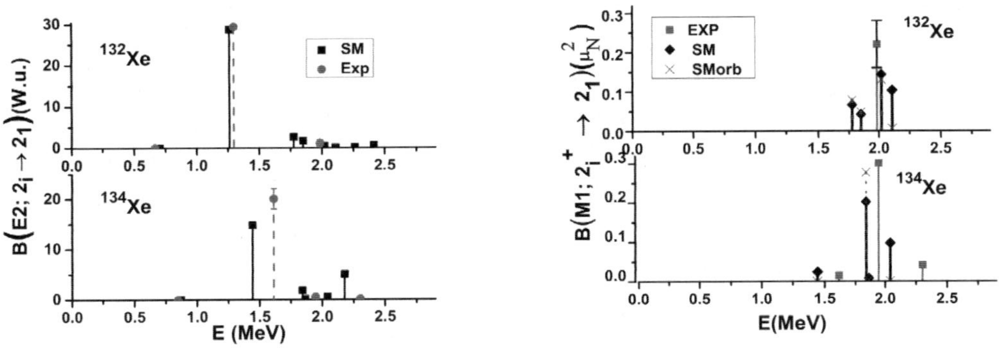

Figure 2. $B(E2; 2_i^+ \to 2_1^+)$ and $B(M1; 2_i^+ \to 2_1^+)$ strengths in ^{132}Xe

mixed symmetry (MS) with respect to the exchange of valence proton and neutron pairs [14]. These MS states were observed unambiguously in ^{94}Mo [15] and, since then, in several other nuclei in the vicinity of the N=50 and N=82 shell closures. Theoretical studies for nuclei in the vicinity of N=82 were carried out within the QPM [16, 17] and in a large scale shell model calculation [18].

The configuration space was determined by the shells $\{2d_{5/2}, 1g_{7/2}, 2d_{3/2}, 3s_{1/2}, 1h_{11/2}\}$. A G matrix [19] derived from the CD-Bonn potential [20] was taken as a two-body interaction. The dimensions of the Hamiltonian matrices range from $N \sim 3.7 \times 10^7$ for ^{132}Xe to $N \sim 0.8 \times 10^9$ for ^{130}Xe.

As shown in Figure 1, a good convergence for the energies is obtained by sampling $\sim 10\%$ of the basis states. The same is true for the strengths $B(E2; 2_1^+ \to 0_1^+)$ of the transitions to the ground state [9]. The computed spectra are in remarkable agreement with the experimental levels [10].

A larger fraction of sampled states is needed in order to reach convergence for the strengths of the transitions between excited states [10]. Since this would be too time consuming for ^{130}Xe, we have considered only the transitions of ^{132,134}Xe [10]. Here, we report some meaningful results.

The strengths of the $E2$ transitions to the ground state are in good agreement with the experiments [10]. The SM 2_1^+ collects by far the largest strength and is shown to correspond to the IBM symmetric one-boson 2^+ or, in microscopic terms, to the QPM isoscalar one-phonon quadrupole vibrational mode. The residual strength goes to the isovector $2_3^+ \to 0_1^+$ which qualifies the 2_3^+ state as MS one-boson state.

Figure 2 (left panel) shows that a very good agreement with experiments was obtained also for the $2_i^+ \to 2_1^+$ transitions. The 2_2^+ is strongly coupled to the 2_1^+ by the isoscalar $E2$ operator. It is therefore strongly collective and corresponds to a proton-neutron symmetric two-boson state.

The same Figure 2 (right panel) shows that the measured $M1$ strength is concentrated mostly, if not solely, on a single 2^+ state, while the SM strength is distributed among few 2^+ states, all clustered around a pronounced peak, lying close to the experimental one. All transitions are almost purely orbital except for the highest in energy which is due to spin. If we exclude this transition, the total SM strength coincides with the experimental value within the errors.

5. Concluding remarks

We have seen that the iterative sampling procedure brings the energy eigenvalues to convergence with about 10% of the basis states. A larger fraction of states is necessary for the convergence of the strengths of the $E2$ and, especially, the $M1$ transitions among excited states. It is, therefore, desirable to explore the possibility of developing a parallel version of the code in order to reduce the execution time.

In its present status, the algorithm can be applied to space of dimensions of several hundred millions. Within these limits, the method yields at once an arbitrary number of orthogonal eigenstates with a given J and provides an exhaustive description of the low-lying spectroscopic properties of nuclei.

Indeed, when applied $^{132-134}$Xe isotopes, the method yielded complete spectra and a full scheme of electromagnetic transitions in good agreement with the experiments. It allowed also to determine the collectivity and the proton-neutron symmetry of the low-lying states.

References

[1] Lanczos C 1950 *J. Res. Natl. Bur. Stand.* **45** 252
[2] Caurier E and Nowacki F 1999 *Acta Phys. Pol. B* **30** 705
[3] Caurier E *et al* 2005 *Rev. Mod. Phys.* **77** 427
[4] Brown B A and Raen W D M 2007 *NUSHELL@MSU, MSUNSCL Report*(unpublished)
[5] Otsuka T *et al* 2001 *Prog. Part. Nucl. Phys.* **47** 319
[6] Koonin S E, Dean D J and Langanke K 1997 *Phys. Rep.* **278** 1
[7] Andreozzi F, Porrino A and Lo Iudice N 2002 *J. Phys. A* **35** L61
[8] Andreozzi F, Lo Iudice N and Porrino A 2003 *J. Phys. G: Nucl. Part. Phys.* **29** 2319
[9] Bianco D, Andreozzi F, Lo Iudice N, Porrino A and Knapp F 2011 *J. Phys. G: Nucl. Part. Phys.* **38** 025103
[10] Bianco D, Andreozzi F, Lo Iudice N, Porrino A and Knapp F 2011 *Phys. Rev.* C **84** (2011) 024310
[11] von Garrel H *et al.* 2006 *Phys. Rev.* C **73** 054315
[12] Ahn T *et al.* 2009 *Phys. Lett.* B **679** 19
[13] Coquard L *et al.* 2010 *Phys. Rev.* C **82** 024317
[14] Arima A, Otsuka T, Iachello F and Talmi I 1977 *Phys. Lett.* B **66** 205
[15] Pietralla N *et al* 1999 *Phys. Rev. Lett.* **83** 1303
[16] Lo Iudice N, Stoyanov Ch and Tarpanov D 2008 *Phys. Rev.* C **77** 044310
[17] Lo Iudice N, Stoyanov Ch and Pietralla N 2009 *Phys. Rev.* C **80** 024311
[18] Sieja K, Martinez-Pinedo G, Coquard L and Pietralla N 2009 *Phys. Rev.* C **80** 054311
[19] Hjorth-Jensen M *et al* 1995 *Physics Reports* **261** 125
[20] Machleidt R 1989 *Adv. Nucl. Phys.* **19** 189

Journal of Physics: Conference Series **366** (2012) 012005 doi:10.1088/1742-6596/366/1/012005

Group contractions and conformal maps in nuclear structure

Dennis Bonatsos

Institute of Nuclear Physics, N.C.S.R. "Demokritos", GR-15310 Aghia Paraskevi, Attiki, Greece

E-mail: `bonat@inp.demokritos.gr`

Abstract. Group contraction is a procedure in which a symmetry group is reduced into a group of lower symmetry in a certain limiting case. Examples are provided in the large boson number limit of the Interacting Boson Approximation (IBA) model by a) the contraction of the SU(3) algebra into the $[R^5]SO(3)$ algebra of the rigid rotator, consisting of the angular momentum operators forming SO(3), plus 5 mutually commuting quantities, the quadrupole operators, b) the contraction of the O(6) algebra into the $[R^5]SO(5)$ algebra of the γ-unstable rotator. We show how contractions can be used for constructing symmetry lines in the interior of the symmetry triangle of the IBA model.

In mathematics, a conformal map is a function which preserves angles. We show how this procedure can be used in the framework of the Bohr Hamiltonian, leading to a Hamiltonian in a curved space, in which the mass depends on the nuclear deformation β, while it remains independent of the collective variable γ and the three Euler angles. This Hamiltonian is proved to be equivalent to that obtained using techniques of Supersymmetric Quantum Mechanics.

1. Group contractions

The SU(3)$\rightarrow$ $[R^5]SO(3)$ contraction [1, 2] is a procedure in which the full SU(3) algebra, consisting of 8 noncommuting generators, is shrinked into an SO(3) algebra (consisting of 3 noncommuting generators), accompanied by 5 mutually commuting operators (the quadrupole operators). This simplification occurs in the limit of large boson number in which, in SU(3), all intrinsic excitations rise in energy, isolating the ground state band so that SU(3) goes over, approximately, into a simple rigid rotator. The resulting algebraic structure is, indeed, known [3] to be the algebra of the rigid rotator.

The SU(3) commutation relations read

$$[\hat{L}_\xi, \hat{L}_\nu] = -\sqrt{2}(1\xi 1\nu|1\xi+\nu)\hat{L}_{\xi+\nu}, \qquad [\hat{L}_\xi, \hat{Q}^{(2)}_{SU(3),\nu}] = -\sqrt{6}(1\xi 2\nu|2\xi+\nu)\hat{Q}^{(2)}_{SU(3),\xi+\nu}, \tag{1}$$

$$[\hat{Q}^{(2)}_{SU(3),\xi}, \hat{Q}^{(2)}_{SU(3),\nu}] = \frac{3}{4}\sqrt{\frac{5}{2}}(2\xi 2\nu|1\xi+\nu)\hat{L}_{\xi+\nu}. \tag{2}$$

The second order Casimir operator is

$$\hat{C}_2[SU(3)] = \frac{2}{3}\left[2\hat{Q}^{(2)}_{SU(3)}\cdot\hat{Q}^{(2)}_{SU(3)} + \frac{3}{4}\hat{L}\cdot\hat{L}\right], \tag{3}$$

while its eigenvalues in the Elliott basis, (λ,μ), are

$$C_2(\lambda,\mu) = \frac{2}{3}(\lambda^2+\mu^2+\lambda\mu+3\lambda+3\mu). \tag{4}$$

If we consider SU(3) irreducible representations (irreps) with large values of $C_2(\lambda,\mu)$, that is for large boson numbers, we can rescale the quadrupole operator as

$$\hat{q}^{(2)}_{SU(3),\xi} = \frac{\hat{Q}^{(2)}_{SU(3),\xi}}{\sqrt{C_2(\lambda,\mu)}}. \tag{5}$$

The first two commutation relations remain unchanged by the rescaling, while the last one becomes

$$[\hat{q}^{(2)}_{SU(3),\xi}, \hat{q}^{(2)}_{SU(3),\nu}] = \frac{3}{4}\sqrt{\frac{5}{2}}(2\xi 2\nu|1\xi+\nu)\frac{\hat{L}_{\xi+\nu}}{C_2(\lambda,\mu)}. \tag{6}$$

Then in the limit of large values of $C_2(\lambda,\mu)$ one gets

$$[\hat{q}^{(2)}_{SU(3),\xi}, \hat{q}^{(2)}_{SU(3),\nu}] = 0. \tag{7}$$

This result, which is obtained for large boson number, is called the contraction of SU(3) to $[R^5]$SO(3), where $[R^5]$SO(3) is the algebra of the rigid rotator [3], generated by the angular momentum operators of SO(3) and the five commuting operators $\hat{q}^{(2)}_{SU(3),\xi}$, $\xi=-2,-1,0,1,2$.

An immediate consequence of Eqs. (6) and (7) is that, in the contraction limit, terms proportional to the angular momentum $\hat{L}$ can be ignored. In the IBA framework, in which $\hat{L}$ is proportional to $(d^\dagger\tilde{d})^{(1)}$, this implies that $(d^\dagger\tilde{d})^{(1)}$ terms can be ignored.

In the limit of large values of $C_2(\lambda,\mu)$ and $\lambda\geq\mu$ the intrinsic quadrupole moments become

$$q_0 = \frac{1}{2\sqrt{2}}(2\lambda+\mu+3), \qquad q_2 = \frac{1}{4}\sqrt{3(\mu-K)(\mu+K+2)}, \tag{8}$$

where K is the eigenvalue of the angular momentum projection on the body-fixed z-axis, for which $K\leq L$ is valid, as one can see from the algorithm of the SU(3)$\supset$SO(3) reduction [2]. For states with $\lambda >> L$ (therefore also $\lambda >> K$) and $\lambda >> \mu$ one then obtains [1] $q_0=\lambda/\sqrt{2}$, while q_2 becomes negligible. Since the ground state band belongs to the $(2N,0)$ irreducible representation (irrep) of SU(3), while other low-lying bands belong to irreps $(2N-4i-6j,2i)$, $i=0,1,2,\ldots$, $j=0,1,2,\ldots$ with relatively low i, j, the contraction occurs in the large N limit. Thus in the case of interest the intrinsic quadrupole moment becomes $q_0=N\sqrt{2}$.

An equivalent statement is that one can approximately replace the operator $\hat{Q}^{(2)}_{SU(3)}$ by the scalar $\lambda/\sqrt{2}$, as one can see from Eqs. (3) and (4), since the terms containing $\hat{L}$ and μ are negligible in this limit, having as a consequence that only the first term in the rhs of these equations survives. A formal justification for this replacement is given in Ref. [4], where matrix elements of the commutators of the relevant parts of the Hamiltonian with the quadrupole operator are properly considered, resulting in the appearance of the intrinsic quadrupole moment.

It should be noticed that the above results have been obtained in irreps with $\lambda >> L$, thus they regard the low lying part of the spectrum.

A similar procedure is followed in the contraction of O(6) to $[R^5]$SO(5) [5, 6]. This is a procedure in which the full O(6) algebra, consisting of 15 noncommuting generators, is shrinked into an SO(5) algebra (consisting of 10 noncommuting generators), accompanied by 5 mutually commuting operators (the quadrupole operators). The resulting algebraic structure is known [6] to be the algebra of the γ-unstable rotator.

The first contraction has been used in Ref. [4] for determining a line of approximate SU(3) symmetry inside the symmetry triangle of the IBA, as described by S. Karampagia in this school [7].

2. Conformal maps

In Ref. [8] it has been proved that the position-dependent effective mass formalism can be equivalently expressed in a curved space. We shall prove here that this connection is possible also in the case of the Bohr Hamiltonian.

Ordering the coordinates as

$$q_1 = \Phi, \quad q_2 = \Theta, \quad q_3 = \psi, \quad q_4 = \beta, \quad q_5 = \gamma, \tag{9}$$

the kinetic energy in the standard Bohr Hamiltonian [9] can be represented as

$$T = \frac{B}{2}\left(\frac{ds}{dt}\right)^2, \qquad ds^2 = g_{ij} dq_i dq_j, \tag{10}$$

the symmetric matrix g_{ij} having the form

$$(g_{ij}) = \begin{pmatrix} g_{11} & g_{12} & g_{13} & 0 & 0 \\ g_{21} & g_{22} & 0 & 0 & 0 \\ g_{31} & 0 & g_{33} & 0 & 0 \\ 0 & 0 & 0 & g_{44} & 0 \\ 0 & 0 & 0 & 0 & g_{55} \end{pmatrix}, \tag{11}$$

with

$$g_{11} = \frac{J_1}{B}\sin^2\Theta\cos^2\psi + \frac{J_2}{B}\sin^2\Theta\sin^2\psi + \frac{J_3}{B}\cos^2\Theta, \quad g_{12} = \frac{1}{B}(J_2 - J_1)\sin\Theta\sin\psi\cos\psi, \tag{12}$$

$$g_{13} = \frac{J_3}{B}\cos\Theta, \quad g_{22} = \frac{J_1}{B}\sin^2\psi + \frac{J_2}{B}\cos^2\psi, \quad g_{33} = \frac{J_3}{B}, \quad g_{44} = 1, \quad g_{55} = \beta^2, \tag{13}$$

where the moments of inertia are

$$J_k = 4B\beta^2 \sin^2\left(\gamma - k\frac{2\pi}{3}\right). \tag{14}$$

The determinant of the matrix is

$$g = \frac{J_1 J_2 J_3}{B^3}\beta^2 \sin^2\Theta = 4\beta^8 \sin^2 3\gamma \sin^2\Theta. \tag{15}$$

The relevant volume element is then

$$dV = 2\beta^4 \sin 3\gamma \sin\Theta d\Phi d\Theta d\psi d\beta d\gamma. \tag{16}$$

The inverse matrix is found to be

$$(g_{ij}^{-1}) = \begin{pmatrix} g_{11}^{-1} & g_{12}^{-1} & g_{13}^{-1} & 0 & 0 \\ g_{21}^{-1} & g_{22}^{-1} & g_{23}^{-1} & 0 & 0 \\ g_{31}^{-1} & g_{32}^{-1} & g_{33}^{-1} & 0 & 0 \\ 0 & 0 & 0 & g_{44}^{-1} & 0 \\ 0 & 0 & 0 & 0 & g_{55}^{-1} \end{pmatrix}, \tag{17}$$

with

$$g_{11}^{-1} = \frac{B}{\sin^2\Theta}\left(\frac{\cos^2\psi}{J_1} + \frac{\sin^2\psi}{J_2}\right), \qquad g_{12}^{-1} = -B\left(\frac{1}{J_1} - \frac{1}{J_2}\right)\frac{\sin\psi\cos\psi}{\sin\Theta}, \qquad g_{44}^{-1} = 1, \tag{18}$$

$$g^{-1}_{13} = -B\left(\frac{\cos^2\psi}{J_1}+\frac{\sin^2\psi}{J_2}\right)\frac{\cot\Theta}{\sin\Theta},\quad g^{-1}_{22} = B\left(\frac{\sin^2\psi}{J_1}+\frac{\cos^2\psi}{J_2}\right),\quad g^{-1}_{55}=\frac{1}{\beta^2}, \tag{19}$$

$$g^{-1}_{23} = B\left(\frac{1}{J_1}-\frac{1}{J_2}\right)\cot\Theta\sin\psi\cos\psi,\qquad g^{-1}_{33} = B\left(\frac{\cos^2\psi}{J_1}+\frac{\sin^2\psi}{J_2}\right)\cot^2\Theta+\frac{B}{J_3}. \tag{20}$$

Using these matrix elements and the value of the determinant from Eq. (15) in the usual Pauli–Podolsky prescription [10]

$$(\nabla\Phi)^i = g^{ij}\frac{\partial\Phi}{\partial x^j},\qquad \nabla^2\Phi = \frac{1}{\sqrt{g}}\partial_i\sqrt{g}g^{ij}\partial_j\Phi, \tag{21}$$

we obtain

$$T = -\frac{\hbar^2}{2B}\left[\frac{1}{\beta^4}\frac{\partial}{\partial\beta}\beta^4\frac{\partial}{\partial\beta}+\frac{1}{\beta^2\sin 3\gamma}\frac{\partial}{\partial\gamma}\sin 3\gamma\frac{\partial}{\partial\gamma}-\frac{1}{4\beta^2}\sum_{k=1,2,3}\frac{Q_k^2}{\sin^2\left(\gamma-\frac{2}{3}\pi k\right)}\right], \tag{22}$$

where Q_k are the components of the angular momentum in the intrinsic frame

$$Q_x = -i\left(-\frac{\cos\psi}{\sin\Theta}\frac{\partial}{\partial\Phi}+\sin\psi\frac{\partial}{\partial\Theta}+\cot\Theta\cos\psi\frac{\partial}{\partial\psi}\right), \tag{23}$$

$$Q_y = -i\left(-\frac{\sin\psi}{\sin\Theta}\frac{\partial}{\partial\Phi}+\cos\psi\frac{\partial}{\partial\Theta}-\cot\Theta\sin\psi\frac{\partial}{\partial\psi}\right),\quad Q_z = -i\frac{\partial}{\partial\psi}. \tag{24}$$

The connection between the position-dependent effective mass and curved spaces has been considered in Ref. [8]. According to the findings of Ref. [8], one expects in the present case all elements of the matrix (11) to be divided by f^2, leading to $g'_{ij} = g_{ij}/f^2$. As a result, the determinant of the matrix will be $g' = g/f^{10}$, and the volume element will be $dV' = dV/f^5$. The elements of the inverse matrix will be $g'^{-1}_{ij} = f^2 g^{-1}_{ij}$.

In Ref. [11] the Bohr equation with a mass depending on the deformation $B(\beta) = B_0/(f(\beta))^2$, where B_0 is a constant, has been considered, leading to a modified Bohr equation of the form

$$\left[-\frac{1}{2}\frac{\sqrt{f}}{\beta^4}\frac{\partial}{\partial\beta}\beta^4 f\frac{\partial}{\partial\beta}\sqrt{f}-\frac{f^2}{2\beta^2\sin 3\gamma}\frac{\partial}{\partial\gamma}\sin 3\gamma\frac{\partial}{\partial\gamma}+\frac{f^2}{8\beta^2}\sum_{k=1,2,3}\frac{Q_k^2}{\sin^2\left(\gamma-\frac{2}{3}\pi k\right)}+v_{eff}\right]\Psi = \epsilon\Psi, \tag{25}$$

where reduced energies $\epsilon = B_0E/\hbar^2$ and reduced potentials $v = B_0V/\hbar^2$ have been used, with

$$v_{eff} = v(\beta,\gamma)+\frac{1}{4}(1-\delta-\lambda)f\nabla^2 f+\frac{1}{2}\left(\frac{1}{2}-\delta\right)\left(\frac{1}{2}-\lambda\right)(\nabla f)^2. \tag{26}$$

According to Ref. [8], in order to obtain the Schrödinger equation in the form of Eq. (25), one has to start with the equation

$$H_g\tilde{\Psi} = \left[-\frac{1}{2}\nabla^2+u_g\right]\tilde{\Psi} = \left[-\frac{1}{2}\frac{1}{\sqrt{g'}}\partial_i\sqrt{g'}g'^{-1}_{ij}\partial_j+u_g\right]\tilde{\Psi},\qquad \tilde{\Psi} = f^{5/2}\Psi, \tag{27}$$

while reduced energies and reduced potentials are used, as in Eq. (25). The exponent in the last equation is related to the dimensionality of the space.

Substituting the g' matrix elements and determinant in Eq. (27), and performing the relevant calculation (which closely resembles the pure Bohr case, except for the 44-term), we see that Eqs. (27) and (25) do coincide with

$$u_g = u_{eff} + f\ddot{f} - 2(\dot{f})^2 + 4\frac{f\dot{f}}{\beta}, \quad \dot{f} = \frac{df}{d\beta}, \quad \ddot{f} = \frac{d^2 f}{d\beta^2}. \tag{28}$$

This result has several important consequences.

1) It becomes clear that solving the Schrödinger equation (25) with deformation dependent mass is equivalent to solving a modified Bohr equation (27) with different metric matrix g' and another effective potential, u_g. Between the two equivalent schemes, one chooses to solve Eq. (25) instead of Eq. (27), just because the former can be solved analytically through the use of SUSYQM techniques.

2) The wave functions $\tilde{\Psi} = f^{5/2}\Psi$ are accompanied by the volume element $dV' = dV/f^5$. As a result

$$\int \tilde{\Psi}^* \tilde{\Psi} dV' = \int (f^{5/2}\Psi^*)(f^{5/2}\Psi)\frac{dV}{f^5} = \int \Psi^*\Psi dV, \tag{29}$$

i.e., the wave functions Ψ of the deformation dependent mass problem correspond to the usual Bohr volume element dV.

3) The simple relation between $\tilde{\Psi}$ and Ψ also shows that the wave functions Ψ satisfy the well-known 24 symmetries of Bohr wave functions [9], which the wave functions $\tilde{\Psi}$ satisfy by construction. If these symmetries were not satisfied, the solutions could not have been used for the description of nuclei.

The solution of Eq. (25) using SUSYQM techniques [11] is described in this school by P. E. Georgoudis [12].

References

[1] Le Blanc R, Carvalho J and Rowe D J 1984 *Phys. Lett. B* **140** 155
[2] Rowe D J 1996 *Prog. Part. Nucl. Phys.* **37** 265
[3] Ui H 1970 *Prog. Theor. Phys.* **44** 153
[4] Bonatsos D, Karampagia S and Casten R F 2011 *Phys. Rev. C* **83** 054313
[5] Meyer-ter-Vehn J 1979 *Phys. Lett. B* **84** 10
[6] Elliott J P, Park P and Evans J A 1986 *Phys. Lett. B* **171** 145
[7] Bonatsos D, Karampagia S and Casten R F, these proceedings.
[8] Quesne C and Tkachuk V M 2004 *J. Phys. A: Math. Gen.* **37** 4267
[9] Bohr A 1952 *Mat. Fys. Medd. K. Dan. Vidensk. Selsk.* **26** no. 14
[10] Podolsky B 1928 *Phys. Rev.* **32** 812
[11] Bonatsos D, Georgoudis P E, Lenis D, Minkov N and Quesne C 2011 *Phys. Rev. C* **83** 044321
[12] Bonatsos D, Georgoudis P E, Lenis D, Minkov N and Quesne C, these proceedings.

Journal of Physics: Conference Series **366** (2012) 012006 doi:10.1088/1742-6596/366/1/012006

Scaling ideas in neutrino scattering reactions: application to the MiniBooNE experiment

J.A. Caballero[1], J.E. Amaro[2], M.B. Barbaro[3], T.W. Donnelly[4], J.M. Udías[5]

[1]Departamento de Física Atómica, Molecular y Nuclear, Universidad de Sevilla, 41080 Sevilla, SPAIN
[2]Departamento de Física Atómica, Molecular y Nuclear, Universidad de Granada, 18071 Granada, SPAIN
[3]Dipartimento di Fisica Teorica, Università di Torino and INFN, Sezione di Torino, Via P. Giuria 1, 10125 Torino, ITALY
[4]Center for Theoretical Physics, Laboratory for Nuclear Science and Department of Physics, Massachusetts Institute of Technology, Cambridge, MA 02139, USA
[5]Grupo de Fíisica Nuclear, Departamento de Física Atómica, Molecular y Nuclear, Universidad Complutense de Madrid, 28040 Madrid, SPAIN

Abstract. An exhaustive analysis of the world data on electron scattering has shown that scaling and superscaling properties are fulfilled with great accuracy. The relativistic impulse approximation (RIA) yields a satisfactory description of experimental data. Not only scaling and superscaling behavior emerge, but also the specific asymmetric shape of the experimental scaling function is reproduced. In this work two relativistic approaches are considered to evaluate the neutrino-nucleus cross sections. One, based on the relativistic impulse approximation, relies on the microscopic description of nuclear dynamics using relativistic mean field theory, and incorporates a description of final state interactions. The second is based on the superscaling behavior exhibited by electron scttering data and its applicability, due to the universal character of the scaling function, to the analysis of neutrino scattering reactions. The role played by the vector meson-exchange currents in the two-particle two-hole sector is also incorporated and the results obtained are compared with the recent data for neutrinos measured by the MiniBooNE Collaboration.

1. Introduction

The data on muon neutrino charged-current quasielastic (CCQE) cross sections recently obtained by the MiniBooNE collaboration [1], and its comparison with several theoretical calculations, have led to an important debate concerning the role played by various ingredients entering in the description of the reaction. Although no definitive conclusions are yet in hand, a detailed study of modeling versus experiment for inclusive quasielastic electron scattering and its extension to neutrino processes can shed light on the different interpretations of the discrepancy between theory and experiment.

When compared with MiniBoone CCQE data, the Relativistic Fermi Gas (RFG) model underestimates the total cross section unless an axial mass M_A of the order of 1.35 GeV/c^2 is employed. This value is considerably larger than the accepted world average value [2]. However, although the RFG incorporates a fully relativistic treatment, its description of the nuclear dynamics is clearly too crude to draw specific conclusions on the value of the anomalous axial mass. At the level of the impulse approximation, a number of much more sophisticated descriptions of the nuclear dynamics also underpredict the measured CCQE cross section. Among the difficulties that one faces when comparing models, is that the effect of the ingredients in the model, such as interactions in the final state (FSI), may differ greatly from model to model.

A systematic analysis of the world inclusive (e,e') data has clearly demonstrated that, for sufficiently large momentum transfers, at energy transfers below the QE peak the property of superscaling works rather well [3, 4, 5, 6]. Moreover, from the longitudinal response a phenomenological scaling function has been extracted that shows a clear asymmetry with a long tail extended to larger energy transfers. Assuming the scaling function to be universal, *i.e.*, valid for electromagnetic and weak interactions, in [7, 8] CCQE neutrino-nucleus cross sections were evaluated by using the scaling function extracted from (e,e') data and multiplying it by the corresponding elementary weak cross section. This approach, denoted simply as "SuSA", provides nuclear-model-independent neutrino-nucleus cross sections, but its reliability rests on a basic assumption: the scaling function (extracted from longitudinal (e,e') data) is appropriate for all of the various weak responses involved in neutrino scattering, and is independent of the vector or axial nature of the nuclear current entering the hadronic tensor.

In a recent paper [9] SuSA predictions have been compared with the MiniBooNE data for the double-differential neutrino cross section showing a systematic discrepancy between theory and experiment. Inclusion of 2p-2h Meson Exchange Current (MEC) contributions yields larger cross sections and accordingly better agreement with the data. However, theory still lies below the data at larger angles where the cross sections are smaller. Before drawing definitive conclusions on the anomalous axial mass, it is important to explore alternative approaches that have been shown to be successful in describing inclusive QE (e,e') processes. This is the case for the Relativistic Mean Field (RMF), where a fully relativistic description of the process is incorporated, and FSI are taken into account by using the same relativistic scalar and vector energy-independent potentials considered in the description of the initial bound states.

2. Analysis of results

In this section we discuss the results obtained with the different approaches considered and compare with the experimental data. Details have been presented in previous works. In particular, the SuSA approach and its extension to CC neutrino reactions can be reviewed in [8], whereas the basic ingredients entering in the RMF model applied to inclusive electron and CCQE neutrino reactions are given in [10, 11, 12, 13].

In Fig. 1 we show the double-differential cross section averaged over the neutrino energy flux as a function of the muon kinetic energy T_μ. In each panel the results have been averaged over the corresponding angular bin of $\cos\theta$. In all cases we use the standard value of the nucleon axial mass, *i.e.*, $M_A = 1.03$ GeV/c^2. We compare the theoretical results evaluated using the three approaches, SuSA (green line), SuSA+MEC (blue) and RMF (red), with the MiniBooNE data [1]. We show that the 2p-2h MEC increase the cross section, yielding results that are closer to experiment, specifically, for data up to $\cos\theta \sim 0.6$. At larger angles, the discrepancy with experiment becomes larger while, on the other hand, the role of MEC is seen to be less significant, that is, the difference between SuSA and SuSA+MEC becomes smaller as the scattering angle increases.

Cross sections evaluated with the RMF model also yield reasonable agreement with data for smaller angles, the discrepancy becoming larger as θ increases. However, some differences emerge from the comparison between the RMF and SuSA predictions. As observed, RMF cross sections are in general larger than the SuSA ones. In particular, in the region close to the peak in the cross section, the RMF result becomes larger than the one obtained with SuSA+MEC. This holds especially for large scattering angles. On the contrary, SuSA and SuSA+MEC get more

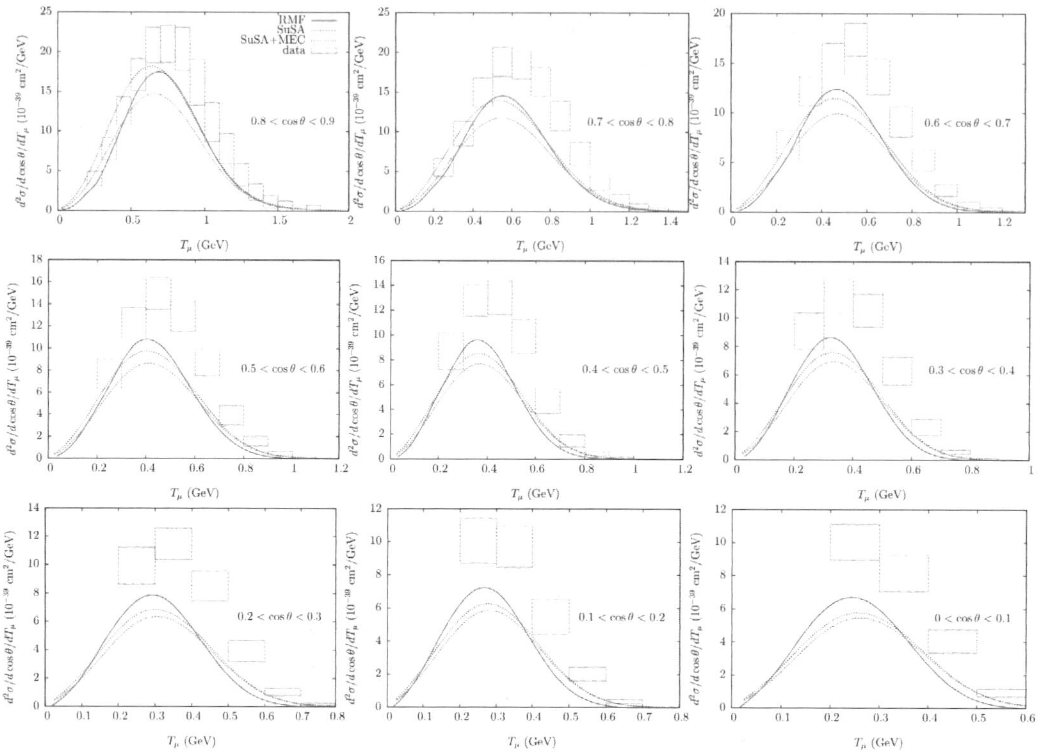

Figure 1. (color online) Flux-integrated double-differential cross section per target nucleon for the ν_μ CCQE process on ^{12}C evaluated in the SuSA (green line), SuSA+MEC (blue) and RMF (red) models and displayed versus the muon kinetic energy T_μ for various bins of $\cos\theta$. The data are from MiniBooNE [1].

strength in the region of high muon kinetic energies. This can be attributed to the breakdown of zeroth-kind scaling in the RMF, i.e., the longitudinal and transverse scaling functions to be not equal.

In Fig. 2 we plot the neutrino-flux-averaged cross section versus the scattering angle at fixed T_μ (averaged over each bin). For low muon momenta the three models tend to underestimate the data, improving the agreement as T_μ increases. As observed, when added to the SuSA results, the 2p-2h MEC yield an enhancement of the cross section whose magnitude increases for more forward scattering angles. With respect to comparison with data, the RMF also underestimates the data at large muon scattering angles, particularly for small T_μ. However, some important differences between RMF and SuSA-based models also emerge. In the six panels presented in Fig. 2 RMF cross sections are the lowest for the smallest values of $\cos\theta$. As we move to more positive $\cos\theta$, the RMF cross section grows faster, lying above the results corresponding to SuSA+MEC in the intermediate region. Finally, for smaller values of the scattering angle, namely $\cos\theta$ approaching 0.9, while RMF inverts its behavior and decreases very rapidly, SuSA and SuSA+MEC approaches to $\cos\theta = 0.9$ show a much softer slope. In fact, this is the region where the discrepancy between RMF and SuSA-based models can be better appreciated. It is very illustrative to point out that the general shape presented by the RMF cross section as a function of $\cos\theta$ fits perfectly well the shape shown by data, although RMF predictions fall below the data for small muon momenta. The different behaviour of the models is partly due to the fact that the RMF is better describing the low-energy excitation region whereas, as already

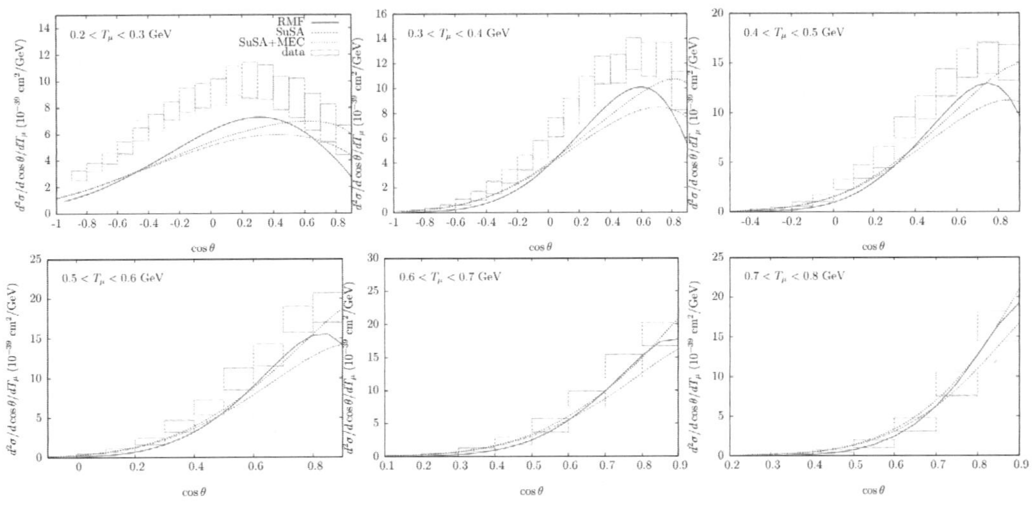

Figure 2. (color online) Double-differential ν_μ CCQE cross section for ^{12}C integrated over neutrino flux versus the outgoing muon scattering angle for various bins of the muon kinetic energy T_μ. Results are given for RMF (red lines), SuSA (green) and SuSA+MEC (blue).

pointed out, the SuSA model has no predictive power at very low angles, where the cross section is dominated by low excitation energies and the superscaling ideas are not supposed to apply.

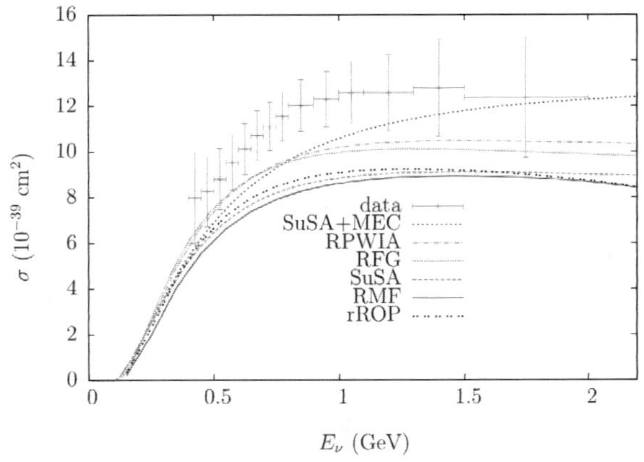

Figure 3. (color online) Total CCQE cross section per neutron versus the neutrino energy. The curves corresponding to different nuclear models are compared with the flux unfolded MiniBooNE data [1].

To conclude, in Fig. 3 we display the total QE cross section per neutron as a function of the neutrino energy and compared with the experimental data. As observed, the discrepancies between the various models tend to be washed out by the integration, yielding very similar results for the models that include FSI (SuSA, RMF and rROP), all of them giving a lower total cross section than the models without FSI (RFG and RPWIA). On the other hand the SuSA+MEC curve, while being closer to the data at high neutrino energies, has a somewhat different shape with respect to the other models, in qualitative agreement with the relativistic calculation of [14].

3. Concluding remarks

In this work we apply the RMF model to CCQE neutrino reactions on ^{12}C corresponding to the kinematics of the MiniBooNE experiment. Results for the flux-averaged double-differential cross sections are compared with data and the predictions given by SuSA and SuSA+MEC models. Generally speaking, the RMF model underestimates the data especially at large muon scattering angles and low muon energies. This was already observed with SuSA and to a somewhat lesser extent with SuSA+MEC. However, the specific behavior shown by RMF clearly differs from that of SuSA and SuSA+MEC; the maximum in $d^2\sigma/d\cos\theta dT_\mu$ as a function of T_μ for various bins of $\cos\theta$ gets higher for RMF, whereas the tail at high T_μ is more pronounced for the SuSA-based models. Also, the general trend shown by the curve corresponding to the double-differential cross section as a function of the scattering angle for bins of T_μ, clearly differs for RMF and SuSA (SuSA+MEC) approaches. Here, it is very interesting to point out that the specific shape followed by RMF predictions fits perfectly well the slope shown by data.

In spite of the discrepancies introduced by the models in the double-differential cross sections, RMF and SuSA approaches provide almost identical results for the single-differential cross section, this being found to lie below the data. Although the inclusion of 2p-2h MEC contributions increases the differential cross section being closer to data, its shape is best reproduced by the RMF and does not improve with the inclusion of the 2p-2h MEC contributions in SuSA. Finally, the impact of the 2p-2h contribution on the total cross section increases with the neutrino energy, suggesting that the data can be explained without the need for a large nucleon axial mass. However more refined calculations taking care of correlation currents, MEC effects in the axial-vector channel, *etc.*, should be performed before definitive conclusions can be drawn.

This work was partially supported by DGI (Spain): FIS2008-01143, FPA2010-1742 FIS2008-04189, by the Junta de Andalucía, by the INFN-MEC collaboration agreement, projects FPA2008-03770-E-INFN, ACI2009-1053, the Spanish Consolider-Ingenio 2000 programmed CPAN (CSD2007-00042), and part (TWD) by U.S. Department of Energy under cooperative agreement DE-FC02-94ER40818.

References

[1] A. A. Aguilar-Arevalo *et al.* [MiniBooNE Collaboration], Phys. Rev. D **81**, 092005 (2010).
[2] V. Bernard, L. Elouadrhiri and U. G. Meissner, J. Phys. G **28**, R1 (2002).
[3] D. B. Day, J. S. McCarthy, T. W. Donnelly and I. Sick, Ann. Rev. Nucl. Part. Sci. **40**, 357 (1990).
[4] T. W. Donnelly and I. Sick, Phys. Rev. Lett. **82**, 3212 (1999).
[5] T. W. Donnelly and I. Sick, Phys. Rev. C **60**, 065502 (1999).
[6] C. Maieron, T. W. Donnelly and I. Sick, Phys. Rev. C **65**, 025502 (2002).
[7] J. E. Amaro, M. B. Barbaro, J. A. Caballero, T. W. Donnelly and C. Maieron, Phys. Rev. C **71**, 065501 (2005).
[8] J. E. Amaro, M. B. Barbaro, J. A. Caballero, T. W. Donnelly, A. Molinari and I. Sick, Phys. Rev. C **71** (2005) 015501.
[9] J. E. Amaro, M. B. Barbaro, J. A. Caballero, T. W. Donnelly and C. F. Williamson, Phys. Lett. B **696**, 151 (2011).
[10] C. Maieron, M. C. Martinez, J. A. Caballero and J. M. Udias, Phys. Rev. C **68**, 048501 (2003).
[11] J. A. Caballero, Phys. Rev. C **74** (2006) 015502.
[12] J. A. Caballero, J. E. Amaro, M. B. Barbaro, T. W. Donnelly and J. M. Udias, Phys. Lett. B **653**, 366 (2007).
[13] J. A. Caballero, J. E. Amaro, M. B. Barbaro, T. W. Donnelly, C. Maieron and J. M. Udias, Phys. Rev. Lett. **95**, 252502 (2005).
[14] J. Nieves, I. R. Simo and M. J. V. Vacas, Phys. Rev. **C 83**, 045501 (2011); arXiv:1106.5374 [hep-ph].

Journal of Physics: Conference Series **366** (2012) 012007 doi:10.1088/1742-6596/366/1/012007

Simple patterns and Symmetries in Nuclei: Correlations of Mass and Spectroscopic Observables

R.F. Casten[1] and R.B. Cakirli[2,3]
[1]A.W. Wright Nuclear Structure Laboratory, Yale University, New Haven, CT 06520 USA
[2]Max-Planck-Institut für Kernphysik, Saupfercheckweg 1, D-69117 Heidelberg, Germany
[3]Department of Physics, University of Istanbul, Istanbul, Turkey

E-mail: richard.casten@yale.edu

Abstract. The relation of collective observables in deformed nuclei is discussed with particular emphasis on the correlations between masses and spectroscopic observables. The role of valence nucleon number in modulating such correlations is introduced and exemplified by data on the Pt isotopes.

1. Introduction

With the accumulation of extensive data on nuclei in the last decades, it is now possible to study correlations between various observables in ways, and to an extent, never before possible. It is also possible in many cases to inspect nuclear data over large stretches of isotopic sequences, often spanning significant fractions of a major neutron shell. In this way, one can study the evolution of structure simultaneously from several perspectives.

2. Correlations of observables

The purpose of this short discussion is to outline some current ideas along these lines. We will focus on a few key observables – namely, nuclear charge radii, two–neutron separation energies, and observables that selectively reveal collective correlations such as 2_1^+, $R_{4/2}$ and $B(E2: 2_1^+ \rightarrow 0_1^+)$ -- as well as on how nuclear structure varies with nucleon number. We start with a simple example, shown in Fig. 1, illustrating the data on $R_{4/2}$ and S_{2n} for the rare earth region, centered on the quantum phase transition around N~90. The left panel shows a clear shape change from spherical ($R_{4/2}$ values well below 3.0) to axial deformed ($R_{4/2}$ values approaching 3.33) as a function of proton number. There is a sudden change from concave to convex between $N = 88$ and 90 that forms a kind of bubble pattern [1] characteristic of the dissolution of a subshell at $Z = 64$ [2], driven by the valence *p-n* interaction. The important point for the present discussion is that this structural change is reflected as well in the S_{2n} values (Fig. 1(right)), which interrupt their characteristic nearly linear downtrend against N near N=90 when the nuclei gain binding energy due to the onset of deformation. We will now extend the discussion of such correlations to a wider set of observables.

Figure 2, taken from Ref. [3], shows the data for the five observables mentioned earlier on the left. They each behave very smoothly and regularly across the region shown. This is typical of nuclear data and reflects the striking simplicity of nuclear phenomenology. This simplicity is all the more remarkable when one considers that heavy nuclei consist of up to a couple hundred nucleons, of two types, interacting with both the strong force and the electromagnetic force, and filling about 60% of the nuclear volume in which they orbit about 10^{21} times per sec. Despite this remarkably regular behavior, the patterns on the left are each different from each other and it is not easy to use the behavior of one observable to predict or anticipate the behavior of another. However, if one takes the *differentials* of these observables, one obtains the panels on the right. Now, we again see regular behavior, but now each of the patterns resembles every other one. Previously, occasional examples of pairwise correlations have been noted (Refs. [4,5], see Ref. [3] for a summary of these) but such an extensive set of correlations as in Fig. 2 (and, for other regions, shown in Ref. [3]) has never heretofore been noted. These correlations are so similar that they have a special use in nuclei far from stability where data will be sparse: data on one or two of them [*e.g.*, S_{2n} and $E(2_1^+)$] can be used to estimate values or ranges for the others, facilitating experiments, guiding theory, and providing an expectation against which deviations can be assessed (such deviations may occur if new physical effects enter in exotic nuclei far from stability).

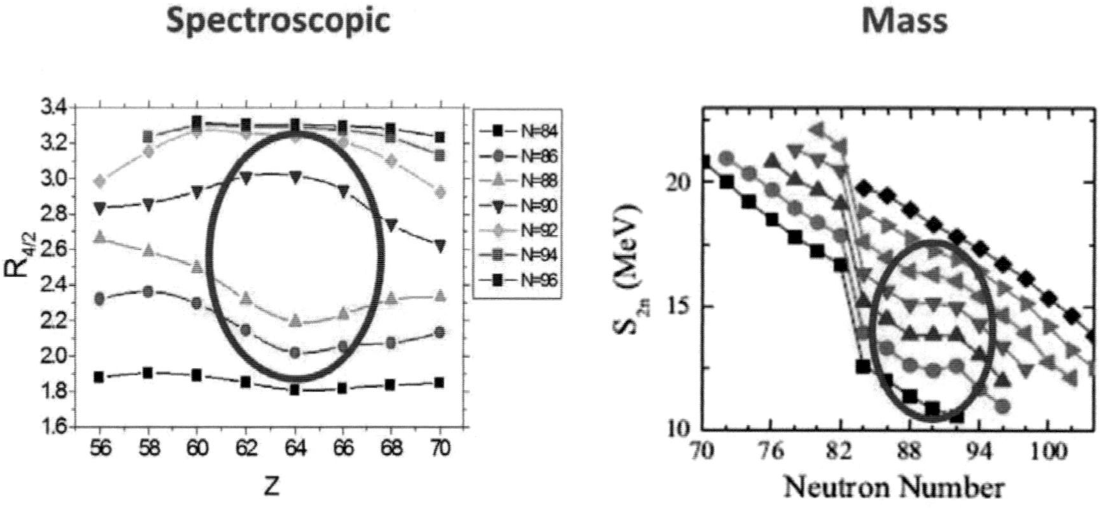

Figure 1. Experimental $R_{4/2}$ and S_{2n} values in the rare earth region showing the correlation of behavior in the N~90 quantum phase transitional region.

3. Sensitivity of masses to collectivity

The correlation of S_{2n} values with collective observables brings up an important point – namely the sensitivity of masses to structure and vice versa. That is, the collective correlations that give rise to the data on observables such as $R_{4/2}$ or $E(2_1^+)$ or $B(E2)$ values also affect the nuclear binding and therefore S_{2n}. Exploiting this linkage can give new insights. The dependence of collectivity on the *number* of valence nucleons [6,7] brings in a third ingredient that can be extremely important. We have explored this in the past [7] but it has again become significant with new data that is emerging [8]. To probe the sensitivity to valence nucleon number, one needs a collective model that depends explicitly on that number. The IBA [9] is the obvious choice. The most common IBA Hamiltonian uses two parameters (plus a scale factor). The Hamiltonian can be written as [10]:

$$H = a[(1 - \zeta)n_d - (\zeta / 4N_B)\, Q \cdot Q] \qquad (1)$$

Here, the first term gives a spherical structure and dominates for $\zeta = 0$, while the second term gives collective, deformed structures and includes a parameter χ in the Q operator. The parameter ζ varies from zero to 1 and χ from zero (γ-soft nucleus) to $-\sqrt{7}/2 = -1.32$ for an axially deformed nucleus. Clearly, the Hamiltonian reflects a competition between spherical-driving and deformation-driving terms. An important point is that the first term of Eq. 1 goes roughly as the number of valence nucleons (2 x N_B where N_B is the boson number) while the second term goes as the square of that number. Therefore, even for constant parameters, the resulting structure will tend towards deformed as the number of valence nucleons increases. Since the essence of collectivity lies in a mixing of non-collective states, with the lowest perturbed state decreasing in energy and exhibiting collective, in-phase, correlations, it is also clear that H(IBA) will involve collective contributions to binding. Thus, in general, as the number of valence nucleons increases, collectivity will increase and so will collective binding. Only the second term in H contributes to this binding, which, therefore, scales as N_B^2. This latter effect is well known but often forgotten.

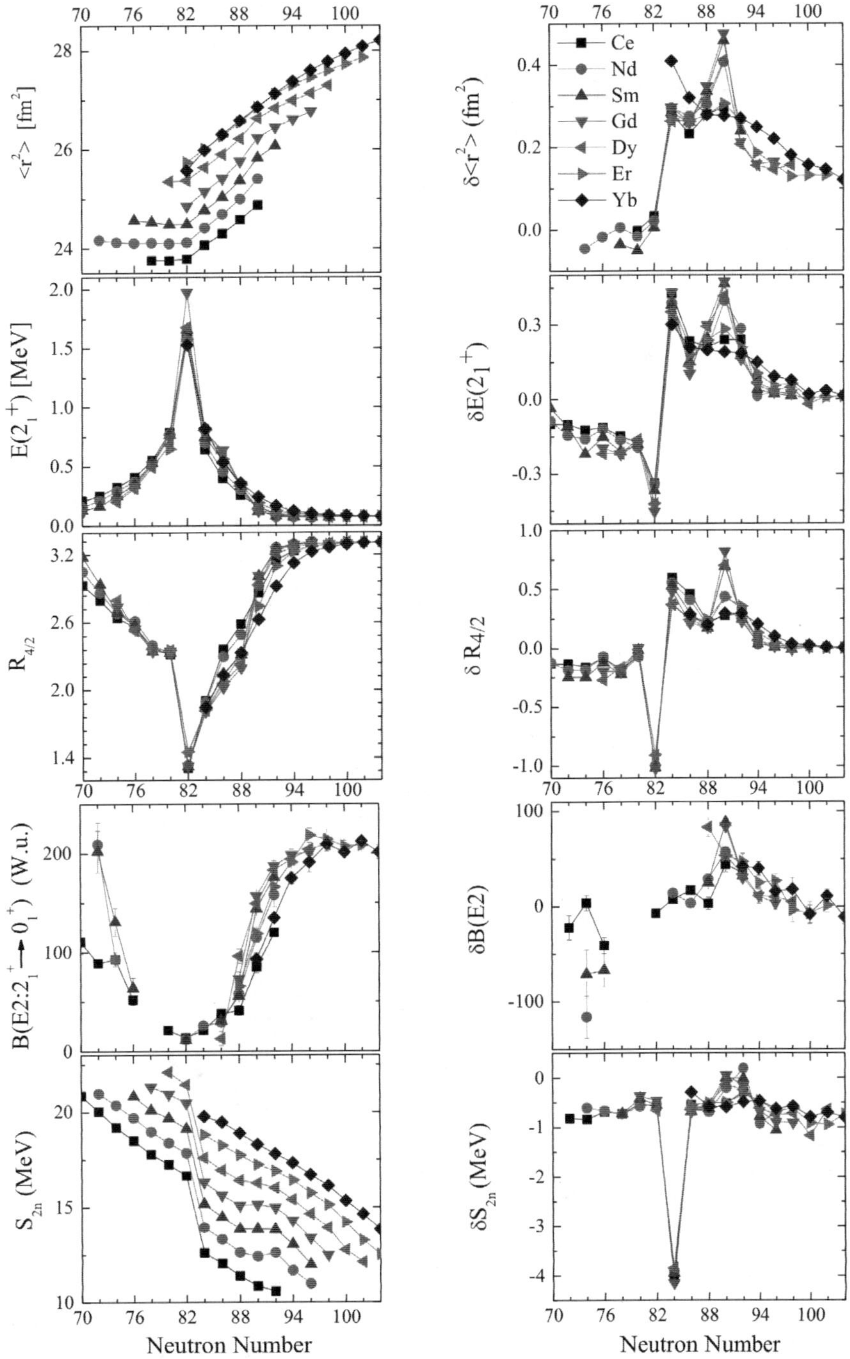

Figure 2. Normal (left) and differential (right) plots for five observables representing integral, collective and single particle aspects of structure (taken from Ref. [3]).

The dependence of collective binding on the boson number and the parameters of the Hamiltonian is illustrated in Fig 3. Note the very strong, nearly quadratic, rise in binding energy with N_B and the strong dependence on ζ and (slightly weaker) on χ as well. It is useful to illustrate these effects in a synoptic way. Since the IBA has three dynamical symmetries, U(5) – a vibrator, SU(3) – an axial rotor, and O(6) – a γ-soft rotor, it is convenient to represent its structures in terms of a symmetry triangle. The one shown in Fig. 4, for 16 bosons, is shaded according to the legend to indicate the collective binding energies. Note that they increase rapidly towards SU(3) from all directions. Similar triangles for lower boson numbers would exhibit similar shading but with a lower scale of binding.

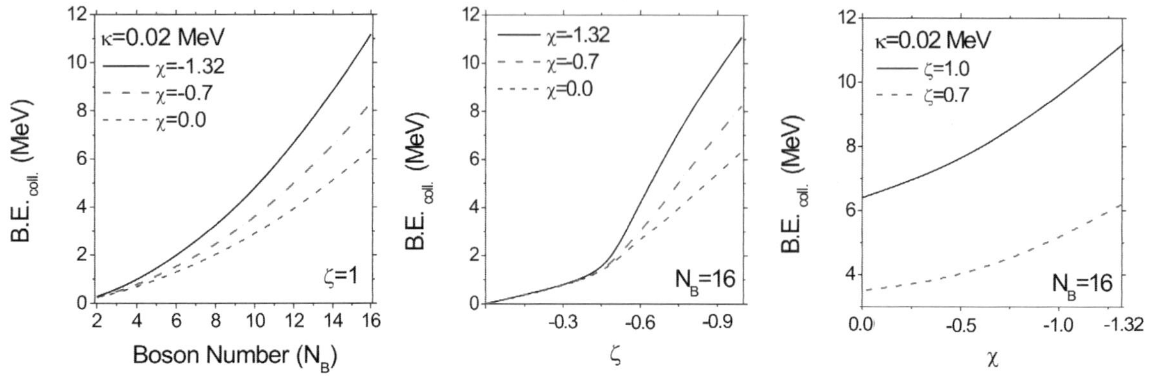

Figure 3. Dependence of the collective contributions to binding energies on boson number, and on ζ and χ in the IBA model. Based on Ref. [7].

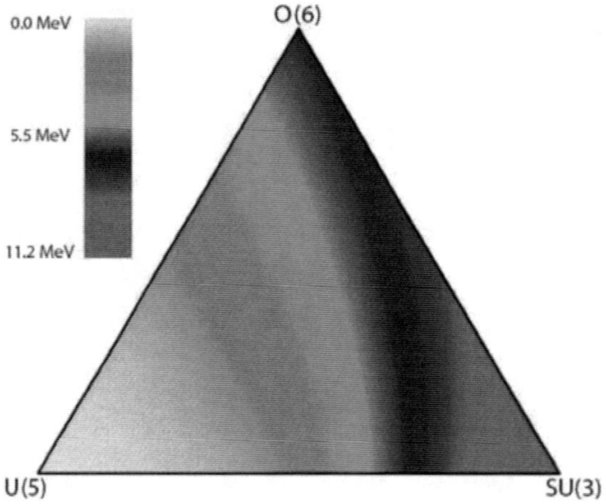

Figure 4. Symmetry triangle of the IBA showing the collective contributions to binding for calculations with 16 bosons. Based on Ref. [7].

These results have an important implication for the correlations between S_{2n} and the other observables in Fig. 2, namely the effect on S_{2n} of changes in structure *depends on the number of valence nucleons.* This is illustrated in Fig. 5 for the Pt isotopes. Let us first consider the 2^+ energies on the left. They show the characteristic behavior one typically sees across a major shell [see also Fig. 2 (left)], namely a drop in $E(2_1^+)$ after the magic number as collectivity begins to set in, then a minimizing near mid-shell and a reversal of trend in the second half of the shell. Notice that rate of the initial decline in energy abruptly changes at $N \sim 100$, then flattens, then abruptly changes again at $N = 110$, and continues upward after that, exhibiting another abrupt change in upward movement at $N = 120$. So, we have three distinct sudden changes in pattern. Now consider the plot of S_{2n}. Two of these kinks in 2^+ energies are reflected in the binding energies, namely those at $N = 98$ and 110. The one at $N = 120$ is accompanied by an almost perfectly smooth behavior in S_{2n}.

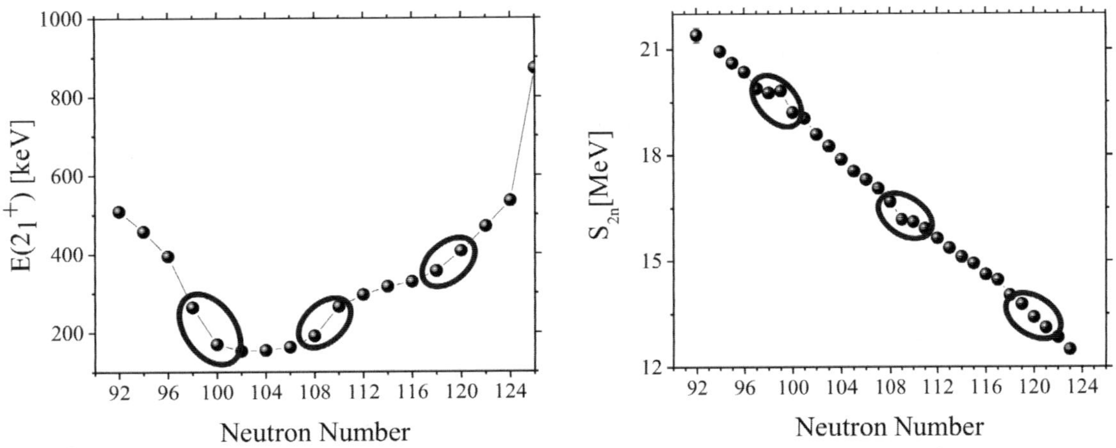

Figure 5. Empirical values of $E(2_1^+)$ and S_{2n} for the Pt isotopes (the ellipses highlight specific regions discussed in the text).

Why does this happen? The answer seems to be the following. The collective binding depends more sensitively on changes in structure for larger boson numbers than for smaller ones, and the increased sensitivity is quadratic in N_B. For $N = 98$, 110 and 120, the Pt isotopes have boson numbers $N_B = 10$, 10, and 5, respectively. So, roughly speaking, this effect alone would make S_{2n} roughly four times *more* sensitive to a given change in $E(2_1^+)$ for the lighter two Pt isotopes than for the heavier. In addition, Pt with N=120 is far less deformed than the Pt isotopes with either $N = 98$ or 110 (as can be seen from the $R_{4/2}$ values, for example) and thus will be positioned further from the SU(3) vertex, still further reducing the collective binding and hence the sensitivity of S_{2n} to structural changes. The combination of these effects will be considerably larger than a factor of four in reduced sensitivity for N=120 and hence it is natural that one would not see the structural change evidenced by the 2^+ energy reflected in S_{2n} in that case.

This type of analysis of the correlations, including explicitly the effects of valence nucleon numbers, is new and would seem to add an important new tool in our arsenal with which to understand the systematic behavior of nuclear data and the role of collectivity in various observables.

Acknowledgments

We are grateful to F. Iachello, K. Blaum, and Yu.A. Litvinov for useful and enlightening discussions. Work supported by the US DOE under Grant Number DE-FG02-91ER-40609, and the Alexander von Humboldt Foundation.

References

[1] R.B. Cakirli and R.F. Casten, Phys. Rev. C **78**, 041301(R) (2008).
[2] R.F. Casten, D.D. Warner, D.S. Brenner and R.L. Gill, Phys. Rev. Lett. **47**, 1433 (1981).
[3] R.B. Cakirli, R.F. Casten and K. Blaum, Phys. Rev. C **82**, 061306(R) (2010).
[4] A. Wolf *et al.*, Phys. Rev. C **49**, 802 (1994).
[5] I. Angeli and R.J. Lombard, Z. Phys. A **324**, 299 (1986).
[6] J.E. García-Ramos, C. De Coster, R. Fossion and K. Heyde, Nucl. Phys. A **688**, 735 (2001).
[7] R.B. Cakirli, R.F. Casten, R. Winkler, K. Blaum and M. Kowalska, Phys. Rev. Lett **102**, 082501 (2009).
[8] D. Shubina, Yu.A.Litvinov, Private Communication.
[9] A. Arima and F. Iachello, *The Interacting Boson Model*, Cambridge University Press, Cambridge, England, 1987.
[10] Pavel Cejnar, Jan Jolie and Richard F. Casten, Rev. Mod. Phys. **82**, 2155 (2010).

Journal of Physics: Conference Series **366** (2012) 012008 doi:10.1088/1742-6596/366/1/012008

One-phonon isovector $2^+_{1,\mathrm{MS}}$ state in the neutron rich nucleus ^{132}Te

M Danchev[1], G Rainovski[1,2], N Pietralla[2], A Gargano[3], A Covello[3], C Baktash[4], J R Beene[4], C R Bingham[5], A Galindo-Uribarri[4], K A Gladnishki[1], C J Gross[4], V Yu Ponomarev[2], D C Radford[4], L L Riedinger[5], M Scheck[2], A E Stuchbery[6], J Wambach[2], C -H Yu[4] and N V Zamfir[7]

[1] Faculty of Physics, St. Kliment Ohridski University of Sofia, 1164 Sofia, Bulgaria
[2] Institut für Kernphysik, TU Darmstadt, 64289 Darmstadt, Germany
[3] INFN, Complesso Universitario di Monte S. Angelo, Via Cintia, I-80126 Napoli, Italy
[4] Physics Division, ORNL, Oak Ridge, Tennessee, 37831, USA
[5] Department of Physics and Astronomy, University of Tennessee, Knoxville, TN 37996, USA
[6] Department of Nuclear Physics, ANU, Canberra, ACT 0200, Australia
[7] Wright Nuclear Structure Laboratory, Yale University, New Haven, CT 06520-8124, USA

Abstract. The 2^+_2 state in ^{132}Te is identified as the one-phonon MSS in a projectile Coulomb excitation experiment presenting a firm example of a MSS in unstable, neutron rich nuclei. The results of shell-model calculations based on the low-momentum interaction V_{low-k} are in good agreement with experiment demonstrating the ability of the effective shell-model interaction to produce states of mixed-symmetry character.

1. Introduction

Neutron-rich nuclei in the vicinity of double-magic shell closures away from the line of stability are currently of great interest. This interest is motivated by the endeavour to understand the nuclear many-body system at extreme neutron excess and fuelled by new experimental data which have become available in the recent years due to the progress made in the production of radioactive ion beams (RIBs). Nuclei with a few nucleons outside doubly closed shells play a special role. They provide direct information on the single-particle energies and the best testing ground for different components of the effective interaction used in nuclear shell model.

The one-phonon 2^+ vibrational states in even-even nuclei are the simplest collective excitations. Due to the two-fluid nature of nuclear matter they appear as a symmetric [the one-phonon 2^+_1 fully symmetric state (FSS)] or an antisymmetric combination of the involved proton and neutron configurations. Because of its isovector nature the $2^+_{1,\mathrm{ms}}$ state decays with a strong $M1$ transition to the one-phonon FSS and with a weak $E2$ transition to the ground state. This unique decay serves as an experimental fingerprint for the $2^+_{1,\mathrm{ms}}$ state. In the framework of IBM-2 [1] the one-phonon FSS and mixed-symmetry state (MSS) are orthogonal states, built on the same microscopic configurations. Moreover, it has been shown that the absolute $B(M1; 2^+_{1,\mathrm{ms}} \rightarrow 2^+_1)$ strength is highly sensitive to the proton-neutron balance of the wave functions through a mechanism dubbed configurational isospin polarization (CIP) [2]. The CIP mechanism and its manifestations have been experimentally confirmed in ^{92}Zr [3]. The case of significant CIP, which can be expected in ^{136}Te, should be manifested by comparatively small absolute $M1$ rates. The opposite case of vanishing CIP can be expected in ^{132}Te which leads to a strong $M1$ transition between the one-phonon MS and FSSs. Apparently, the CIP of one-phonon MS and FSSs in ^{132}Te and ^{136}Te allows the proton-neutron balance in their wave functions to be studied by measuring and comparing their absolute $B(M1; 2^+_{1,\mathrm{ms}} \to 2^+_1)$ values. The new experimental data on MSSs in the stable $N = 80$ isotones have revealed their evolution with increasing proton number which allows a prediction for the energy of the one-phonon MSS in the ^{132}Te to be made [4]. The fit procedure [5] used in Ref. [4] suggests the 2^+_2 level in ^{132}Te at 1665 keV as a candidate for the one-phonon MSS. Shell-model calculations [6] have corroborated this prediction.

In this study we report on the first firm experimental identification of a one-phonon MSS in the neutron-rich unstable nucleus ^{132}Te, demonstrating that the projectile Coulomb excitation of RIBs is the proper experimental technique to study the MSSs in exotic nuclei. The observed strong $B(M1; 2^+_{1,\mathrm{ms}} \to 2^+_1)$ value is in agreement with the expectations for vanishing CIP which is also confirmed by the performed shell model calculations.

2. Experimental details and discussion

^{132}Te is one of the first radioactive neutron-rich nuclei in hich the $B(E2; 2^+_1 \to 0^+_1)$ value was measured in projectile Coulomb excitation reactions of a RIB [7]. The excited states of ^{132}Te have been identified in a $\gamma - \gamma$ measurement following the β^- decay of ^{132}Sb [8]. The second excited state at 1665.3 keV discovered in the latter study decays predominantly to the 2^+_1 state by a 690.9 keV transition and to the ground state by a very weak 1665.3 keV transition $[I_\gamma(691\mathrm{keV})/I_\gamma(1665\mathrm{keV}) = 100(52)]$. The observed intensity dominance of the decay to the 2^+_1 state relative to the high-energy decay to the 0^+_1 ground state is typical for the decay of a $2^+_{1,\mathrm{ms}}$ state due to its large $B(M1; 2^+_{1,\mathrm{ms}} \to 2^+_1)$ value. The magnetic moment of the 2^+_1 state in ^{132}Te has been determined experimentally by using the technique of recoil in vacuum (RIV) after a projectile Coulomb excitation reaction on a carbon target [9]. Providing that states above the 2^+_1 are also populated in this experiment, their relative γ-ray yields with respect to the 2^+_1 state measure the relative Coulomb excitation cross-sections. This experimental information, in combination with the known $B(E2; 0^+_1 \to 2^+_1)$ value [7] can give the decay strength. This information eventually could reveal the character of the 2^+_2 state in ^{132}Te. With this idea in mind we have re-evaluated the data from the experiment described in [9].

The experiment was carried out at the HRIBF Facility at Oak Ridge National Laboratory. The ^{132}Te RIB was accelerated to 3 MeV/u and Coulomb-excited on a thick self-supporting ^{12}C target. The data were acquired for 64 hours with a beam intensity of about 3×10^7 pps. γ rays resulting from Coulomb excitation and decay of ^{132}Te nuclei were detected with the CLARION array [10]. ^{12}C ions scattered out of the target were detected in the HyBall array [10]. The used rings in HyBall covered carbon scattering angles between 7° and 44° [9]. γ-rays observed in coincidence with the ^{12}C ions detected in HyBall were corrected for Doppler shift. In order to reduce the background from uncorrelated CLARION-HyBall coincidences, the time difference between CLARION and HyBall was projected and the γ spectrum from the gate set on uncorrelated events was subtracted from the γ spectrum from the gate set on correlated events. This procedure yields the γ-ray spectrum shown in figure 1. In this spectrum all γ-rays from the radioactive decays of the beam are completely subtracted. Besides the 974-keV transition, which dominates the spectrum and represents the decay of the 2^+_1 state, we also have observed a γ-ray with energy of 691 keV (see the inset in figure 1). This peak is sharp and well pronounced, which indicates that it is emitted in flight with a speed and direction equal to those of the excited ^{132}Te ions. From this observation, together with the coincidence conditions and the background subtraction procedure, it is clear that the 691-keV γ-line represents the decay of a low-lying Coulomb-excited state of ^{132}Te. Indeed, a γ-ray with this energy is known [8] in the decay scheme of ^{132}Te. It represents the $(2^+_2) \to 2^+_1$ transition. As seen in figure 1 the 691 keV line is certainly not the known 697 keV $4^+ \to 2^+$ transition [8]. The 1665-keV level was tentatively assigned spin-parity 2^+ in Ref. [8] on the basis of observed decay to the 2_1^+ and 0_1^+ states. Given the observation of the 1665-keV level under the present experimental conditions, the only possible spin-parity assignment for it is 2^+.

From the peak areas of the 974-keV [N = 30400(500)] and the 691-keV [N = 354(102)] transitions obtained from the spectrum in figure 1 and the reported branching ratio for the decay of the 2_2^+ state [8] the relative population of the 2_2^+ state with respect to the population of the 2_1^+ state is $1.01(28) \times 10^{-2}$. It measures the relative Coulomb excitation (CE) cross sections. Erroneous target thickness were reported in Ref. [7, 9]. The correct thickness of the target in both references was 1.13(6) mg/cm^2. As a consequence, the values for the $B(E2; 0_1^+ \to 2_1^+)$ in ^{132,134,136}Te given in Table II of Ref. [7] should read 0.216(22) e^{2}b^2, 0.114(13) e^{2}b^2, 0.122(18) e^{2}b^2, respectively. The new value for the $B(E2; 0_1^+ \to 2_1^+)$ in ^{132}Te influences the adopted value for the lifetime of the 2_1^+ state and consequently the value for its magnetic moment deduced in [9, 11]. From the re-evaluated lifetime of the 2_1^+ state (τ = 2.2(5)ps) and using the RIV calibration from [11] we find $g(2_1^+) = (+)0.46(5)$. We stress that these new values do not affect the phenomenon of lowering the $B(E2)$ value in ^{136}Te and its consequences as discussed above and in Ref. [7].

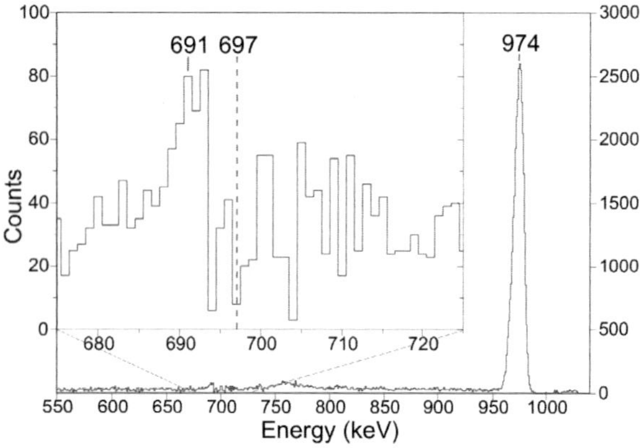

Figure 1. (Color online) Doppler corrected, background subtracted spectrum of all γ rays observed in coincidence with ^{12}C recoils detected in HyBall. For details see the text. The inset shows the existence of 691-keV transition.

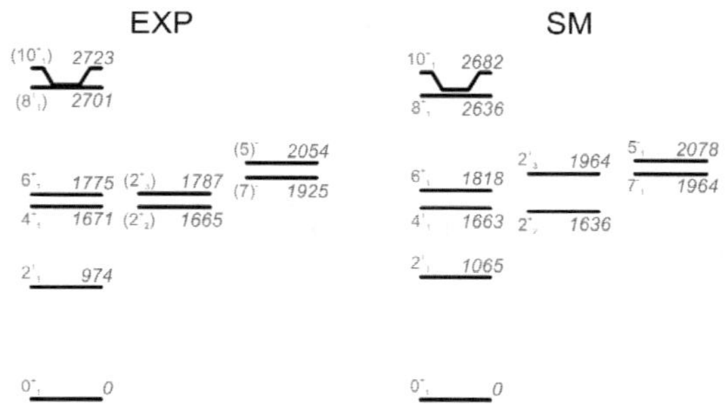

Figure 2. (Color on-line) The experimental [8] (EXP) and the calculated (SM) spectra of low-lying states of ^{132}Te.

The experimental relative population of the 2_2^+ state was fitted to the Winther-de Boer theory using a multiple CE code [12] and taking into account the energy loss of the beam in the target ($\sim$ 80 MeV). Absolute cross sections were derived using the new value $B(E2; 0_1^+ \to 2_1^+)$=0.216(22) and the branching ratio for the decay of the 2_2^+ state [8].

The combination of the Coulomb excitation yields, the known decay branching ratio, and the fact that the $B(E2; 2_2^+ \to 2_1^+)$ value is extremely unlikely to exceed the vibrational estimate of twice the $B(E2; 2_1^+ \to 0_1^+)$ implies that the 691-keV transition is predominantly a $M1$ transition. Variation of unknown $E2$ strength between 0 and 20 W.u. introduces an uncertainty in the final matrix elements of less than 8%. Unknown quadrupole moments of the 2_1^+ and the 2_2^+ states were varied between the extreme rotational limits which introduces additional uncertainties for the matrix elements of about 1%. The sizes of the resulting matrix elements are insensitive to the choice of their signs within the statistical uncertainties. The mean values of the final results are derived assuming a pure $M1$ $2_2^+ \to 2_1^+$ transition and vanishing quadrupole moments, while the estimated uncertainties account for the variations of these quantities. The final results are

Journal of Physics: Conference Series **366** (2012) 012008 doi:10.1088/1742-6596/366/1/012008

Observable	Unit	Experiment	Shell Model
$B(E2;2_1^+ \to 0_1^+)$	[W.u.]	$10(1)^a$	7.8
$\mu(2_1^+)$	$[\mu_N]$	$+0.92(10)^b$	0.68
$B(E2;2_2^+ \to 0_1^+)$	[W.u.]	$0.5(1)^c$	0.21
$B(E2;2_2^+ \to 2_1^+)$	[W.u.]	$0 \div 20^c$	0.24
$B(M1;2_2^+ \to 2_1^+)$	$[\mu_N^2]$	$5.4(3.5)^c$ ($>0.23^d$)	0.20
$\mu(2_2^+)$			0.69

Table 1. Comparison of the available experimental data on the electromagnetic properties of the 2_1^+ and the 2_2^+ states in ^{132}Te with results of the shell model calculations.
[a] From Ref. [7] and the present work.
[b] From Ref. [11, 22] and the present work.
[c] From the Coulomb excitation analysis in this work and the branching ratio of the decay of 2_2^+ from Ref. [8].
[d] From the Coulomb excitation analysis and the detectability limit for 1665-keV transition in this work.

$B(E2;2_2^+ \to 0_1^+) = 0.5(2)$W.u. and $B(M1;2_2^+ \to 2_1^+) = 5.4(3.5)\mu_N^2$.

The extremely large $B(M1)$ value has a large uncertainty which is dominated by the uncertainty of the branching ratio [8]. On the other hand, due to the use of a low-Z target the excitation processes are predominantly one step. Therefore, the result obtained for the $B(E2;2_2^+ \to 0_1^+)$ value, which is also the primary fitting parameter in the Coulomb excitation code, is more reliable. We have not observed the 1665-keV transition in our data but from the detection limit (see e.g. [13]) at 1665 keV we obtained lower limit for $I_\gamma(691\text{keV})/I_\gamma(1665\text{keV}) > 4.2$. Replacing the branching ratio of 2_2^+ state from Ref. [8] with the calculated lower estimate we establish the lower limit of $B(M1;2_2^+ \to 2_1^+) > 0.23\mu_N^2$. This value was obtained entirely on the basis of the current data. Even this lower limit for the $B(M1;2_2^+ \to 2_1^+)$ value clearly shows that the 2_2^+ state of ^{132}Te at 1665 keV is the one-phonon MSS.

^{132}Te nucleus, with two proton and two neutron holes away from the double magic nucleus ^{132}Sn, is a natural candidate to be studied within the realistic shell-model framework, which has proved to be a valuable tool to investigate nuclei in this region (see Ref. [14] and references therein). Calculations have been performed along these lines, focusing on the structure of the 2^+ states. We consider ^{132}Sn as a closed core and let the valence protons and neutron holes occupy the five levels $0g_{7/2}$, $1d_{5/2}$, $1d_{3/2}$, $2s_{1/2}$, and $0h_{11/2}$ of the 50–82 shell. The single-particle and single-hole energies have been taken from the experimental spectra of ^{133}Sb and ^{131}Sn [15], respectively. The only exception is the proton $\epsilon_{s_{1/2}}$ which has been taken from Ref. [16] the $h_{11/2}$ neutron-hole level which has been taken from Ref. [17].

The two-body component of the effective Hamiltonian has been derived within the framework of perturbation theory [18] starting from the CD-Bonn NN potential [19] renormalized by way of the V_{low-k} approach [20] with a cutoff momentum of $\Lambda = 2.2$ fm^{-1}. The shell-model calculations were performed using the OXBASH computer code [21]. The calculated energy spectrum of low-lying states of ^{132}Te is compared with experimental data in figure 2. The shell model calculations reproduce the ordering and the excitation energies of the states. The largest deviations between the calculated energies and the experimental data are 91 keV for the 2_1^+ state and 177 keV for the 2_3^+ state while for the other states the deviations are less than 50 keV. The calculated and the experimental electromagnetic properties of 2_1^+ and 2_2^+ states are summarized in Table 1.

All the experimental values in Table 1 are well reproduced except the $B(M1;2_2^+ \to 2_1^+)$ strength for which the calculations reproduce only the lower limit. The MS character of the calculated 2_2^+ state of ^{132}Te is evident from the structure of its wave function too. In terms of the basic 2^+ proton and neutron excitations the shell-model wave functions of the ground and the two lowest lying 2^+ states can be presented as follows:

$$|0_1^+\rangle = 0.94|0_1^+\rangle_\nu|0_1^+\rangle_\pi + ... \quad (1)$$

$$|2_1^+\rangle = 0.66|0_1^+\rangle_\nu|2_1^+\rangle_\pi + 0.62|2_1^+\rangle_\nu|0_1^+\rangle_\pi + ... \quad (2)$$

$$|2_2^+\rangle = 0.58|0_1^+\rangle_\nu|2_1^+\rangle_\pi - 0.63|2_1^+\rangle_\nu|0_1^+\rangle_\pi + ... \quad (3)$$

where $\pi(\nu)$ denote the respective excitations in ^{134}Te (^{130}Sn) and "..." means minor components. Equations (2,3) indicate almost equal proton and neutron contributions to the 2_1^+ and the 2_2^+ states of ^{132}Te, *i.e.* no CIP is present. This is due to almost equal energies of the basic 2^+ proton [$E_x(2_1^+$; ^{134}Te)= 1279 keV] and neutron [$E_x(2_1^+$; ^{130}Sn)= 1221 keV] configurations which even in the case of very weak pn interaction leads to one-phonon 2^+ states with a balanced proton-neutron character. The main difference between Eqs. (2) and (3) is the opposite sign of the neutron and proton components of the wave function of the 2_2^+ state [see Equation (3)] which makes it antisymmetric with respect to interchanges of proton and neutron components in the wave function. Eq.(3) represents a shell-model wave function which describes a MSS. The isovector character of this wave function leads to the relatively large $B(M1; 2_2^+ \rightarrow 2_1^+)$ value. The shell-model calculations confirm the nature of the two lowest lying 2^+ states of ^{132}Te as FS and MS states, respectively with balanced neutron and proton components.

3. Summary

In summary, by using the data from a projectile Coulomb excitation experiment we have identified the 2_2^+ state of ^{132}Te as the one-phonon MSS. This is the first case of a MSS of an unstable, neutron rich nucleus identified on the basis of a large absolute $M1$ transition strength. The experimental results prove that projectile Coulomb excitation experiments on light targets are an appropriate technique to study MSSs of exotic nuclei. The performed shell model calculations based on the V_{low-k} interaction successfully reproduce the experimental data and the isovector character of the 2_2^+ state of ^{132}Te. The shell-model wave functions of the one-phonon states have a balanced proton-neutron character as expected from the evolution of collectivity in the neutron rich tellurium isotopes around the $N = 82$ shell closure.

Acknowledgments

G.R. is a research fellow of the Alexander von Humboldt Foundation. This work was supported by the DFG (Grants No. Pi 393/2-2 and No. SFB 634), by the German-Bulgarian exchange program (Grant No. PPP 50751591), by the BgNSF (Grant No. DO 02-219), by the HIC for FAIR, by the Office of Nuclear Physics, U. S. Department of Energy Grant No. DE-FG02-96ER40983, and by the ARC grant DP0773273.

References

[1] Iachello F 1984 *Phys. Rev. Lett.* **53** 1427
[2] Holt J D, Pietralla N, Holt J W, Kuo T T S and Rainovski G 2007 *Phys. Rev.* C **76** 034325
[3] Werner V *et al.* 2008 *Phys. Rev.* C **78** 031301(R)
[4] Ahn T *et al.* 2009 *Phys. Lett.* B **679** 19
[5] Sau J and Heyde K 1981 *Phys. Rev.* C **23** 2315
[6] Sieja K, Martinez-Pinedo G, Coquard L and Pietralla N 2009 *Phys. Rev.* C **80** 054311
[7] Radford D C *et al.* 2002 *Phys. Rev. Lett.* **88** 222501
[8] Hughes R O *et al.* 2005 *Phys. Rev.* C **71** 044311
[9] Stone N J *et al.* 2005 *Phys. Rev. Lett.* **94** 192501
[10] Gross C J *et al.* 2000 *Nucl. Instrum. Methods Phys. Res.* A **450** 12
[11] Stuchbery A E and Stone N J 2007 *Phys. Rev.* C **76** 034307
[12] Alder K and Winther A 1966 *Coulomb Excitation* (New York: Academic press)
[13] Knoll G F 2000 *Radiation Detection and Measurement* 3ed., (New York: John Wiley & Sons Inc.)
[14] Covello A, Coraggio L, Gargano A and Itaco N 2011 *J. Phys.: Conf. Series* **267** 012019
[15] NNDC On-line Data Service from the ENSDF database, file revised as of May 23, 2011.
[16] Andreozzi F, Coraggio L, Covello A, Gargano A, Kuo T T S and Porrino A 1997 *Phys. Rev.* C **56** R16
[17] Fogelberg B *et al.* 2004 *Phys. Rev.* C **70** 034312
[18] Coraggio L *et al.* 2009 *Prog. Part. Nucl. Phys.* **62** 135 and references therein
[19] Machleidt R 2001 *Phys. Rev.* C **63** 024001
[20] Bogner S, Kuo T T S, Coraggio L, Covello A and Itaco N 2002 *Phys. Rev.* C **65** 051301
[21] Brown B A, Etchegoyen A and Rae W D M *The computer code OXBAH, MSU-NSCL, Report 534*
[22] Benczer-Koller N *et al.* 2008 *Phys. Lett.* B **664** 241

Journal of Physics: Conference Series **366** (2012) 012009 doi:10.1088/1742-6596/366/1/012009

Dipole and quadrupole collectivity in atomic nuclei

J. Darai[1], J. Cseh[2], N.V. Antonenko[3], G.G. Adamian[3], A. Georgieva[4]

[1] Institute of Experimental Physics, University of Debrecen, Debrecen, Pf. 105, Hungary-4010
[2] Institute of Nuclear Research, Hung. Acad. Sci., Debrecen, Pf. 51, Hungary-4001
[3] Bogoliubov Laboratory, Joint Insitute for Nuclear Research, 141980 Dubna, Russia
[4] Institute for Nuclear Research and Nuclear Energy, Bulg. Acad. Sci., Sofia 1784, Bulgaria

E-mail: darai@atomki.hu

Abstract. The interrelation of the dipole, i.e. cluster and quadrupole collectivity of atomic nuclei is studied in the examples of the superdeformed and hyperdeformed sates of $N = Z$ nuclei. Our method is largely based on symmetry-considerations.

1. Introduction

The atomic nuclei are typical examples of mesoscopic systems, in which the number of building blocks is not too small, and not too large. It is a longstanding problem how to find the relevant degrees of freedom, which are few enough such that a correct theoretical description can be based on them, yet the important features of the system are involved. Actually this task is as old as the nuclear research, and the methods which were developed for its solution are exported to other fields of research as well.

Our present understanding is that nuclear collectivity serves as a useful guide in this complex problem. There are different types of collective motion in nuclei, and they can be most easily characterized by their multipole nature.

The dipole collectivity is related to the clusterization. In this picture a nucleus is considered to be built up from two (or more) smaller nuclei. The relative motion of these clusters is characterized by a dipole vector. This kind of molecule-like (or in simple cases dumbbell-like) structure-model is a very old one. It was invented by G. Gamow [1] (as an alpha-particle model), even before the discovery of the neutron, i.e. before realizing that the nucleus consists of protons and neutrons.

Another important type of collectivity is the quadrupole one, which describes the deformation of the nuclear shape from spherical to prolate or oblate. When dealing with the quadrupole collectivity the nucleus is considered as a microscopic liquid drop, which can vibrate and rotate. This behaviour turned out to be very general, almost all nuclei show this feature.

The relation of the different structure models, or that of the different collective phenomena, like. e.g. clusterzation and quadrupole deformation is of utmost interest. Without exploring this connection we do not have a coherent theoretical understanding.

In this contribution we investigate the interrelation of the quadrupole deformation and clusterization in some examples of superdeformed and hyperdeformed states of $N = Z$ nuclei.

2. Methods of investigation

The superdeformed (SD) and hyperdeformed (HD) states of nuclei are defined by their quadrupole shape. They correspond to ellipsoids with ratios of main axes of 2:1:1, and 3:1:1, respectively. The investigation of their clusterization is important from two aspects. First it contributes to our understanding of the nuclear structure. Second it gives valuable information on the nuclear reactions, which can populate the state in question. In particular, a binary cluster configuration with the clusters in their ground intrinsic states is uniquely related to a reaction channel, like target and projectile in their ground states.

Both in our determination of the shape isomers, and in the investigation of their possible clusterizations symmetry considerations play an important role. In particular, the quasidynamical (or effective) $U(3)$ symmetry is used [2]. It is a generalization of the concept of the real $U(3)$ symmetry, known to be approximately valid in light nuclei [3]. The quasi-dynamical symmetry is more general in the following sense. The Hamiltonian breaks the symmetry in such a way that the $U(3)$ quantum numbers are not valid for its eigenvectors either (contrary to the case of the real $U(3)$ dynamical symmetry). In other words neither the operator is symmetric (i.e. it is not a $U(3)$ scalar), nor its eigenvectors (i.e. they do not transform according to a single irreducible representation) [4]. Yet, the symmetry is present is some sense.

An asymptotic Nilsson-state serves as an intrinsic state for the quasi-dynamical $SU(3)$ representation. Thus the effective quantum numbers are determined by the Nilsson states in the regime of large deformation [5]. When the deformation is not large enough, then we can expand the Nilsson-states in the asymptotic basis, and calculate the effective quantum numbers based on this expansion [6].

The $SU(3)$ quantum numbers uniquely determine the quadrupole shape of the nucleus [7], thus we obtain the shape isomers from them. In particular, a self-consistency calculation is performed with respect to the quadrupole shape of nucleus. It is based on the application of the quasidynamical $U(3)$ quantum numbers [8], and in those cases when a detailed comparison can be made with the more traditional energy-minimum calculations, the results are very similar [8, 9, 10].

Once the shape isomers have been found, the next question is how they are related to cluster configurations. To find their connection we use the Harvey prescription [11] and the $U(3)$ selection rule [10]. They can incorporate the effects of the exclusion principle, only in an approximate way, of course. But it is a well-defined way, and its validity can be checked by making a comparison with the results of the fully microscopic description, where they are available. In geometrical terms the $U(3)$ selection rule expresses the similarity of the quadrupole-deformation of the cluster configuration and the shell-model (or collective model) state.

Energetic preference represents a complementary viewpoint for the selection of clusterization. We incorporate it in two different ways: i) by applying simple binding energy arguments [12], and ii) by performing double-folding calculations, according to the dinuclear system model [13, 14].

3. Case studies

In this section we discuss some cases, which are in the focus of recent theoretical and experimental investigations.

3.1. ^{36}Ar

The SD band of the ^{36}Ar nucleus was reported in [15]. Following the experimental observation a considerable theoretical effort has been concentrated on this band. In [16] e.g. the possible binary clusterizations of this state was studied systematically. Similar studies have been done also for the ground, and the hyperdeformed bands. The latter one had been predicted from alpha-cluster model calculations [17]. One of the interesting conclusions of this work [16], was that the HD state of the ^{36}Ar nucleus could be populated in the ^{24}Mg+^{12}C and ^{20}Ne+^{16}O reactions.

A recent analysis of the ^{24}Mg+^{12}C elastic scattering [18] revealed the fact that the cross section can be described only by supposing resonances on top of the potential scattering. This very careful analysis incorporated phase-shift study, as well as Regge-pole and energy-dependent resonance calculations. The existence of five resonances have been proved, which have angular momenta 2, 4, 6, 7, 8. These states together with the resonances from the ^{20}Ne+^{16}O reactions seem to establish a rotational band [18]. Its moment of inertia is in a very good agreement with that of the HD shape predicted from alpha-cluster model [17]. The similarity of the (predicted and observed) moments of inertia, and the fact that the resonances were seen in exactly those reactions, which define the preferred cluster-configurations of the HD shape suggest that the recently observed band is a good candidate for the hyperdeformed shape isomer of the ^{36}Ar nucleus.

Since a candidate for the HD state showed up, the exciting question arises if such a shape can be seen in shell model calculation as well. In [8] we have carried out a Nilsson-model + quasi-dynamical $SU(3)$ calculation in order to find the answer. This calculation gives a HD state which has exactly the same symmetry as that from the alpha-cluster model. The moment of inertia from these models agrees well with the one indicated by the experiment.

3.2. ^{56}*Ni*

The stable elongated shapes of the ^{56}Ni nucleus, which are relevant for clusterization, have been determined from a similar Nilsson-model + quasi-dynamical $U(3)$ symmetry calculation in [19], and are shown in Figure 1. In this figure it is not the minima, rather the horizontal plateaus, which correspond to the stable shapes. (They are insensitive to the small changes of the input parameter. Furthermore, these deformations fulfill the selfconsistency argument between the input and output deformation-parameters to some approximation.)

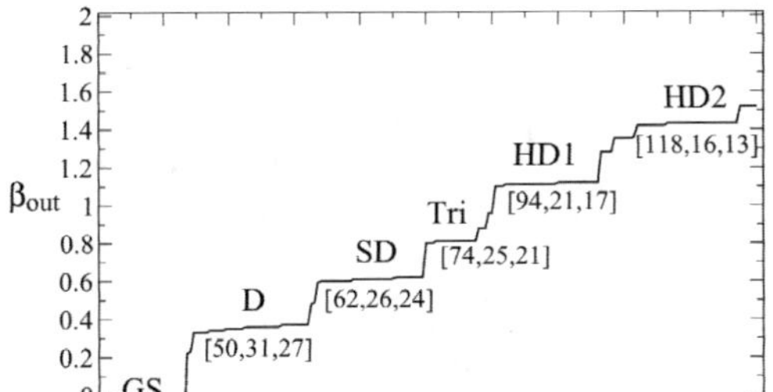

Figure 1. Quadrupole deformation of the ^{56}Ni nucleus from the Nilsson model with effective U(3) quantum numbers.

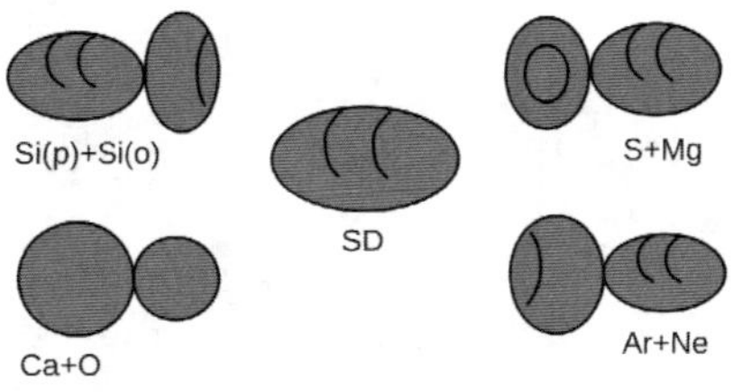

Figure 2. The superdeformed state of the ^{56}Ni nucleus (central part), and its allowed alpha-like binary cluster configurations. The energetics prefer ^{28}Si+^{28}Si and ^{40}Ca+^{16}O configurations to ^{32}S+^{24}Mg and ^{36}Ar+^{20}Ne.

The triaxial ground-state (for which the experimental deformation is $\beta_2 = 0.173$) is followed by a prolate-like deformed state of $0\hbar\omega$ excitation. The next region of stability corresponds to the superdeformed shape. This state represents 4 nucleon excitation, being very much in line with [20, 21]. Then appears an even more deformed state with triaxial shape, and two pronounced hyperdeformed shapes.

We have found that the ground state of ^{56}Ni prefers asymmetric cluster configurations, from among the alpha-like clusterization only ^{4}He+^{52}Fe is allowed. The deformed, superdeformed and largely deformed triaxial states match with several clusterizations. The allowed alpha-like binary configurations of the SD state are shown in Fig. 2. Structure considerations suggest that the correlated ^{28}Si+^{28}Si and ^{40}Ca+^{16}O resonances correspond to the superdeformed state of ^{56}Ni, but not to the hyperdeformed one. In the latter case the ^{40}Ca+^{16}O configuration has a strong structural forbiddenness [19]. The ^{24}Mg+^{32}S cluster configuration on the other hand, which is determined by the entrance channel of the ternary fission experiment [22] matches both with the SD and HD states, and with the largely deformed triaxial state in between.

The triaxial state is of special interest, because it is thought to be related to the molecular resonances of two ground-state-like (oblate) ^{28}Si clusters in their equator-to-equator configuration. If the equator-to-equator configuration is not exactly parallel, then other alpha-like binary clusterizations, like e.g. ^{24}Mg+^{32}S, are also possible.

The energetic preference of the cluster configurations were obtained from binding-energy arguments [12], and from calculations based on the Dinuclear System Model [13]. The latter ones were performed both for the pole-to-pole configurations and for the ones derived from the microscopic considerations. The ^{4}He+core configuration turned out to be the most preferred one, followed by the ^{8}Be+core one. Then a group of the ^{12}C, ^{28}Si, and ^{16}O clusters come, with close-lying values, but in different order from different calculations. The ^{24}Mg and ^{20}Ne turned out to be the least-preferred alpha-like clusters.

More detailed investigations on this nucleus is presented in [19]. In this work the results of the energetic calculation with three different methods are compared: in addition to the binding energy, and double folding, mentioned above, there the extended liquid drop model [23] was also applied.

3.3. ^{28}*Si*

Concerning the superdeformed state of the ^{28}Si nucleus, there has been an experimental candidate from [24]. Recent AMD calculations [25], as well as our Nilsson-model studies [26] predict a moment of inertia, which is different from that of the experimental data of [24]. The alpha-cluster model also gave a SD state [27] in line with these recent theoretical studies. Both AMD and our binary-cluster studies indicate the importance of the ^{24}Mg+^{4}He and ^{16}O+^{12}C clusterizations.

These circumstances initiated a critical overview of the available experimental information, and their extension [26]. It turned out that some states were populated in (α, γ) reactions, but not in (p, γ) [28]. These states represent another candidate for the rotational band with superdeformed shape. The analysis of a recent in-beam gamma-ray measurement [26] gives further support for this idea. The new candidate for the SD state has a moment of inertia very close to the theoretical predictions.

4. Summary and conclusions

In this paper we have considered the connection between the dipole and quadrupole collectivity of some states in $N = Z$ nuclei. In particular elongated shape isomers, e.g. superdeformed and hyperdeformed states and their possible binary clusterizations have been investigated. Both in finding the stable shapes and in determining their relations to cluster configurations symmetry considerations were applied extensively.

We have determined the shape isomers from the quasidynamical $U(3)$ symmetry, obtained from Nilsson-calculations. In searching for the possible binary clusterizations of the shape isomers we have taken into account both natural laws which govern the building up of a nucleus from smaller constituents: the exclusion principle and the energy-minimum principle.

This kind of investigations contribute to the understanding of the structure of the specific states, and give information on the possible reaction channels, which can populate them. Furthermore, they deepen our understanding on the connection between the different collective modes of the atomic nucleus.

Acknowledgments

This work was supported by the OTKA (Grant No. K72357), TAMOP (4.2.1./B-09/1/KONV-2010-0007/IK/IT), MTA-JINR (EAI-2009/001) and MTA-BAS (No. 7) bilateral projects.

References

[1] Gamow G 1930 *Proc. Roy. Soc.* A **126** 632
[2] Rochford P, Rowe D J 1988 *Phys. Lett.* B **210**, 5;
Rowe D J, Rochford P, Repka J 1988 *J. Math. Phys.* **29**, 572
[3] Elliott J P 1958 *Proc. Roy. Soc.* A **245** 128 and 562
[4] Cseh J 2006 *Proc. IV Int. Symp. on Quantum Theory and Symmetries (Varna)* (Sofia: Heron Press) p. 918
[5] Jarrio M, Wood J L and Rowe D J 1991 *Nucl. Phys.* A **528** 409
[6] Hess P O, Algora A, Hunyadi M, Cseh J 2002 *Eur. Phys. J.* A **15** 449
[7] Rowe D J 1985 *Rep. Prog. Phys.* **48** 1419
[8] Cseh J, Darai J, Sciani W, Otani Y, Lepine-Szily A, Benjamin E A, Chamon L C, Lichtenthaler R 2009 *Phys. Rev.* C **80** 034320
[9] Cseh J, Darai J, Algora A, Yepez-Martinez H, Hess P O 2008 *Rev. Mex. Fis.* S. **54** (3) 30
[10] Cseh J, Darai J 2008 *AIP Conf. Proc., Fusion08 (Chicago)* 1098 p 225, and references therein
[11] Harvey M 1975 *Proc. 2nd Int. Conf. on Clustering Phenomena in Nuclei* (College Park) USDERA report ORO-4856-26 p 549
[12] Buck B, Merchant A C, Perez S M 2000 Few-Body Systems **29** 53
[13] Volkov V V 1978 *Phys. Rep.* **44** 93; *Deep inelastic nuclei reactions* (Energoizdat,Moscow, 1982)
[14] Schneidman T M, Adamian G G, Antonenko N V Jolos R V, Scheid W 2002 *Phys. Lett.* **B526** 322
[15] Svensson C E, Machiavelli A O, Juodagalvis A, Poves A, Ragnarsson I, Aberg S, Appelbe D E, Austin R A E, Baktash C, Ball G C et al 2000 *Phys. Rev. Lett.* **85** 2693
[16] Cseh J, Algora A, Darai J, Hess P O 2004 *Phys. Rev.* C **70** 034311
[17] Rae W D M, Merchant A C 1992 *Phys. Lett.* **B279** 207
[18] Sciani W, Otani Y, Lepine-Szily A, Benjamin E A, Chamon L C, Lichtenthaler R, Darai J, Cseh J 2009 *Phys. Rev.* C **80** 034319
[19] Darai J, Cseh J, Antonenko N V, Royer G, Algora A, Hess P O, Jolos R V, Scheid W 2011 *Phys. Rev.* C **84** 024302
[20] Rudolph D et al 1999 *Phys. Rev. Lett.* **82** 3763
[21] Zhang J, Merchant A C, Rae W D M 1994 *Phys. Rev.* C **49** 562
[22] Von Oertzen W et al 2008 *Eur. Phys. Jour.* A **36** 279
[23] Royer G, Remaud B 1985 *Nucl. Phys.* A **444** 477
[24] Kubono S et al 1986 *Nucl. Phys.* A **457** 461
[25] Taniguchi Y, Kanada-En'yo Y, Kimura M 2009 *Phys. Rev.* C **80** 044316
[26] Jenkins D G, Darai J, Cseh J et al, in preparation
[27] Zhang J, Rae W D M, Merchant A C 1994 *Nucl. Phys.* A **575** 61
[28] Brennensein J et al 1995 *Z. Phys.* A **352** 149; ibid. p.279; ibid. p.403.

Journal of Physics: Conference Series **366** (2012) 012010 doi:10.1088/1742-6596/366/1/012010

Richardson-Gaudin description of pairing in atomic nuclei

Stijn De Baerdemacker

Ghent University, Department of Physics and Astronomy, Proeftuinstraat 86, 9000 Gent, Belgium

Department of Physics, University of Toronto, Toronto, Ontario M5S 1A7, Canada

Department of Physics, University of Notre Dame, Notre Dame, IN 46556-5670, USA

E-mail: `stijn.debaerdemacker@ugent.be`

Abstract. The present contribution discusses a connection between the exact Bethe Ansatz eigenstates of the reduced Bardeen-Cooper-Schrieffer (BCS) Hamiltonian and the multi-phonon states of the Tamm-Dancoff Approximation (TDA). The connection is made on the algebraic level, by means of a deformed quasi-spin algebra with a bosonic Heisenberg-Weyl algebra in the contraction limit of the deformation parameter. Each exact Bethe Ansatz eigenstate is mapped on a unique TDA multi-phonon state, shedding light on the physics behind the Bethe Ansatz structure of the exact wave function. The procedure is illustrated with a model describing neutron pairing in ^{56}Fe.

1. Introduction

Pairing correlations comprise an important part of the low-energy structure of atomic nuclei [1, 2]. Due to the short-range nucleon-nucleon interaction, two individual nucleons have a tendency to couple to a zero total angular momentum pair [3]. In analogy with the Bardeen-Cooper-Schrieffer (BCS) theory of superconductivity in metals [4], it was suggested that the ground state of an even-even atomic nucleus is a condensate of zero total angular momentum recoupled Cooper pairs [5]. This mechanism is capable of explaining the systematic $J^\pi = 0^+$ spin assigment of the ground state in even-even atomic nuclei, as well as the energy required to excite these nuclei from the ground state to the first excited state, the so-called pairing gap. In the nuclear shell model [6, 7], the pairing interaction is effectively modelled by means of the BCS Hamiltonian. The standard approach to solve the corresponding Schrödinger equation is to employ the BCS approximation, which assumes that the ground state is a coherent state of Cooper pairs [4]. While this assumption is valid in the thermodynamic limit, certain issues arise for finite-size systems. First, the total number of particles is not a conserved quantity in the BCS wavefunction, so pair-number fluctuations cause significant uncertainties in the ground-state energy. Second, BCS theory predicts a sharp phase transition to the superconducting phase at a critical interaction strength, in contrast with the smooth transitions obverved in finite-size systems [8]. For sufficiently small finite-size systems, the basis of Slater determinants is small enough to extract exact spectroscopic information in the full basis [9]. However, with the recent developments in superconducting nanograins [10] and exotic nuclei far from the valley of stability, there is a need for exact canonical techniques to diagonalise the BCS Hamiltonian.

Journal of Physics: Conference Series **366** (2012) 012010 doi:10.1088/1742-6596/366/1/012010

Richardson & Sherman have shown that the reduced, or level-independent, BCS Hamiltonian is exactly solvable with a Bethe Ansatz wavefunction, provided the parameters in the wavefunction are the solution of a set of non-linear algebraic equations, the so-called Richardson-Gaudin equations [11, 12]. The significance of the RG approach is that the diagonalisation of a Hamiltonian living in an exponentially scaling Hilbert space, is reduced to solving a set of algebraic equations scaling linearly with the system size. However the RG equations have proven to be very difficult to solve numberically [13], due to the presence of singular values in the interaction strengths, where the RG variables become degenerate. Recent innovative approaches [14, 15, 16, 17] have removed this obstacle, however the physical interpretation of the RG variables, including their singular behaviour, remained enigmatic.

The present contribution will discuss how the RG variables of the reduced BCS Hamiltonian are related to the elementary excitation modes of the Tamm-Dancoff Approximation (TDA), and how this can shed light on the physical nature of the RG variables.

2. Richardson's solution of the reduced BCS Hamiltonian

Consider one family of fermions, living in a single-particle Hilbert space composed of n single-particle levels. We assume rotational symmetry, so the good quantum numbers of every level are given by the set $(n_i, l_i, s_i, j_i, m_i)$, and we abbreviate the set (n_i, j_i, l_i, s_i) with the index i $(i = 1 \dots n)$ [6]. The single-particle energies and degeneracies of the levels are given by respectively ε_i and $\Omega_i = 2j_i + 1$. The reduced BCS Hamiltonian reads

$$\hat{H} = \sum_{i=1}^{n} \varepsilon_i \hat{n}_i + g \sum_{i,k=1}^{n} \hat{S}_i^\dagger \hat{S}_k, \tag{1}$$

with $\hat{S}_i^\dagger$ and $\hat{S}_i$ respectively the pair creation/annihilation operator, and $\hat{n}_i$ the number operator

$$\hat{n}_i = \sum_{m_i=-j_i}^{j_i} \hat{a}_{i,m_i}^\dagger \hat{a}_{i,m_i}, \quad \hat{S}_i^\dagger = \sum_{m_i=-j_i}^{j_i} (-)^{j_i+m_i} \hat{a}_{i,m_i}^\dagger \hat{a}_{i,-m_i}^\dagger, \quad \hat{S}_i = (\hat{S}_i^\dagger)^\dagger. \tag{2}$$

The set of operators $\{\hat{S}_i^\dagger, \hat{S}_i, \hat{S}_i^0\}$ (with $\hat{S}_i^0 = \frac{1}{2}\hat{n}_i - \frac{1}{4}\Omega_i$) spans an $su(2)$ quasi-spin algebra for the level i. The allowed $su(2)_i$ representations are $|d_i, \mu_i\rangle$ with $d_i = \frac{1}{4}\Omega_i - \frac{1}{2}v_i$ and $\mu_i = \frac{1}{2}n_i - \frac{1}{4}\Omega_i$, where v_i denotes the seniority of the level i, *i.e.* the number of particles not pairwize coupled to zero [18].

The Hamiltonian (1) can be diagonalised by means of the Bethe Ansatz product wavefunction

$$|\psi\rangle = \prod_{\alpha=1}^{N} \left(\sum_{i=1}^{n} \frac{\hat{S}_i^\dagger}{2\varepsilon_i - E_\alpha} \right) |\theta\rangle, \tag{3}$$

with N the number of pairs and $|\theta\rangle$ the particle vaccuum, provided the RG variables E_α $(\alpha = 1 \dots N)$ are a solution of the RG equations

$$1 + 2g \sum_{i=1}^{n} \frac{d_i}{2\varepsilon_i - E_\alpha} - 2g \sum_{\beta \neq \alpha}^{N} \frac{1}{E_\beta - E_\alpha} = 0 \quad (\forall \alpha = 1..N). \tag{4}$$

The associated energy of the eigenstate (3) is then given by $E = \sum_{\alpha=1}^{N} E_\alpha$.

3. The RG equations and the TDA

The TDA assumes that the eigenstates of a quantum many-body Hamiltonian can be approximated at first order by means of harmonic multi-phonon states of an elementary excitation modes [6, 19]. The elementary TDA modes of the BCS Hamiltonian are generalised pair creation operators

$$\hat{b}_k^\dagger = \sum_{i=1}^{n} Y_{ki} \hat{S}_i^\dagger. \tag{5}$$

Note that there are a total of n different TDA modes, with the lowest one ($k = 1$) refered to as the *collective* mode. Thanks to the $su(2)$ symmetry of the BCS Hamiltonian, the elementary TDA modes (5) coincide with the exact wavefunctions for one single pair (eq. (3) with $N = 1$), and consequently, the energy of the elementary TDA mode can be interpret as a RG variable for a system with 1 pair. Therefore, the multi-phonon states of the TDA seem as a good approximations for the Bethe Ansatz wave functions of the reduced BCS Hamiltonian (3). This is confirmed by calculating the overlaps of the multi-phonon states with the exact wavefunctions [20]. Depending on the interaction strength g, one finds multi-phonon states overlapping maximally with the exact ground state or first excited state of the system. For weak interaction strengths g, the overlap is largest with multi-phonon states reminiscent of the single-particle fermionic structure of the non-interacting system, whereas for strong g, the largest overlap is found for fully collective TDA states. The physical underpinning of these results is provided by the Pauli principle. By assuming that the wavefunction can be approximated by means of a multi-phonon state, we have transformed a fermion pair into a genuine boson, hence violating the Pauli principle. For a better understanding of this process, it would be interesting to adiabatically switch off the Pauli principle in the exact eigenstate (3), and investigate which TDA multi-phonon state we will recover. For this purpose, we can employ a deformation of the $su(2)$ quasi-spin [17], such that we recover the standard quasi-spin $su(2)$ in one limit of the deformation parameter and the bosonic Heisenberg-Weyl algebra $hw(1)$ in the other. This algebra is given by

$$[\hat{S}_i^0, \hat{S}_i^\dagger] = \hat{S}_i^\dagger, \quad [\hat{S}_i^0, \hat{S}_i] = -\hat{S}_i, \quad [\hat{S}_i^\dagger, \hat{S}_i] = \xi 2\hat{S}_i^0 + (\xi - 1)\frac{1}{2}\Omega_i, \tag{6}$$

with ξ acting as the deformation parameter. It can be verified that $hw(1)$ is obtained for $\xi = 0$, whereas we retain $su(2)$ for $\xi = 1$. Remarkably, if we employ the deformed algebra (6) instead of the standard $su(2)$, the reduced BCS Hamiltonian (1) remains exactly solvable along the whole path of deformation ($\xi \in [0, 1]$). The deformed RG equations are given by

$$1 + g\sum_{i=1}^{n} \frac{\frac{1}{2}\Omega_i - \xi v_i}{2\varepsilon_i - E_\alpha} - 2g\xi \sum_{\beta \neq \alpha} \frac{1}{E_\beta - E_\alpha} = 0, \quad (\forall \alpha = 1 \ldots N). \tag{7}$$

Again, for $\xi = 1$, we obtain the standard RG equations (4), whereas for $\xi = 0$, we obtain N copies of the seqular equation of TDA. So, we can use the deformed RG equations to adiabatically switch off the Pauli principle and observe to which TDA multi-phonon we converge. This has been done for the model describing neutron superfluidity of ^{56}Fe, as discussed in [14]. In Figure 1, the evolution of the ground state RG variables for an interaction strength of $g = -0.500$MeV are plotted as a function of the deformation parameter ξ. The $\xi = 1$ RG variables are given by open circles, the $\xi = 0$ TDA solutions by open squares and the reference single-particle energies by filled circles. The followed paths of every RG variable are plotted as full lines starting at the RG variables and ending at the TDA solutions. From this figure, one can deduce that the RG variables evolve towards a $[3, 3, 0, 5, 0, \ldots]$ TDA multi-phonon structure, where the notation $[\nu_1, \nu_2, \ldots, \nu_n]$ refers to the TDA multi-phonon decomposition, *i.e.* the number ν_k of elementary

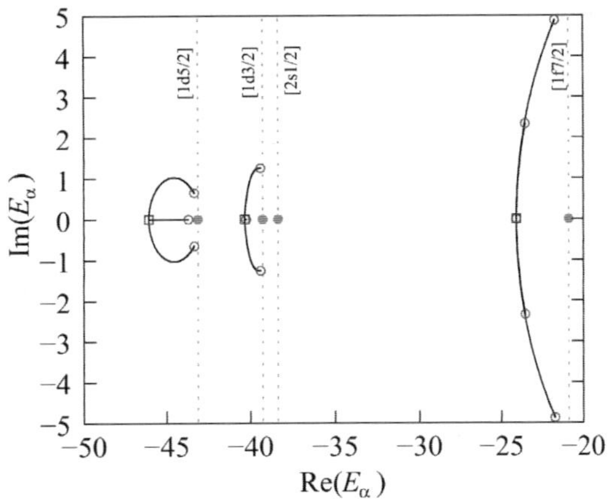

Figure 1. The RG variables (open circles) of the neutron superfluidity model of ^{56}Fe with $g = -0.5$MeV [14]. The corresponding TDA modes (open squares) are connected with the RG variables by means of full lines. The single-particle energies (full circles) are given as reference.

TDA modes k (see eq. (5)) in the TDA multi-phonon state [17]. We can repeat this procedure for every interaction strength g of the model. The results are given in Figure 2 and Table 1. Figure 2 plots the RG variables (solid lines) of the ground state as a function of the interaction strength $|g|$, as well as the 5 lowest TDA elementary modes (dashed lines). The singular values are accentuated by vertical dotted lines. The main result from the calculations is that every time the interaction strength crosses a critical point, the TDA multi-phonon decomposition changes. For instance, we can read from the Table that at $g = -0.923$MeV, the TDA decomposition changes from $[3, 3, 0, 5, 0, \ldots]$ (see Figure 1) to $[11, 0, 0, 0, 0, \ldots]$. As the latter decomposition is solely built from the collective $k = 1$ TDA mode, we can label this regime as the collective regime for the ground state.

$\lvert g_c \rvert$ [MeV]	1	2	3	4	5	6	...	10
0.000	3	2	1	4	1	0	...	0
0.245	3	3	0	4	1	0	...	0
0.272	3	3	0	5	0	0	...	0
0.923	11	0	0	0	0	0	...	0

Table 1. The TDA decomposition $[\nu_1, \nu_2, \ldots, \nu_{10}]$ of the ground state as a function of the interaction strength. The singular interaction strengths $|g_c|$ are given in the first column, and the rows give the TDA decomposition for interaction strengths larger than $|g_c|$.

4. Conclusions

In conclusion, a clear-cut connection is made between the exact Bethe Ansatz wavefunctions of the reduced BCS Hamiltonian and the multi-phonon predictions of the TDA. This is done by means of a deformation of the quasi-spin algebra, allowing us to switch off the Pauli principle in an adiabatic way. The procedure is illustrated with a model describing neutron pairing in ^{56}Fe. The singular interaction strengths mark the points where the TDA decomposition of the Bethe Ansatz wavefunction changes.

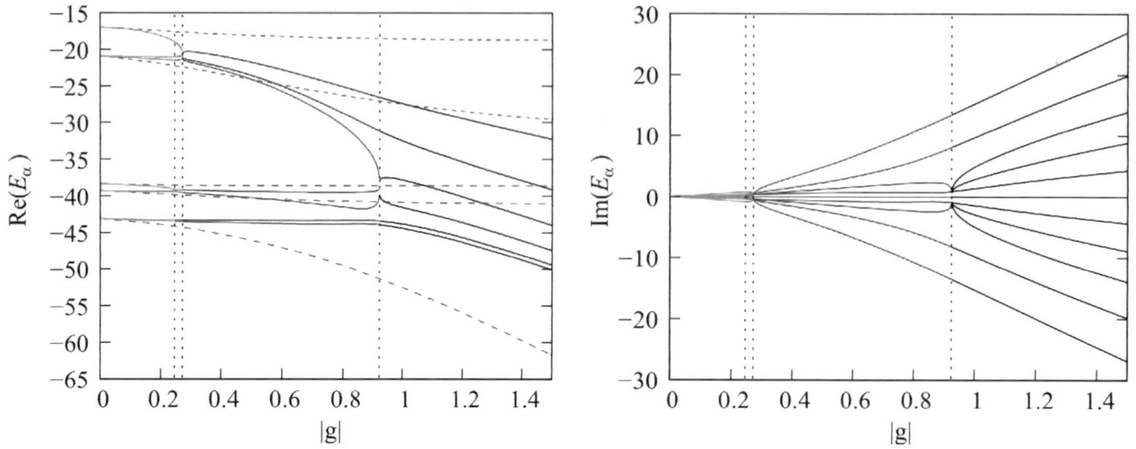

Figure 2. The Real and Imaginary part of the RG variables (solid lines) of the ground state of the neutron super fluidity model of ^{56}Fe [14] as a function of the interaction strength $|g|$. The solutions of the 5 lowest elementary TDA modes (dashed lines) are also plotted.

Acknowledgments
It is a pleasure to acknowledge P. Van Isacker, S. Rombouts, D. J. Rowe, D. Van Neck, P. Johnson, V. Hellemans and K. Heyde for many interesting discussions and suggestions. Financial support comes from the "FWO-Vlaanderen" as well as 2 scholarships for a "Long stay abroad" to support my stay at the University of Toronto and the University of Notre Dame.

References

[1] Dean D J and Hjorth-Jensen M 2003 *Rev. Mod. Phys.* **75** 607
[2] Brink D and Broglia R 2005 *Nuclear superfluidity, pairing in finite systems* (Cambridge: Cambridge University Press)
[3] Mayer M G 1950 *Phys. Rev.* **78** 22
[4] Bardeen J, Cooper L N and Schrieffer J 1957 *Phys. Rev.* **108** 1175
[5] Bohr A, Mottelson B R and Pines D 1958 *Phys. Rev.* **110** 936
[6] Heyde K 1994 *The nuclear shell model* (Berlin: Springer-Verlag)
[7] Rowe D J and Wood J 2010 *Fundamentals of Nuclear Models* (Singapore: World Scientific)
[8] Sandulescu N and Bertsch G F 2008 *Phys. Rev. C* **78** 064318
[9] Volya A, Brown B A and Zelevinsky V 2001 *Phys. Lett. B* **509** 37
[10] von Delft J and Ralph D C 2001 *Phys. Rep.* **345** 61
[11] Richardson R W 1963 *Phys. Lett.* **3** 277
[12] Richardson R W and Sherman N 1964 *Nucl. Phys.* **52** 221
[13] Richardson R W 1966 *Phys. Rev.* **141** 949
[14] Rombouts S M A, Van Neck D and Dukelsky J 2004 *Phys. Rev. C* **69** 061303(R)
[15] Faribault A, El Araby O, Sträter C and Gritsev V 2011 *Phys. Rev. B* **83** 235124
[16] Pan F, Guan X, Xie M, Bao L and Draayer J P 2011 nucl-th/1106.5237v1
[17] De Baerdemacker S 2011 *Proceedings of the 28th International Colloquium on Group Theoretical Methods in Physics (GROUP28)* (*J. Phys. Conf. Ser.* vol 284) ed Angelova M, Zakrzewski W, Hussin V and Piette B p 012020
[18] Talmi I 1993 *Simple models of complex nuclei* (Chur: Harwood academic publishers)
[19] Ring P and Schuck P 2004 *The Nuclear Many-Body Problem* 3rd ed (Berlin: Springer)
[20] Sambataro M 2007 *Phys. Rev. C* **75** 054314

Journal of Physics: Conference Series **366** (2012) 012011 doi:10.1088/1742-6596/366/1/012011

Topics on Nuclear Structure with Electroweak Probes

E. Moya de Guerra[1,2], O. Moreno[1], P. Sarriguren[2], M. Ramon[3]

[1] Departamento de Física Atómica, Molecular y Nuclear, Universidad Complutense de Madrid, Ciudad Universitaria, E-28040 Madrid, Spain

[2] Instituto de Estructura de la Materia, Consejo Superior de Investigaciones Científicas, Serrano 123, E-28006 Madrid, Spain

[3] Departamento de Física Teórica I, Universidad Complutense de Madrid, Ciudad Universitaria, E-28040 Madrid, Spain

E-mail: emoyaval@fis.ucm.es

Abstract. We study some relevant aspects of complex nuclei structure using electroweak probes within the framework of self-consistent mean field theories with Skyrme density-dependent two-body interactions, including pairing and spin-isospin RPA correlations where necessary. We apply the formalism to the study of single and double beta decays as normal modes of the system, as well as to the analysis of parity-violating electron scattering by nuclei. Finally, we profit from the studied processes to draw some conclusions on the neutrino nature (eigenstates mixing).

1. Introduction

We present in this work some of our recent theoretical results on nuclear structure using electroweak probes, including double beta decay (^{76}Ge - ^{76}Se), parity-violating elastic electron scattering by nuclei (^{208}Pb, ^{12}C) and beta particle spectra in single beta decays (^{187}Re, ^{3}H). We examine various nuclear structure properties such as active-shell spectroscopic factors, nuclear deformation, neutron distributions in nuclei or Coulomb and weak interactions between nuclei and scattered or emitted beta particles, among others. The processes under study set stringent tests on the capability of theory to make reliable predictions. The theoretical scheme that we follow in describing the nuclear structure properties lies on the adiabatic time-dependent Hartree-Fock approximation which provides a unified framework for a large variety of nuclear properties: sizes and shapes, density profiles, fission barriers, excitation energies, transition strengths or even energy transfers between single-particle and collective modes [1]. This theory has not yet been fully implemented to explore, for instance, the coupling between different nuclear modes or the dilution of multiphonon states, but many applications have been made to the study of electric excitations (mainly dipole, quadrupole and octupole) around equilibrium, as well as of spin-isospin excitations. Many applications of the latter have been put forward by our group in the last years.

In the following section we focus on the spectroscopic factors of the double-beta decay partners ^{76}Ge and ^{76}Se that have been recently measured. We show that a modification of the spin-orbit terms of the Skyrme energy density functional and the Hartree-Fock potential are needed to reproduce the measured spectroscopic factors. We also show that fair agreement with experiment is obtained for the single beta Gamow-Teller branches and for the double-beta decay

Journal of Physics: Conference Series **366** (2012) 012011 doi:10.1088/1742-6596/366/1/012011

matrix elements. In the next section we discuss parity-violating elastic electron scattering by nuclei and its potential application to the precise determination of neutron densities in nuclei. We study nuclear isovector and isoscalar densities for $N > Z$ and for $N = Z$ stable nuclei obtained within the self-consistent mean field approximation. We compare the values of the parity-violation asymmetry (PVA) at low and intermediate q-values for different N/Z and A values. Distorted wave calculations of PVA are shown and are compared to plane wave impulse approximation. In the last section we show our results on the energy spectrum of the electron emitted in the beta decay of ^{187}Re, including the distortion of its wave function due to the Coulomb interaction with the daughter nucleus. We will exploit this information to analyze the effect on the spectrum of a hypothetical keV-mass contribution to the emitted neutrino.

2. Active-shell spectroscopic factors in beta decaying nuclei: ^{76}Ge→^{76}Se case and the single and double beta decay nuclear matrix elements

The search for the neutrino mass is pushing strongly experimental and theoretical work on nuclear double-beta decay because of the crucial role of a possible detection of the zero-neutrino mode. However to settle the absolute scale of the neutrino mass, the precise theoretical calculation of the nuclear matrix element is equally important as the detection of the zero-neutrino mode. An important ingredient of those calculations are the spectroscopic factors.

The best studied case of two-neutrino double-beta decay is that of ^{76}Ge going to ^{76}Se. The relevant protons and neutrons involved in the double-beta process are those in the valence shells $1p_{3/2}$, $1p_{1/2}$, $0f_{5/2}$, $0g_{9/2}$, whose occupation probabilities have been recently measured by Schiffer *et al.* (for neutrons) and by Kay *et al.* (for protons) [2]. These data are not reproduced by previous mean-field calculations using Woods-Saxon potential, and although shell-model calculations have also been carried out with occupation probabilities in good agreement with experimental ones, the corresponding single particle energies are not available in the literature. In the deformed Hartree-Fock calculation of this work we allow the Skyrme force parameters to vary and search for a parametrization that provides better agreement with the experimental occupations in the valence shells of ^{76}Ge and ^{76}Se. We focus on the effect of the two-body spin-orbit strength, W_0, that appears to be of great relevance to describe properly the proton and neutron valence shells occupations. In addition, the use of the experimental ground-state deformations contributes to the agreement between the calculated occupations and the experimental ones [3].

The new single-particle energies and occupation probabilities obtained with this procedure have an effect on the matrix element of the two-neutrino double-beta decay ^{76}Ge→^{76}Se and on the GT strength distributions of both nuclei going to the intermediate nucleus ^{76}As. Theoretical results are shown in Fig. 1 together with experimental data (light lines). A first attempt of pnQRPA calculation using the Sk3 Skyrme force for spherical ground states is presented in the dashed black lines, whereas the solid black lines correspond to deformed ground states (experimental deformations) and using an increased value of the two-body spin-orbit strength for neutrons, $W_n = 200$ MeV fm^5. The deformed calculation agrees better with the experiment, as shown in the low energy regions of the GT (plus and minus) accumulated strength distributions (to the left) and in the running sum of the double-beta decay matrix element (to the right). The final value of the double-beta decay matrix element, reached by the running sum beyond 18 MeV, lies within the experimental range (shaded region) for the improved calculation, whereas the initial one (spherical, Sk3) is very far.

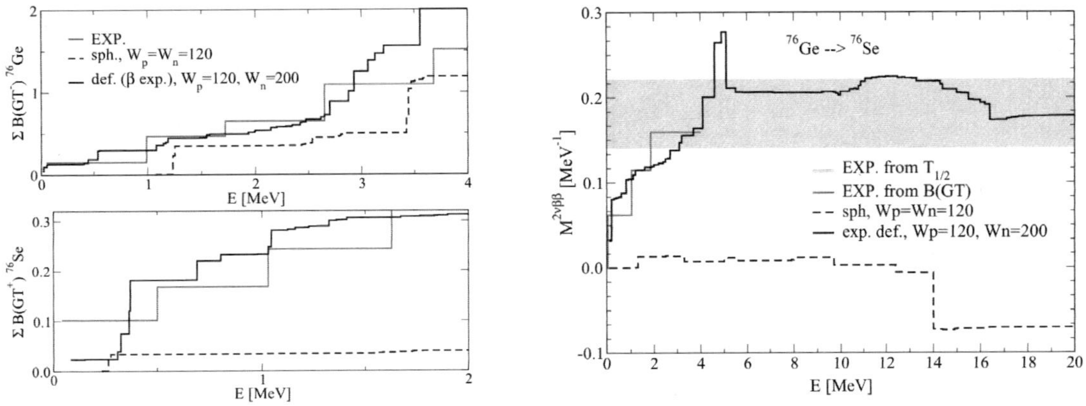

Figure 1. Right-hand panel shows the $2\nu2\beta$ matrix element accumulated in energy as a function of the excitation energy of the intermediate nucleus ^{76}As calculated with the standard Sk3 force ($W_n = W_p = 120$) and with the modified Sk3 energy density functional ($W_n = 200, W_p = 120$) and taking deformation into account. The detailed distribution of the accumulated single beta branches GT$^-$ and GT$^+$ at low energies is shown in the left-hand panels.

3. Parity-violation asymmetry in electron-nucleus scattering: ^{208}Pb$(\vec{e}, e)$ and ^{12}C$(\vec{e}, e)$ cases and the precise determination of neutron distributions

We show in Fig. 3 results on PV asymmetry in elastic scattering of polarized electrons by stable $N > Z$ and $N = Z$ nuclear targets, namely ^{208}Pb and ^{12}C respectively [4]. Their ground-state structures have been obtained from a Skyrme Hartree-Fock mean field with pairing interactions in BCS approximation. The curves in the figure correspond to the parity-violation asymmetry, which is proportional to the difference between the cross-sections of the scattered electrons polarized parallel and antiparallel to their momenta. Three curves (dark lines) are presented for ^{208}Pb, and two for ^{12}C (light lines). In the case of ^{208}Pb, two calculations are in plane-wave (PW) Born approximation, one (solid line) corresponding to the actual proton and neutron form factors and the other one (dashed line) corresponding to the same function for both the proton and the neutron form factors, both normalized to 1. The latter shows a simple proportionality to q^2, whereas the former shows a complex pattern for which the actual nuclear structure is responsible. The divergences in this last curve are filled in or smoothed out when the distorted-wave (DW) calculation is carried out (dotted line). This calculation, which is the one to be compared to experimental data, takes into account the distortion of the incoming and outgoing electron's wave function due to the presence of the nuclear Coulomb field, which is very important in this case (and not so much in the case of ^{12}C).

The results for ^{12}C in the same figure are for plane-wave calculations using the actual ground-state form factors (solid line), and the same proton and neutron form factors (dashed line). The latter shows again a simple q^2 dependence whereas the structure of the former is related to the isospin mixing in the nuclear ground state (nominally the isospin is $T = 0$ since it has $N = Z$).

4. Charged lepton spectrum in single-beta decay: ^{187}Re case and the search for sterile neutrinos

The differential decay rate with respect to the energy of the electron emitted in a single beta minus decay depends on the nuclear transition matrix element and on several kinematic factors of the emitted leptons. The latter come from the differential density of asymptotic final states

Journal of Physics: Conference Series **366** (2012) 012011 doi:10.1088/1742-6596/366/1/012011

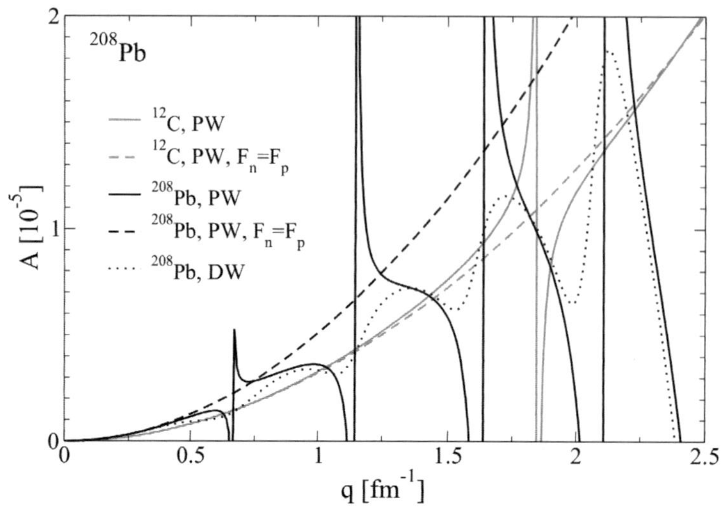

Figure 2. Parity-violation asymmetry of elastic electron scattering by ^{208}Pb and by ^{12}C. For each nucleus we show results of PW calculations (full lines) and also assuming equal proton and neutron form factors (dashed lines). For ^{208}Pb we also show the result of a DW calculation using realistic proton and neutron densities (dotted line).

and from the shape factor:

$$\frac{d\Gamma^{\chi}_{l_j}}{dE_e} = K\ B\ p_e\ p_{\nu_\chi}\ E_e\ (E_0 - E_e)\ F_l(Z', E_e)\ S_l(p_e, p_{\nu_\chi})\ \theta(E_0 - E_e - m_{\nu_\chi}) \quad (1)$$

where K is a constant ($\sim 2 \cdot 10^{-36}$ keV^{-4}), B is the squared nuclear reduced matrix element (dimensionless), p_e and p_{ν_χ} are the lepton momenta, and E_e and E_0 are the electron total energy and maximum total energy respectively. The relativistic Fermi function $F_l(Z', E_e)$ accounts for the Coulomb interaction between the residual nucleus (charge Z') and the emitted electron with an angular momentum l with respect to the nucleus. The shape factor is a kinematic dependence coming from the total transition amplitude, which takes different forms depending on the nature of the decay (allowed or forbidden). For instance, for an allowed decay its value is simply 1, whereas for a unique first forbidden decay it depends on the orbital angular momentum quantum number l of the emitted electron:

$$S_0(p_e, p_\nu) = \frac{1}{3}\, p_\nu^2 \quad \text{and} \quad S_1(p_e, p_\nu) = \frac{1}{3}\, p_e^2 \quad (2)$$

The shape and range (the latter given by the step function θ) of the differential decay rate, Eq. 1, strongly depend on the mass m_{ν_χ} of the considered neutrino χ [5]. In figure 3 left-panel we show the differential decay rate of the electron emitted in the beta decay of ^{187}Re (Q_β = 2.47 keV) for different masses of the emitted neutrino. Only the leading contribution (electron p-wave) is shown. The massless case is a very good approximation to the usual light active neutrino (with mass $m_{\nu_a} \sim 1$ eV), whereas the keV-cases account for hypothetical heavy sterile neutrinos, which have been recently proposed as dark matter candidates [5]. The neutrino emitted in a beta decay should obviously be composed mainly of the light active neutrino, with only a very small mixing of the heavy sterile one. We will consider the following mixing of active-sterile neutrinos or antineutrinos:

$$|\nu_e\rangle = \cos\theta\,|\nu_a\rangle + \sin\theta\,|\nu_s\rangle \quad (3)$$

Journal of Physics: Conference Series **366** (2012) 012011 doi:10.1088/1742-6596/366/1/012011

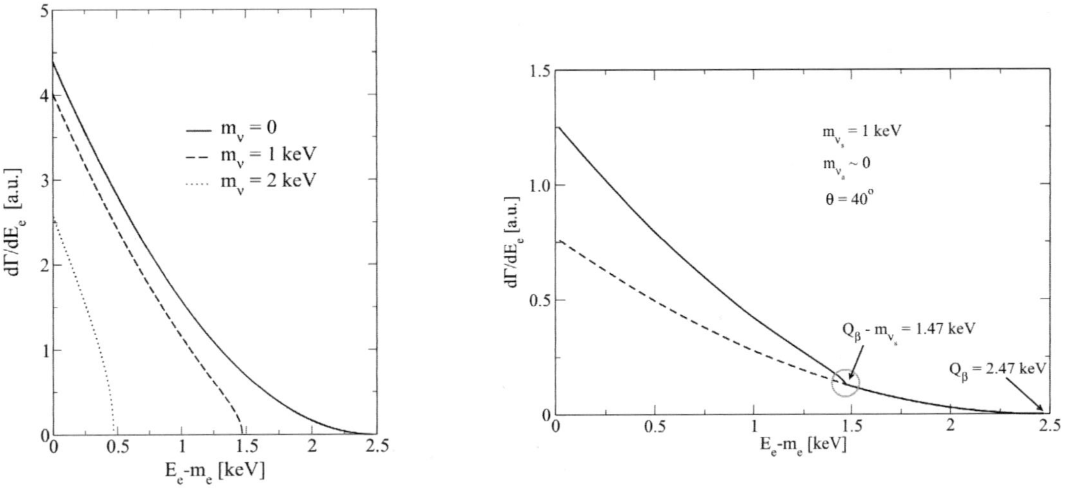

Figure 3. Left-hand panel shows the differential beta decay rate (in arbitrary units of the order 10^{-37}) of ^{187}Re for different masses m_ν of the emitted neutrino. Right-hand panel shows the differential decay rate when the emitted neutrino is a mixing of a light neutrino ($m_{\nu a} \sim 0$) with a heavy neutrino ($m_{\nu_s} = 1$ keV), the mixing angle being much too large in this plot for the sake of clarity.

In figure 3 right panel we show the effect on the ^{187}Re differential decay rate when the neutrino emitted in association with the electron contains a mixing of a 1 keV sterile neutrino [5]. The mixing angle in this plot is unrealistically large in order to clearly show the kink produced in the electron spectrum when the neutrino has enough energy for the heavy neutrino to contribute ($E_\nu = Q_\beta - m_{\nu_s} = 1.47$ keV).

Acknowledgments

We acknowledge our collaborators Dr. Udías, Prof. Faessler, Prof. Donnelly for fruitful dicussions. This work was supported by Ministerio de Ciencia e Innovación (Spain), Contracts FIS2008-01301 and CSPD-2007-00042@Ingenio2010, and UCM 'Grupo de Física Nuclear'-910059.

References

[1] Villars F 1977 *Nucl. Phys. A* **285** 269
Moya de Guerra E and Villars F 1977 *Nucl. Phys. A* **285** 297
Moya de Guerra E and Villars F 1978 *Nucl. Phys. A* **298**, 109
Moya de Guerra E 1982 *Phys. Rev. Lett.* **48** 922
Moya de Guerra E 2001 *The limits of the mean field, in An Advanced Course* in *Modern Nuclear Physics (Ed. Arias, J.M. and Lozano M.)* Springer.

[2] Schiffer J P *et al.* 2008 *Phys. Rev. Lett.* **100** 112501
Kay B P *et al.* 2009 *Phys. Rev. C* **79** 021301(R)

[3] Moreno O, Moya de Guerra E, Sarriguren P and Faessler A 2010 *Phys. Rev. C* **81** 041303(R)

[4] Moreno O, Moya de Guerra E, Sarriguren P and Udias J M 2010 *J. Phys. G: Nucl. Part. Phys.* **37** 064019
Moreno O, Sarriguren P, Moya de Guerra E, Udias J M, Donnelly T W and Sick I 2009 *Nucl. Phys. A* **828** 306

[5] de Vega H J, Moreno O, Moya de Guerra E, Ramon Medrano M, Sanchez N G 2011 arXiv:1109.3452v2 [hep-ph].

Journal of Physics: Conference Series **366** (2012) 012012 doi:10.1088/1742-6596/366/1/012012

Systematic Study of the Pygmy Dipole Resonance

V. Derya[1], J. Endres[1], M. N. Harakeh[2], D. Savran[3,4], H. J. Wörtche[2], and A. Zilges[1]

1 Institut für Kernphysik, Universität zu Köln, 50937 Köln, Germany
2 KVI, University of Groningen, 9747 AA Groningen, The Netherlands
3 ExtreMe Matter Institute EMMI and Research Division, GSI, 64291 Darmstadt, Germany
4 Frankfurt Institute for Advanced Studies FIAS, 60438 Frankfurt, Germany

E-mail: `derya@ikp.uni-koeln.de`

Abstract. In recent years, a systematic comparative study of the pygmy dipole resonance (PDR) using the (γ,γ') and the $(\alpha,\alpha'\gamma)$ reactions on the nuclei ^{140}Ce, ^{138}Ba, ^{124}Sn, ^{94}Mo, and ^{48}Ca was performed. The aim of this systematic study is to explore the structure of the PDR by the usage of two complementary probes and several different nuclei. The real-photon scattering experiments were performed at the S-DALINAC in Darmstadt, Germany, while the $(\alpha,\alpha'\gamma)$ coincidence experiments were performed at the Big-Bite Spectrometer (BBS) at KVI in Groningen, The Netherlands. The comparison of the results from both kinds of experiments showed an energetic splitting of the PDR into two parts that might be due to the different isospin character of these states.

First experimental results of the $(\alpha,\alpha'\gamma)$ experiments on the nuclei ^{94}Mo and ^{48}Ca will be presented.

1. Introduction

The main component of the electric dipole strength in atomic nuclei is the isovector giant dipole resonance (IVGDR) which exhausts almost the total energy-weighted sum rule [1]. Moreover, in neutron-rich nuclei electric dipole strength to low-lying $J^\pi = 1^-$ states around the neutron separation threshold, the electric pygmy dipole resonance (PDR), has been observed [2].

The most direct way to induce dipole transitions from the ground state, i.e. excite 1^- states in even-even stable nuclei, is the (γ,γ') reaction, because real photons are highly selective to low multipolarities [3, 4, 5, 6]. Experiments using the (γ,γ') reaction are thus a commonly used method to study electric dipole excitations. Most of the available data on the PDR in many nuclei were established in this kind of experiments. Observables like excitation energy, spin, parity, and $B(E1)$ strength of excited states can be determined using the (γ,γ') reaction in a model-independent way. In the experiments, the emitted γ rays were detected by HPGe detectors which provide an excellent energy resolution that is essential because of the high density of excited levels.

Additional information about the structure of the PDR can be deduced by a combination with a complementary excitation mechanism. The interaction of an α particle with an atomic nucleus, at intermediate energies, includes strong interaction mainly with the surface besides electromagnetic interaction with the whole nucleus. Furthermore, the strongly interacting α-particle is a scalar-isoscalar probe which therefore excites states of natural parity and isoscalar nature. The experimental method which was applied to allow for using α particles as a probe

Journal of Physics: Conference Series **366** (2012) 012012 doi:10.1088/1742-6596/366/1/012012

of electric dipole strength is the α-γ coincidence method. The $(\alpha, \alpha'\gamma)$ coincidence experiments at E_α=136 MeV and forward scattering angles were performed at KVI in Groningen, The Netherlands. The beam was provided by the AGOR cyclotron and the scattered α particles were detected by the QQD-type magnetic Big-Bite Spectrometer which was positioned at 3.5° or 5° (depending on the nucleus) with respect to the beam axis. Around the target chamber an array of HPGe detectors was placed for γ spectroscopy with an excellent energy resolution. A hardware coincidence between HPGe detectors and the BBS was used as a main trigger to reduce the count rates and the dead time of the system. This setup enables the determination of the singles α-scattering cross sections by integrating over the full solid angle of γ detection. More details about the experimental setup are given in Ref. [7].

One result of the performed systematic study of the PDR in $(\alpha, \alpha'\gamma)$ and (γ, γ') experiments is the observation of an energetic splitting in a low-energy part, which is observed in both types of experiments, and a high-energy part, which is observed only in the (γ, γ') experiments. It was first discovered in ^{140}Ce [8] and it has been shown, by the study of ^{138}Ba and ^{124}Sn [9, 10], that this is not an exceptional case but a general behaviour of the low-lying electric dipole strength (see Fig. 1).

In the following sections, results for the nuclei ^{94}Mo and ^{48}Ca experiments will be presented.

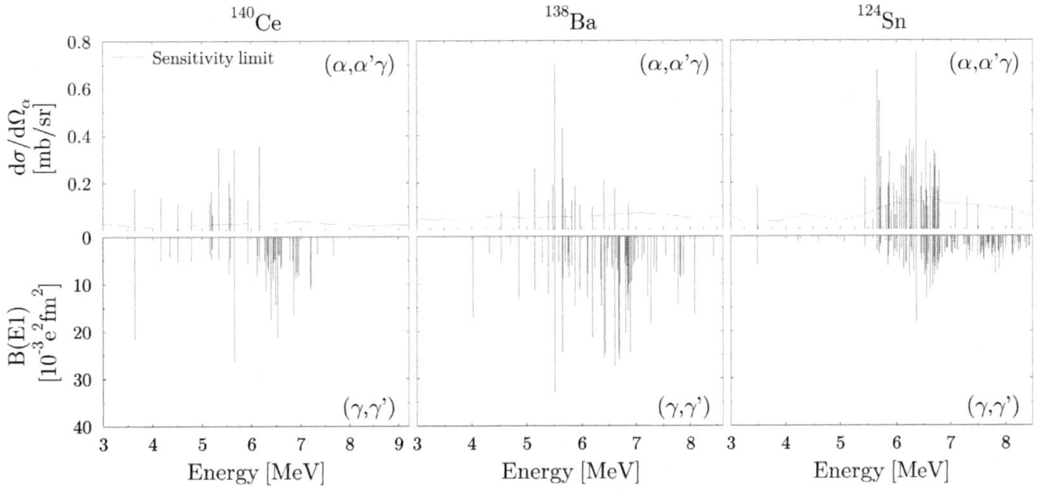

Figure 1. Experimental results of the systematic study of the PDR in ^{140}Ce, ^{138}Ba, and ^{124}Sn [3, 8, 9, 10, 11]. The singles α-scattering cross sections were determined in the $(\alpha, \alpha'\gamma)$ experiment and the $B(E1) \uparrow$ transition strengths were determined in the (γ, γ') experiment. The dashed lines correspond to the sensitivity limit in the $(\alpha, \alpha'\gamma)$ experiments.

2. The non-magic nucleus ^{94}Mo

In order to extend the systematic study, the nucleus ^{94}Mo was measured with the α-γ coincidence method. The BBS was placed at 3.5° with respect to the beam axis and the HPGe detector array consisted of seven HPGe detectors.

In contrast to the other nuclei studied, this nucleus is non-magic but still spherical and close to a sub-shell closure. The low-lying 1^- states, which were excited by the α particles, could be identified by a comparison to results from a (γ, γ') experiment. In Fig. 2, the singles cross sections are shown. They were determined in this experiment from the summed γ spectrum with a gate on the excitation energy $E_x = E_\gamma$, which is the condition for a de-excitation to the ground state. A branching to an excited state with a level energy E_{level} becomes visible in the γ spectrum by gating on $E_x = E_\gamma - E_{level}$. First results indicate a branching of low-lying 1^- states to the first 2_1^+ state, which is included in the calculation of the cross section. If no branching was observed, a branching ratio of $\Gamma_0/\Gamma = 1$ was assumed.

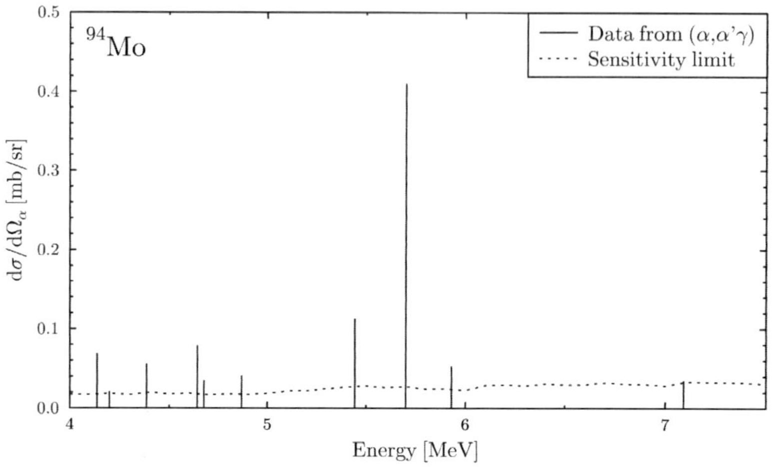

Figure 2. Singles α-scattering cross sections of dipole excitations in ^{94}Mo. Branching ratios are included. If no branching was observed, a branching ratio of $\Gamma_0/\Gamma = 1$ was assumed. The dashed line corresponds to the experimental sensitivity limit.

3. The doubly-magic nucleus ^{48}Ca

Whereas in heavy neutron-rich nuclei the occurrence of the PDR has been shown, the existence of the PDR mode in lighter nuclei is still an open question. Light nuclei which have a neutron halo are known (for example ^{11}Li [12]), but in these cases only single weakly-bound nucleons contribute to the excitation, so it is a different phenomenon than the PDR in heavy nuclei. Further investigations are necessary to clarify in which mass region and at which neutron-proton asymmetries the development of a neutron skin and the PDR starts. Dipole transitions in the Calcium chain have been studied systematically in (γ, γ') experiments at the Darmstadt High-Intensity Photon Setup [13] at the S-DALINAC (see Refs. [14, 15, 16]). Furthermore, the summed $B(E1)$ strength can be reproduced by theoretical calculations [16]. Knowledge of the development of electric dipole-strength in the Calcium chain, with increasing neutron number and therefore increasing N/Z ratio, may provide access to some unanswered questions concerning the low-lying dipole strength.

The doubly-magic isotope ^{48}Ca, which is known to have a neutron skin (Refs. [17, 18]), was studied by real photon scattering at the S-DALINAC [14] and by the α-γ coincidence method at the BBS at KVI. The setup for the $(\alpha, \alpha'\gamma)$ experiment was the same as for ^{94}Mo except that the BBS was positioned at 5° and in total six HPGe detectors were arranged around the target chamber. Again, the α-beam energy in the experiment was $E_\alpha = 136$ MeV. First results of this experiment show that almost all dipole excitations which were observed in the (γ, γ') experiment are also visible in the summed γ spectrum with a gate on ground-state transitions (see Fig. 3), which was obtained in the α-γ coincidence experiment. Surprisingly, the state at

7.3 MeV which has the highest $B(E1)\uparrow$ value of $16.5(19)\cdot10^{-3}e^2fm^2$ (see Fig. 4) of in total $55.7(41)\cdot10^{-3}e^2fm^2$ was not excited by the α particles. In contrast, the state at 7.6 MeV, which was strongly excited by α particles, has a small $B(E1)\uparrow$ transition strength and therefore seems to have a different structure compared to the other observed transitions. The analysis is still in progress.

Recently, in addition a direct parity measurement in ^{48}Ca has been performed by means of the $(\vec{\gamma},\gamma')$ reaction. The polarized and nearly mono-energetic photons were produced at the High Intensity Gamma-Ray Source (HI$\vec{\gamma}$S) facility at the Duke University in Durham, USA. First results assign negative parity to all observed dipole excitations.

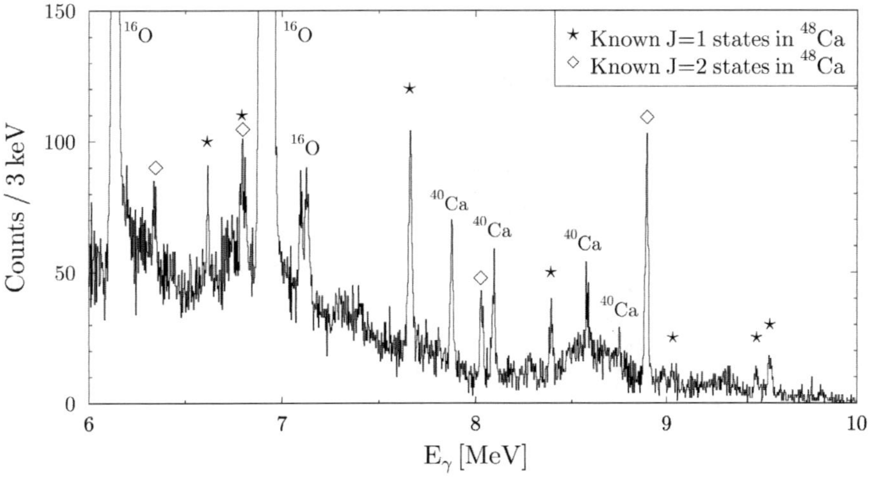

Figure 3. Summed γ spectrum with a gate on $E_x = E_\gamma$ measured in the α-γ coincidence experiment on ^{48}Ca. Dipole transitions are marked with stars. Besides the transitions of excited states in ^{48}Ca, also transitions in ^{40}Ca and ^{16}O are visible.

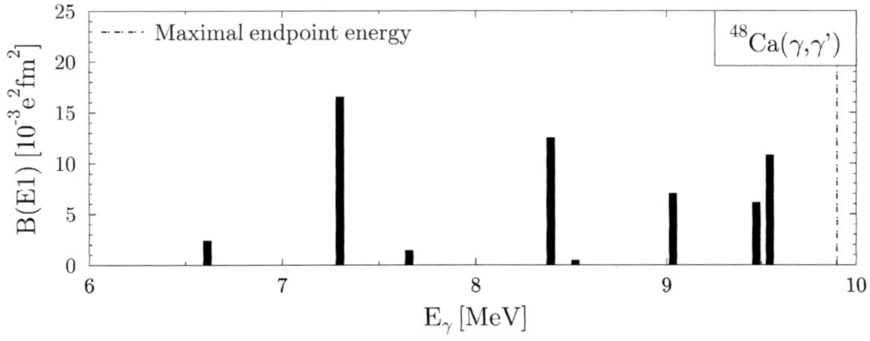

Figure 4. The $B(E1)$ strength distribution was determined in a (γ,γ') experiment at the S-DALINAC [14, 15].

4. Summary and outlook

The α-γ coincidence method, which was performed to study the low-lying electric dipole strength, was introduced and first results for two nuclei, ^{94}Mo and ^{48}Ca, which were obtained by using this method were presented. Singles α-scattering cross sections were determined for ^{94}Mo from the summed γ spectrum with a gate on ground-state transitions. There are clear indications for decay of two states into the first 2_1^+ state in addition to decay to the ground state.

Dipole excitations in the doubly-magic ^{48}Ca are visible in the summed γ spectrum with a gate on ground-state transitions and will allow the determination of absolute singles cross sections. The comparison to results from a (γ, γ') experiment once again emphasizes the importance of using different excitation mechanisms to get a deeper insight into the structure of low-lying 1^- states, especially their isospin character.

Acknowledgment
This work was supported by the DFG (ZI 510/4-1 and SFB 634), by the EU under EURONS Contract No. RII3-CT-2004-506065 in the 6th framework programme, and by the Alliance Program of the Helmholtz Association (HA216/EMMI).

References
[1] Harakeh M N and Woude A van der 2001 *Giant Resonances* (Oxford University Press, Inc.)
[2] Bartholomew G A *et al.* 1973 *Adv. Nucl. Phys.* **7** 229
[3] Zilges A *et al.* 2002 *Phys. Lett. B* **542** 43
[4] Schwengner R *et al.* 2008 *Phys. Rev. C* **78** 064314
[5] Tonchev A P *et al.* 2010 *Phys. Rev. Lett.* **104** 072501
[6] Savran D *et al.* 2011 *Phys. Rev. C* **84** 024326
[7] Savran D *et al.* 2006 *Nucl. Instr. Meth. Phys. Res. A* **564**
[8] Savran D *et al.* 2006 *Phys. Rev. Lett.* **97** 172502
[9] Endres J *et al.* 2009 *Phys. Rev. C* **80** 034302
[10] Endres J *et al.* 2010 *Phys. Rev. Lett.* **105** 212503
[11] Govaert K *et al.* 1998 *Phys. Rev. C* **57** 2229
[12] Tanihata I *et al.* 1985 *Phys. Rev. Lett.* **55** 2676
[13] Sonnabend K *et al.* 2011 *Nucl. Instr. Meth. Phys. Res. A* **640** 6
[14] Hartmann T *et al.* 2002 *Phys. Rev. C* **65** 034301
[15] Hartmann T *et al.* 2004 *Phys. Rev. Lett.* **93** 192501
[16] Isaak J *et al.* 2011 *Phys. Rev. C* **83** 034304
[17] McCamis R H *et al.* 1986 *Phys. Rev. C* **33** 1624
[18] Gambacurta D *et al.* 2011 *Phys. Rev. C* **84** 034301

Journal of Physics: Conference Series **366** (2012) 012013 doi:10.1088/1742-6596/366/1/012013

The Inverse Kinematics Thick Target scattering method as a tool to study cluster states in exotic nuclei.

A Di Pietro[1], D Torresi [1,2,3], M Zadro[4], L Cosentino[1], C Ducoin[5], P Figuera[1], M Fisichella[1,6], M Lattuada[1,2], C Maiolino[1], A Musumarra[1,2], M Papa[5], MG Pellegriti[1,2], M Rovituso[2], D Santonocito[1], G Scalia[1,2], V Scuderi[1,2] and E Strano[1,2]

[1]INFN Laboratori Nazionali del Sud, Catania, Italy
[2]Dipartimento di Fisica ed Astronomia, Universitá di Catania, Catania, Italy
[3]CSFNSM, Catania Italy
[4]Ruder Boškovič Institute, Zagreb, Croatia
[5]INFN Sezione di Catania Catania, Italy
[6]Dipartimento di Fisica, Università di Messina, Messina, Italy

E-mail: dipietro@lns.infn.it

Abstract. The inverse kinematics thick target scattering method (TTIK) was used to measure the ^{8}Li resonant elastic scattering on a ^{4}He target in order to investigate ^{8}Li-α cluster configurations in excited states of ^{12}B. A ^{8}Li beam at E_{beam} = 30.6 MeV, provided by the radioactive beam facility EXCYT of LNS in Catania, while passing through a thick Helium target decreases continuously its energy. Thus, elastic scattering can be induced starting from the initial beam energy down to zero. Time of flight measurement between beam particles passing through a MCP detector and α-particles impinging on the ΔE stage of a Si telescope detector allows the discrimination of elastic scattering from nuclear reaction events, thus representing an improvement of the TTIK method. In this paper the used experimental technique and preliminary results will be described.

1. Introduction

It is well known that cluster structures occur in many light nuclei in their excited states and in some cases also in their ground states. Due to their very large binding energy, α-particles have been considered to be the ideal cluster. Excited states of light N=Z nuclei can indeed be described as having α-cluster configurations [1]. Recently more exotic description of cluster configurations have been introduced in order to describe the structure of n-rich nuclei. For example, theoretical studies of cluster structures on neutron-rich Boron isotopes were performed by Kanada En'jo [2] using the Antysimmetrized Molecular Dynamics. These calculations show how cluster structure become more evident and more exotic when the neutron number of the isotopes increases. For instance, the ^{19}B ground state can be described by a ^{11}Li-^{8}He configuration, while the less neutron rich ^{13}B, due to the N=8 shell closure, exhibits a single particle behavior. The existence of a cluster structure above the α-particle emission threshold is predicted for the excited states of ^{12}B. Calculations using the Generator Coordinate Method

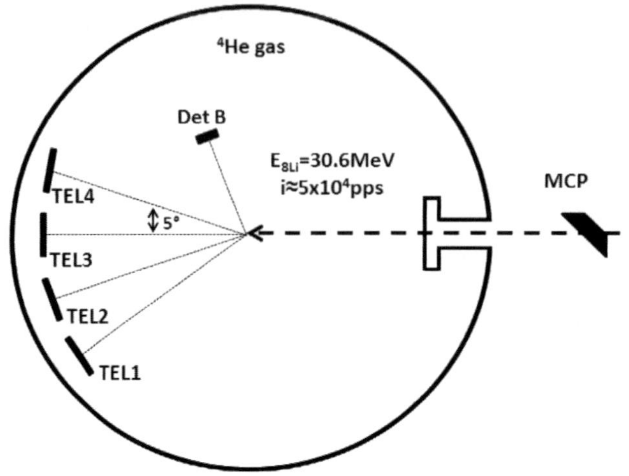

Figure 1. Sketch of the experimental set-up used for the experiment. Si-detector telescope B was used to measure the stopping power.

(GCM) have been performed by [3]. In this study the presence of a positive parity band with a large reduced width for α-decay, located just above the α-decay threshold, supports the picture of α-^{8}Li cluster configurations for excited states in ^{12}B.
With the aim of investigating such peculiar cluster configurations in ^{12}B we performed the measurement of ^{8}Li+^{4}He resonant elastic scattering.

2. Experimental technique

The ^{8}Li beam was produced by the radioactive beam facility EXCYT at INFN-Laboratori Nazionali del Sud in Catania [5]. EXCYT uses the ISOL technique to produce the beam and it uses a K-800 superconducting cyclotron as a driver and of a 14 MV tandem as post-accelerator for the radioactive beams. The energy of the ^{8}Li beam delivered by the accelerator was E_{lab}= 30.6 MeV. Before entering the scattering chamber the beam passes through the aluminized Mylar foil (0.8 μm thick) of a Micro Channel Plate (MCP) detector. This detector provides a reference signal (stop signal) for the time of flight measurement (TOF) and, at the same time, it allows for counting the beam particles. The average beam intensity was $\sim 5 \cdot 10^{4}$ pps. Downstream the MCP detector it is placed a kapton foil (13 μm thick) that separates the in-vacuum beam line from the scattering chamber CT2000. This latter was filled with ^{4}He enriched gas at a pressure P$\sim$700 mbar.
An array of four ΔE-E Silicon detector telescopes is placed at forward angles at a distance of $\approx$180 cm from the kapton foil. The ΔE stage is made by four quadrant Silicon Pad detectors, 50-60 μm thick whereas, the E stage consists of Double Sided Silicon Strip Detectors, 16+16 strips, 500-1000 μm thick. The ΔE detector provides also the start signal for the TOF measurement. In figure 1 a sketch of the used experimental set-up is shown.

When passing through the ^{4}He gas the beam reduces its energy, hence the elastic scattering is induced in a continuous range of energies starting from the initial beam energy (30.6 MeV) down to zero; this is called TTIK method [4]. With this technique the ^{4}He gas target has a threefold role. It acts at the same time as target, as degrader and as beam stopper. The ^{8}Li beam is in fact stopped in the gas before reaching the detectors, while the recoiling ^{4}He particles, owing to their lower stopping power, reach the detectors which are placed at forward angles (backward angles in CM system). An experimental limitation of using the TTIK method is

Journal of Physics: Conference Series **366** (2012) 012013 doi:10.1088/1742-6596/366/1/012013

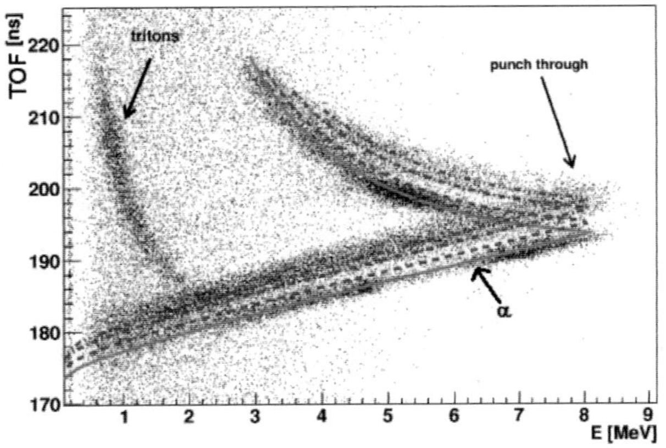

Figure 2. TOF versus ΔE plot for the detector at 0°. The scatter plot represents the experimental data. The full line represents the calculations for elastic scattering α particles events, the dashed lines represent the calculations for the inelastic scattering of the first three states of ^{8}Li E_x=0.9808, 2.255, 3.21 MeV.

represented by α-particles coming from other possible reaction mechanisms. Only by analyzing the particle energy deposition spectra it not possible to identify the reaction mechanisms for the α-particle production. For this reason we decided to improve the TTIK method published in [4] by introducing the TOF measurement. The measured energy of a detected α-particle depends on the reaction process in which it was produced, as well as on the position in the gas target, where the reaction has occurred. Therefore, the TOF measurement allows us to discriminate elastic from reaction events. When using the TTIK method, the excitation energy of the compound nucleus is reconstructed from the energy of the detected α-particle by computing the energy loss in the ^{4}He gas of both the beam and the recoiling α. Therefore, stopping power knowledge is of paramount importance in this type of measurements. Stopping power measurements of ^{4}He gas for Li-ions have also been performed. This was done using a ^{7}Li beam. The incorrect knowledge of the stopping power affects, not only on the position of a peak in the extracted excitation function but also on the extracted cross-section, as it has been discussed in [6].

3. Results

In the following, results obtained for the detector placed at θ_{lab}=0° will be discussed. As mentioned above, TOF and ΔE-E measurement were performed. The TOF measurement, beside allowing the identification of elastic scattering from other reaction processes it allowed also to identify in charge. It was possible, in fact, for particles stopping on the ΔE detector, to discriminate H from He. Particles punching-through the ΔE detector were identified via the telescope technique.

In figure 2, the TOF is shown as a function of the energy deposited in the ΔE detector. Since inverse time logic was used, large TOF ($\sim$ 220 ns) corresponds to events which occur at the highest energies (entrance of the chamber), whereas smaller TOF ($\approx$175 ns) corresponds to events which originated closer to the ΔE detectors. The α-particle punch-through energy is around 8 MeV. The lines drawn upon the experimental data show calculated values of the TOF as a function of the detected energy both for elastic and inelastic scattering, corresponding to the first three excited levels of ^{8}Li. The calculations include the kinematics, energy losses of

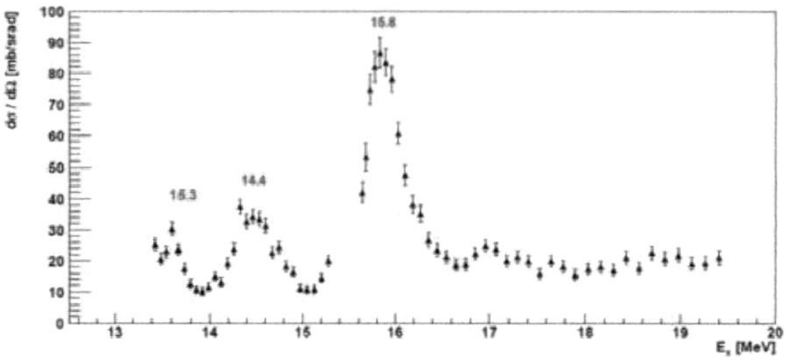

Figure 3. Preliminary ^{12}B excitation function at 0°.

^{8}Li and α-particles in ^{4}He gas and energy losses of α-particles in the entrance dead layer of the detector. On this plot it is evident the separation of elastic contribution from inelastic or break-up events. In figure 2 it can also be observed that α-particles are well separated from hydrogen isotopes at energies above $\approx$2 MeV. The elastic events, which stop in the ΔE detector, were selected by graphical cuts on the TOF-ΔE plot; the ones that punch-through were selected by graphical cut on the ΔE-E plot. For these events the excitation energy spectrum of the ^{12}B compound nucleus was reconstructed.
In figure 2 one can also see events, at E$\leq$5 MeV, which are uncorrelated in time. These events are coming from the α decay of ^{8}Be* produced in the ^{8}Li radioactive decay (^{8}Li $(\beta^-)^8$Be$^* \rightarrow 2\alpha$) and they are uncorrelated in time since the $T_{1/2}(^8\mathrm{Li})$=840 ms. These events are source of background for the elastic events and they have been subtracted by shifting the graphical cut of the elastic events on a different time region of the TOF-ΔE plot.
In order to reconstruct the ^{12}B excitation function from the detected α-particle energies a precise knowledge of the stopping power of ^{4}He for ^{8}Li and ^{4}He is required. For this reason, as mentioned above, during the same experiment a measurement of the stopping power was also performed. The results were found in agreement with SRIM calculations.
The elastic scattering excitation function extracted for the telescope placed at 0° degree is shown in figure 3. Three different peaks are clearly observed at 13.6, 14.4 and 15.8 MeV. Evidences of the peaks at $E_x(^{12}\mathrm{B})$= 13.6 and 15.8 MeV have already been reported in the literature [7], whereas this is the first evidence for a state in ^{12}B at 14.4 MeV.

4. Conclusions

In this contribution the measurement of the ^{8}Li-α elastic scattering excitation function performed by using the TTIK method is reported. In this measurement the technique reported in [4] was improved by introducing the time of flight measurement. The TOF allows for the discrimination of elastic scattering events from the ones produced by other reaction processes. The elastic scattering excitation function in the range 13.5 MeV$\leq E_x \leq$ 19.5 MeV was measured. Three different peaks are observed at ^{12}B excitation energies of 13.6, 14.4 and 15.8 MeV. These results are in agreement with previous measurements reported in [7] for E_x=13.6 and 15.8 MeV whereas no information is present in the literature for the peak at 14.4 MeV.

Before extracting resonance parameters with R-matrix calculations, evaluation of the experimental resolution, for which one needs to take into account beam size, angular and energy straggling for beam particles and recoiling α-particles, is required.

5. References

[1] K. Ikeda et al. Progr. Theo. Phys. Suppl. Extra Number (1968) 464.
[2] Y. Kanada En'yo et al., *Prog. Theor. Phys.* **93** (1995) 115.
[3] P. Descouvemont, *Nucl. Phys. A* **596** (1996) 285.
[4] V.Z. Goldberg et al., *Phys. Rev. C* **69** (2004) 024602.
[5] G. Ciavola et al., *Nucl. Phys. A* **701** (2002) 54c.
[6] M. Zadro et al, *NIM B* **259** (2007) 836.
[7] N.Soic et al., *Europhys. Lett.* **63** (2003) 524.

Journal of Physics: Conference Series **366** (2012) 012014 doi:10.1088/1742-6596/366/1/012014

Symmetry-Adapted *Ab Initio* Shell Model for Nuclear Structure Calculations

J P Draayer[1], **T Dytrych**[1], **K D Launey**[1] **and D Langr**[2]

[1]Department of Physics and Astronomy, Louisiana State University, Baton Rouge, LA 70803, USA

[2] Department of Computer Systems, Czech Technical University in Prague, Prague, Czech Republic

E-mail: draayer@lsu.edu

Abstract. An innovative concept, the symmetry-adapted *ab initio* shell model, that capitalizes on partial as well as exact symmetries that underpin the structure of nuclei, is discussed. This framework is expected to inform the leading features of nuclear structure and reaction data for light and medium mass nuclei, which are currently inaccessible by theory and experiment and for which predictions of modern phenomenological models often diverge. We use powerful computational and group-theoretical algorithms to perform *ab initio* CI (configuration-interaction) calculations in a model space spanned by SU(3) symmetry-adapted many-body configurations with the JISP16 nucleon-nucleon interaction. We demonstrate that the results for the ground states of light nuclei up through $A = 16$ exhibit a strong dominance of low-spin and high-deformation configurations together with an evident symplectic structure. This, in turn, points to the importance of using a symmetry-adapted framework, one based on an LS coupling scheme with the associated spatial configurations organized according to deformation.

1. Introduction

With considerable progress made in recent years in the development of realistic two- and three-nucleon interactions (e.g., [1, 2, 3, 4]), *ab initio* approaches [5, 6, 7], which build on fundamental principles, have successfully described properties of light nuclei while establishing a link between few-nucleon interactions, linked to Quantum Chromodynamics (QCD) considerations, and many-body nuclear dynamics. The predictive power of these approaches to nuclear structure and reactions depends critically on the choice of the realistic interaction and on the ability of the underlying theory, while restricted to available computational resources, to describe multifaceted properties of atomic nuclei from single-particle effects to collective phenomena. For example, light nuclei with equal numbers of protons and neutrons ($N = Z$) and total particle numbers that are multiples of four ($4n$ systems) display a complex pattern in their low-lying energy spectra with certain states continuing to remain a challenge for *ab initio* techniques. An important example of such a configuration is the first excited 0^+ state in ^{12}C (with impact on stellar evolution and nucleosynthesis); another is the first excited 0^+ state in ^{16}O.

The no-core shell model (NCSM) [5] is an *ab initio* configuration-interaction (CI) method that has achieved a good description of low-lying states and nuclear reactions for nuclei up through the p shell [8, 9, 10, 11]. This method uses the Lanczos algorithm to compute the lowest few eigenvalues and associated eigenstates of a realistic Hamiltonian matrix whose elements are calculated in an m-scheme basis, i.e. a basis of Slater determinants constructed from single-particle wavefunctions of the harmonic oscillator. The main limitation of this approach, and its predictive power, is inherently coupled with the combinatorial growth of the m-scheme basis with increasing nucleon number and maximal number of total harmonic oscillator quanta $N_{\max}$.

We have developed a next-generation *ab initio* model, the *ab initio* symmetry-adapted no-core shell model (SA-NCSM), that adopts the first-principle concept, and capitalizes on exact and approximate symmetries of nuclei. Our goal is, using the SA-NCSM model, to reproduce and predict properties of heavy as well as light nuclei, starting with and building upon QCD-informed and inspired interactions. This study focuses on the following current and near-future developments:

- Exploit the existing NCSM technology to prove the method efficacy, revealing (or not) any inherent limitations.
- Explore the need (or not) for interaction renormalization, as well as for winnowing of the model space to physically relevant and tractable subspaces.
- Evaluate the extensibility of the theory and its characteristics vis-à-vis the current as well as emerging computational resources.
- Study the emergence of collective phenomena and tracking their evolution to fundamental features of the interaction.
- Apply the theory to studies of extreme phenomena known to be important to understanding nuclear systems.
- Extend the theory to include coupling to the continuum, and apply the results to studies of nuclear reactions.
- Develop a user friendly desktop version of the SA-NCSM code for simple applications as well as educational and training purposes.

The significance of the SU(3) group, used in the SA-NCSM to construct the many-body basis, for a microscopic description of the nuclear collective dynamics can be seen from the fact that it is the symmetry group of the Elliott model [12], and a subgroup of the $\mathrm{Sp}(3,\mathbb{R})$ symplectic model [13]. Hence, the SA-NCSM holds promise to expand the reach as well as the impact of current *ab initio* approaches toward describing heavier mass nuclei together with collective, deformed, and cluster substructures. This is achieved by recognizing that the choice of coordinates, especially when deformed nuclear shapes dominate, is crucial, and that the SA-NCSM affords a solution in terms of coordinates that reflect symmetries inherent to the nuclear system. While the SA-NCSM states can be obtained through a unitary transformation from the m-scheme basis used in the NCSM, and hence span the entire space, the growth of the model space within the SA-NCSM framework can be managed by winnowing to only physically relevant states as determined through symmetry considerations.

2. SU(3) Symmetry-Adapted Theoretical Framework

The basis states of the *ab initio* SA-NCSM are constructed in the proton-neutron formalism and are labeled by the physical $\mathrm{SU}(3)\supset\mathrm{SO}(3)$ subgroup chain quantum numbers $(\lambda\,\mu)\kappa L$, and by proton, neutron, and total intrinsic spins S_π, S_ν, and S. The orbital angular momentum L is coupled with S to the total orbital momentum J and its projection M_J. Each basis state in this scheme is labeled schematically as $|\vec{\gamma}\,(\lambda\,\mu)\kappa L; (S_\pi S_\nu)S; JM_J\rangle$, where the set of deformation-related $(\lambda\,\mu)$ quantum numbers labels irreducible representations (irreps) of SU(3) that bring forward important information about nuclear shapes and deformation. For example, (00), $(\lambda\,0)$ and $(0\,\mu)$ describe spherical, prolate and oblate shapes, respectively. The label κ distinguishes multiple occurrences of the same L value in the parent irrep $(\lambda\,\mu)$, and $\vec{\gamma}$ distinguishes among configurations carrying the same $(\lambda\,\mu)$ and $(S_\pi S_\nu)S$ labels.

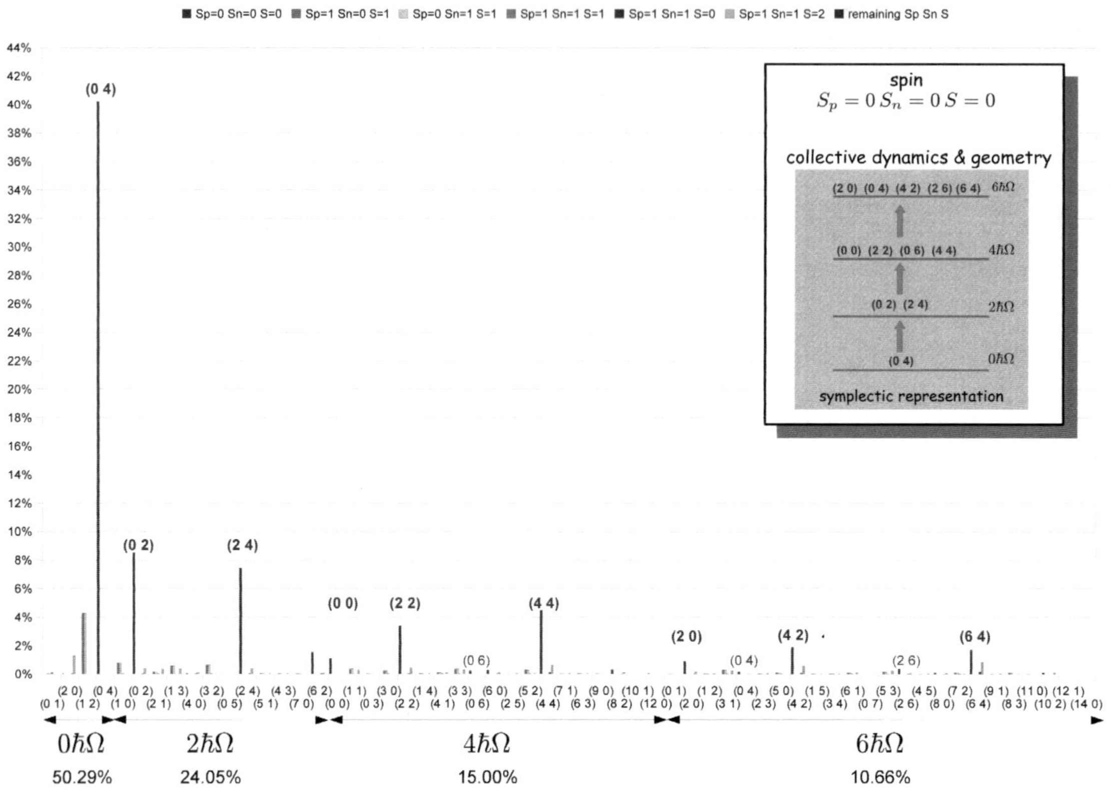

Figure 1. Probability distribution of the lowest calculated 0^+ state for ^{12}C over deformed subspaces labeled by $(\lambda\,\mu)$ for six of the most important spin components $\{S_\pi, S_\nu, S\} = \{0,0,0\}, \{1,0,1\}, \{0,1,1\}, \{1,1,1\}, \{1,1,0\}$ and $\{1,1,2\}$. Blue labels denote SU(3) quantum labels of basis states that belong to the leading $(0\,4)$ symplectic $\mathrm{Sp}(3,\mathbb{R})$ irrep. The wavefunction was obtained using the $N_{\max} = 6$ SA-NCSM with the JISP16 bare interaction and $\hbar\Omega = 10$ MeV.

The SA-NCSM implements a set of powerful algorithms [14, 15], which facilitate calculations of matrix elements of arbitrary (currently up to two-body, but expandable to higher-rank) operators in the SU(3)-scheme basis. The underlying principle behind the SA-NCSM computational kernel is a SU(3)-type Wigner-Eckhart theorem that enables a factorization of the problem into reduced matrix elements (*rmes*) and SU(3) coupling/recoupling coefficients. While the latter can be computed using the publicly available library [16], the former is calculated from a set of single-shell *rmes* by the repetitive application of a SU(3)-type reduction formula for *rmes* of operators acting on two independent subsystems. The resulting algorithm is very general; it allows for the evaluation of the Hamiltonian matrix elements and the use of the resulting eigenvectors to determine other experimental observables.

3. Results and Discussions

We calculated low-lying wavefunctions and some associated physical observables for ^{6}Li, ^{7}Li, ^{12}C, and ^{16}O, using the bare JISP16 interaction [1] for $N_{\max} = 6$ and $\hbar\Omega = 10$ MeV, and compared our results with those obtained using the m-scheme based NCSM approach. To demonstrate the efficacy of the symmetry-adapted selection scheme of the SA-NCSM, *ab initio* SA-NCSM calculations for the ground state of ^{6}Li and ^{7}Li were carried out. The model space included all

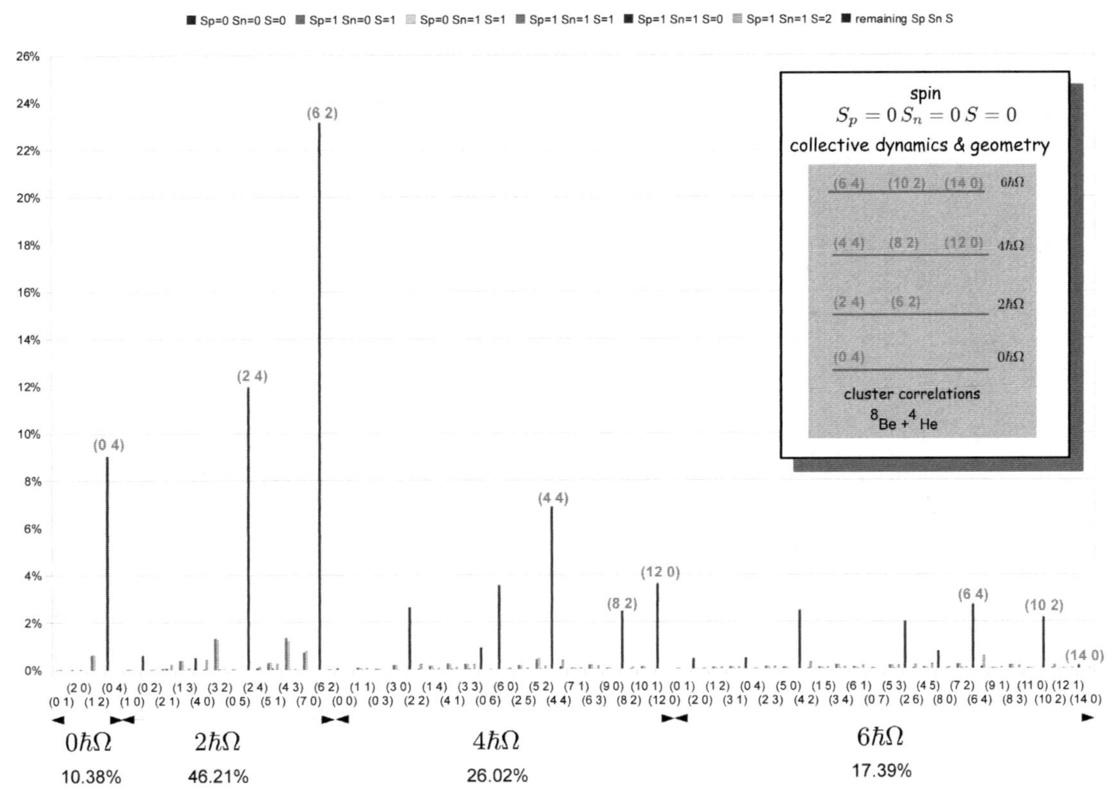

Figure 2. Probability distribution of the calculated 0_6^+ state for ^{12}C over deformed subspaces labeled by $(\lambda\,\mu)$ for six of the most important spin components $\{S_\pi, S_\nu, S\} = \{0,0,0\}, \{1,0,1\}, \{0,1,1\}, \{1,1,1\}, \{1,1,0\}$ and $\{1,1,2\}$. Green SU(3) labels denote physically allowed configurations of the ^{8}Be+^{4}He cluster basis states. The wavefunction was obtained using the $N_{\max} = 6$ SA-NCSM with the JISP16 bare interaction and $\hbar\Omega = 10$ MeV.

the configurations up through $N_{\max} = 4$ (full space) with the $N_{\max} = 6$ subspace restricted to only a few SU(3) and S configurations. The outcome points to the remarkable result that only a small fraction of the model space (19 SU(3) configurations for ^{6}Li and 15 SU(3) configurations for ^{7}Li) yields high overlap with the corresponding NCSM wavefunction and furthermore, most ($\sim 99\%$ for ^{6}Li and $\sim 97\%$ for ^{7}Li) of the corresponding binding energy.

In addition, the *ab initio* $N_{\max} = 6$ SA-NCSM results for the 0^+ ground state of ^{12}C reveal the dominance of the $0\hbar\Omega$ component with the foremost contribution coming from the leading $(0\,4)$ $S = 0$ irrep (Fig. 1). We find that important SU(3) configurations are typically organized into structures with $\mathrm{Sp}(3,\mathbb{R})$ symplectic symmetry. For example, the $(0\,4)$ symplectic irrep gives rise to $(0\,2)$ and $(2\,4)$ configurations in the $2\hbar\Omega$ subspace, as shown in the insert of Fig. 1, and those configurations indeed realize the major components of the wavefunction in this subspace. This further confirms the significance of the symplectic symmetry to nuclear dynamics. Similar results are observed for the 0^+ ground state of ^{16}O. The outcome points to the fact that the relevant model space can be systematically selected and further expanded to higher $N_{\max}$.

Another interesting result is related to the 'Hoyle' state, the second 0^+ state in ^{12}C that lies at 7.65 MeV and is of particular astrophysical interest. However, its *ab initio* description still remains beyond the reach of traditional NCSM calculations. Due to the presence of alpha-cluster structures, achieving convergence of this state requires ultra-large NCSM basis spaces for $N_{\rm max}$ considerably higher than the cutoffs currently considered. Nonetheless, in the SA-NCSM framework, it is now possible to identify the 'Hoyle' state and to study its structure. In particular, the lowest ten 0^+ states in ^{12}C as obtained in $N_{\rm max} = 6$ SA-NCSM calculations using the JISP16 interaction were examined for characteristic SU(3) configurations that could point to a cluster structure. These configurations, given in the insert of Fig. 2, can be easily understood in the framework of a microscopic cluster model, which builds upon ^{8}Be+^{4}He clusters and yields physically allowed deformations of the compound ^{12}C system. Among all excited 0^+ states, only the sixth 0^+ state reveals a pronounced pattern predominantly comprised of the characteristic deformations (Fig. 2). Hence, it is this state that is expected to converge to the 7.65-MeV 0^+ of ^{12}C for large $N_{\rm max}$ spaces, with higher-lying configurations, such as (12 0), (14 0) and (16 0) expected to grow in importance.

4. Conclusion

The SA-NCSM advances an extensible microscopic framework for studying nuclear structure and reactions that capitalizes on advances being made in *ab initio* methods while exploiting symmetries – exact and partial, known to dominate the dynamics. We have developed a symmetry-adapted *ab initio* shell-model code with the view toward exploring the properties of nuclei far from stability using externally provided realistic interactions derived from Quantum Chromodynamics (QCD) considerations. This, in turn, allows us to begin to address, in a focused way, the origin of collective degrees of freedom in nuclei, that is, the emergence of collectivity from fundamental principles.

Acknowledgements

Supported by the U.S. National Science Foundation (PHY-0500291 & OCI-0904874), the U.S. Department of Energy (DOE-0904874), and the Southeastern Universities Research Association. This research used resources of the National Energy Research Scientific Computing Center, which is supported by the Office of Science of the U.S. Department of Energy under Contract No. DE-AC02-05CH11231.

References

[1] Shirokov A M, Vary J P, Mazur A I, and Weber T A 2007 *Phys. Letts. B* **644** 33
[2] Machleidt R 2001 *Phys. Rev. C* **63** 024001
[3] Coon S A and Han H K 2001 *Few-Body Syst.* **30** 131
[4] Entem D R and Machleidt R 2003 *Phys. Rev. C* **68** 041001
[5] Navrátil P, Vary J P, and Barrett B R 2000 *Phys. Rev. Lett.* **84** 5728; 2000 *Phys. Rev. C* **62** 054311
[6] Wiringa R B and Pieper S C 2002 *Phys. Rev. Lett.* **89** 182501
[7] Hagen G, Papenbrock T, Dean D J, and Hjorth-Jensen M,2008 *Phys. Rev. Lett.* **101** 092502
[8] Maris P, Shirokov A M, and Vary J P 2010*Phys. Rev. C* **81** 021301(R)
[9] Maris P, Vary J P, Navrátil P, Ormand W E, Nam H., and Dean D J 2011*Phys. Rev. Lett.* **106** 202502
[10] Navrátil P 2004 *Phys. Rev. C* **70** 054324
[11] Quaglioni S and Navrátil P 2008*Phys. Rev. Lett.* **101** 092501
[12] Elliott J P 1958 *Proc. Roy. Sot. A* **245** 128
[13] Rosensteel G and Rowe D J 1977 *Phys. Rev. Lett.* **38** 10
[14] Draayer J P, Leschber Y, Park S C, and Lopez R 1989 *Comput. Phys. Comm.* **56** 279
[15] Bahri C and Draayer J P 1994 *Comput. Phys. Comm.* **83** 59
[16] Draayer J P and Akiyama Y 1973 *J. Math. Phys.* **14** 1904

Extended (pseudo-)SU(3) models for nuclear structure

K. P. Drumev[1], A. I. Georgieva[1] and J. P. Draayer[2]

[1] Institute for Nuclear Research and Nuclear Energy,
Bulgarian Academy of Sciences, Sofia, Bulgaria;
[2] Department of Physics and Astronomy,
Louisiana State University, Baton Rouge, LA 70803, USA

Abstract. As a simpler version of an extended (pseudo-)SU(3) model which acts in a model space of more than one oscillator shells, an extended pairing-plus-quadrupole model is introduced. The Hamiltonian consists of quadrupole-quadrupole plus pp-, nn- and pn-isovector and isoscalar pairing terms as well as terms describing the pair-scattering between two shells. Energy spectra and shape parameters for two $N = Z$ nuclear systems having the same nucleon content as the nuclei ^{20}Ne and ^{60}Zn which belong to two different areas of the nuclear chart ($ds - fp$ or $upper - fp - gds$ shells) are calculated and the role of the two parts in the interaction, depending on the parameter strengths, is discussed. Both extended models provide for a microscopic description of nuclei while being selective on the choice of the relevant model space.

keywords: models based on group theory, neutron-proton pairing, pairing-plus-quadrupole model, SU(3) model, SO(8) model

1. Introduction

Until recently, SU(3) shell-model calculations - real SU(3) [1] for light nuclei and pseudo-SU(3) [2] for heavy nuclei - have been performed in either only one space (protons and neutrons filling the same shell, e.g. the ds shell) or two spaces (protons and neutrons filling different shells, e.g. for rare-earth and actinide nuclei). Number of interesting and important results for low-energy features like energy spectra, shape description and electromagnetic transition strengths, have been published over the years [3, 4, 5].

Up to now, SU(3)-based methodologies have not been applied to nuclei with mass numbers $A = 56$ to $A = 100$, which is an intermediate region where conventional wisdom suggests the break down of the assumptions that underpin their use in the other domains. In particular, the $g_{9/2}$ intruder level that penetrates down from the shell above due to the strong spin-orbit splitting appears to be as spectroscopically relevant to the overall dynamics as the normal-parity $f_{5/2}, p_{3/2}, p_{1/2}$ levels. Specifically, in this region the effect of the intruder level cannot be ignored or mimicked through a "renormalization" of the normal-parity dynamics which is how it has been handled to date. Moreover, both protons and neutrons occupy the same oscillator shells which suggests strong proton-neutron correlations. These facts should be addressed with the appropriate choice of a Hamiltonian.

In order to build a complete shell-model theory, opposite-parity levels need to be included in the model space especially if experimentally observed high-spin or opposite-parity states are to be described. After we shortly introduce the basics of the extended SU(3) shell model which, for the first time, explicitly includes particles from the complete unique-parity sector, we present some results in the case of a simpler Hamiltonian consisting of only the pairing plus quadrupole-quadrupole terms.

2. Extended (pseudo-)SU(3) model

It is a microscopic theory in the sence that both SU(3) generators - the angular momentum and the quadrupole operators - are given in terms of individual nucleon coordinate and momentum variables. Results for the quality of the pseudo-SU(3) symmetry for the nuclei ^{64}Ge, ^{68}Se [6] with realistic interactions suggest that one can perform symmetry-adapted calculations thus reducing significantly the size of the model space.

The many-particle basis states

$$|\{a_\pi; a_\nu\}\rho(\lambda,\mu)\kappa L, \{S_\pi, S_\nu\}S; JM\rangle \tag{1}$$

are built as $SU(3)$ proton (π) and neutron (ν) strongly-coupled configurations with well-defined particle number and good total angular momentum J. Here, the proton and neutron quantum numbers are indicated by $a_\sigma = \{a_{\sigma N}, a_{\sigma U}\}\rho_\sigma(\lambda_\sigma, \mu_\sigma)$, where the $a_{\sigma\tau} = N_{\sigma\tau}[f_{\sigma\tau}]\alpha_{\sigma\tau}(\lambda_{\sigma\tau}, \mu_{\sigma\tau})$ are the basis-state labels for the four spaces in the model (σ stands for π or ν, and τ stands for normal (N) or unique (U) parity levels). In the last expression, $N_{\sigma\tau}$ denotes the number of particles in the corresponding space, $[f_{\sigma\tau}]$ - the spatial symmetry label and $(\lambda_{\sigma\tau}, \mu_{\sigma\tau})$ - the SU(3) irrep label. Multiplicity indices $\alpha_{\sigma\tau}$ and ρ_σ count different occurrences of $(\lambda_{\sigma\tau}, \mu_{\sigma\tau})$ in $[f_{\sigma\tau}]$ and in the product $\{(\lambda_{\sigma N}, \mu_{\sigma N}) \times (\lambda_{\sigma U}, \mu_{\sigma U})\} \to (\lambda_\sigma, \mu_\sigma)$, respectively. First, the particles from the normal and the unique spaces are coupled for both protons and neutrons. Then, the resulting proton and neutron irreps are coupled to a total final set of irreps. The total angular momentum J results from the coupling of the total orbital angular momentum L with the total spin S. The ρ and κ are, respectively, the multiplicity indices for the different occurences of (λ, μ) in $\{(\lambda_\pi, \mu_\pi) \times (\lambda_\nu, \mu_\nu)\}$ and L in (λ, μ). The basis states (1) used in the model calculations are by construction directly linked to the shell-model Lie algebra $U(4\Omega)$, which contains the $SU(3)$ quantum numbers (λ_τ, μ_τ) for the proton and neutron systems.

The most general Hamiltonian that has been used until now for SU(3) model calculations is of the form

$$H = H_{s.p.} - \frac{\chi}{2} Q.Q - G\sum_{\sigma,\tau}(S^+)^{\sigma\tau}(S^-)^{\sigma\tau} + aJ^2 + bK_J^2. \tag{2}$$

It includes single-particle terms $H_{s.p.}$ as well as the quadrupole-quadrupole $Q.Q$ and the pairing interaction within a shell, represented by the standard pair-creation $(S^+)^{\sigma\tau} = \frac{1}{2}\sum_{\eta l j m_j}(-)^{l+j-m_j}(a^\dagger_{\eta l j m_j})^{\sigma\tau}(a^\dagger_{\eta l j -m_j})^{\sigma\tau}$ and annihilation $(S^-)^{\sigma\tau} = ((S^+)^{\sigma\tau})^\dagger$ operators. The two rotor-like terms J^2 and K_J^2 (the square of the total angular momentum and its projection on the intrinsic body-fixed axis) are used to "fine tune" the energy spectra, adjusting the moment of inertia of the ground band and the position of the gamma $K = 2^+$ bandhead, respectively. Other interactions that have been used are the Casimir operators of SU(3) of second and third order C_2 and C_3. Because of the specifics of the systems studied in the past, proton-neutron terms in the interaction have not been needed. The single-particle terms together with the pairing interactions mix the SU(3) basis states, which allows for a realistic description of the energy spectra of the nuclei.

3. Pairing-plus-quadrupole model: results in two shells

Pairing-plus-quadrupole model, introduced by Bohr, Mottelson and Pines [7], and Belyaev [8] is an important tool for the description of the nuclear structure for a couple of reasons but mostly because it uses a simple Hamiltonian which includes both short-range and long-range terms. Although various calculations in this model can be found [9, 10, 11], it has never been applied for full-space calculations in more than one of the low-lying oscillator shells or for restricted number of particles in higher-lying shells. SU(3) realization of the model up to two (proton and neutron) spaces has been developed [12, 13] and the effects of the quadrupole-quadrupole, identical-particle pairing, and even single-particle interactions have been studied.

When proton-neutron pairing terms are included as well, we consider simultaneously isovector ($S = 0, T = 1$) and isoscalar ($S = 1, T = 0$) pairing which have the SO(8) dynamical symmetry. The problem of the classification of the states for the case of the total pairing is solved in [14] for particles in one shell. It is also based on chain of the shell-model Lie algebra $U(4\Omega) \supset [U(\Omega) \supset SO(\Omega)] \otimes [SU_{ST}(4) \supset SU_S(2) \otimes SU_T(2)]$ and its complementarity to the chains of subalgebras typical for the SO(8) model. For example, the chain $SO(8) \supset SO_{ST}(6) \supset SO_S(3) \otimes SO_T(3)$ has been identified as complementary to the latter and used in the classification of the states for a given number of nucleons. For a certain nucleus (good T_z value), these pairing eigenstates can also be expressed as linear combinations of the SU(3) basis states (1). Solution to the scattering of pairs of particles between two orbitals has also been provided in the special case of seniority $\nu = 0$ and 1 [15].

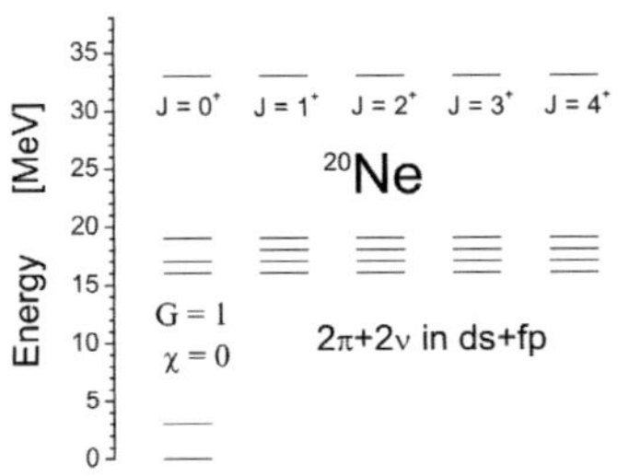

Figure 1. Isovector pairing for ^{20}Ne in the dsfp shell.

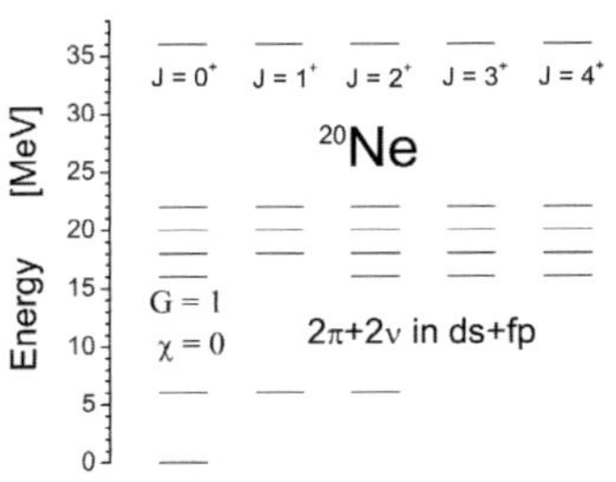

Figure 2. Total pairing for ^{20}Ne in the dsfp shell.

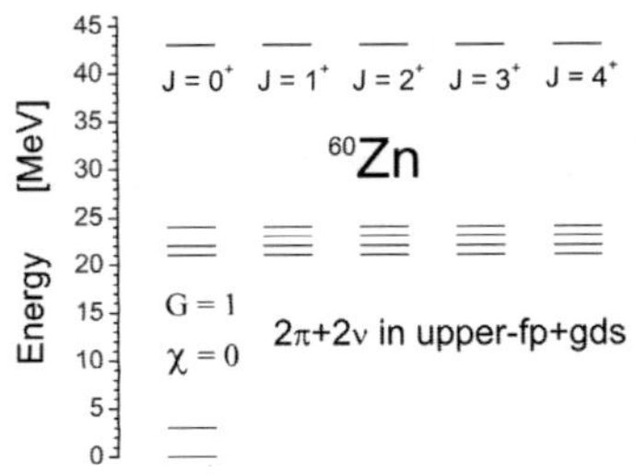

Figure 3. Isovector pairing for ^{60}Zn in the upper-fp+gds shell.

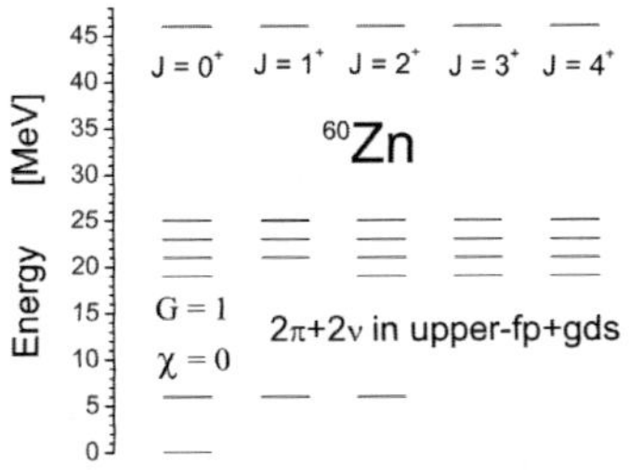

Figure 4. Total pairing for ^{60}Zn in the upper-fp+gds shell.

We use a Hamiltonian that is more general by considering additionally identical-particle pair-scattering terms ($\tau \neq \tau'$), proton-neutron terms ($\pi\nu$) and isoscalar terms which is obviously a more sophisticated form of the well-known pairing-plus-quadrupole model. It has the form

$$H = -\frac{\chi}{2}Q.Q - G\sum_{(JT)}\left\{\sum_{\sigma,\tau}(S^+)^{JT}_{\sigma\tau}(S^-)^{JT}_{\sigma\tau} + \sum_{\sigma,\tau\neq\tau'}(S^+)^{JT}_{\sigma\tau}(S^-)^{JT}_{\sigma\tau'} + \sum_{\tau,\tau'}(S^+)^{JT}_{\pi\nu,\tau}(S^-)^{JT}_{\pi\nu,\tau'}\right\} \quad (3)$$

where, for simplicity, all pairing terms are taken with the same strength. The sum over (JT) describes the isovector $(J = 0, T = 1)$ and the isoscalar $(J = 1, T = 0)$ both at $L = 0$ identical-particle pairing (first term) and pair-scattering (second term), and proton-neutron pairing and pair-scattering (third term in the braces).

Next, we present the results from calculations performed in two shells for two $N = Z$ nuclear systems which we, although not entirely accurate, dub ^{20}Ne and ^{60}Zn. There are no imposed restrictions on the model space and we do not compare with experimental data because the interaction (3) we use could not pretend to be complete and realistic.

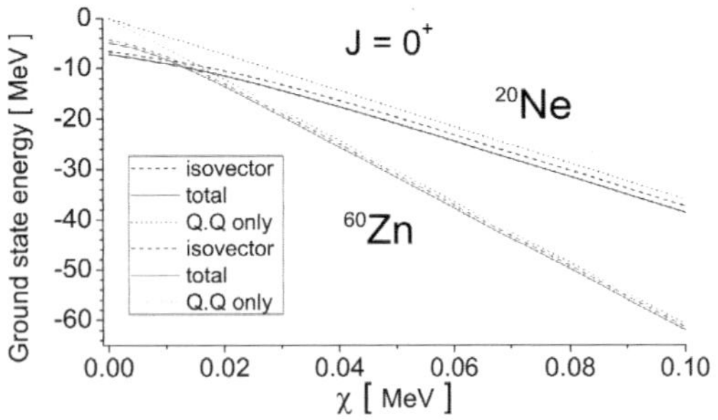

Figure 5. Ground-state energy for ^{20}Ne and ^{60}Zn in the pairing-plus-quadrupole model in two shells.

In Figures 1 and 2 (3 and 4), we can see the energy spectrum of the isovector- and the total-pairing interaction for the systems ^{20}Ne (^{60}Zn). By definition, the isovector pairing spectrum does not depend on the spin S and the isoscalar part does not depend on the isospin T which makes the difference between the Figures 1 and 2 (correspondingly 3 and 4) almost non observable. The states are clustered in groups for the different values of the total seniority quantum number which can be $\nu = 0, 2, 4$ for $J = 0$ but only $\nu = 2, 4$ for the remaining J values. While the first excited state in the isovector results is due to developing rotational features in isospace, the one in the total pairing results just reflects the pairing gap in the lower-lying (namely ds) shell of the model space.

Results from calculations with the Hamiltoinan (3) of the ground-state energy for the nuclei are presented in Figure 5. The quadrupole parameter χ takes different values while the values $G = 0.2$ MeV and $G = 0.1$ MeV for the pairing interaction are fixed as the ones for a typical ds-shell (fp-shell) nucleus. For low values of the quadrupole parameter χ the isovector part has prominent contribution which becomes comparable with the effect from the isoscalar part for values of χ bigger than 0.04 MeV.

Finally, we calculate the expectation values of the two shape parameters - the deformation parameter β and the triaxiality parameter γ, given by the following formulae [16]:

$$k\beta = \frac{2}{3}(C_2 + 3)^{\frac{1}{2}} \quad (4)$$

$$cos(3\gamma) = \frac{C_3}{2(C_2 + 3)^{\frac{3}{2}}}, \quad (5)$$

Journal of Physics: Conference Series **366** (2012) 012015 doi:10.1088/1742-6596/366/1/012015

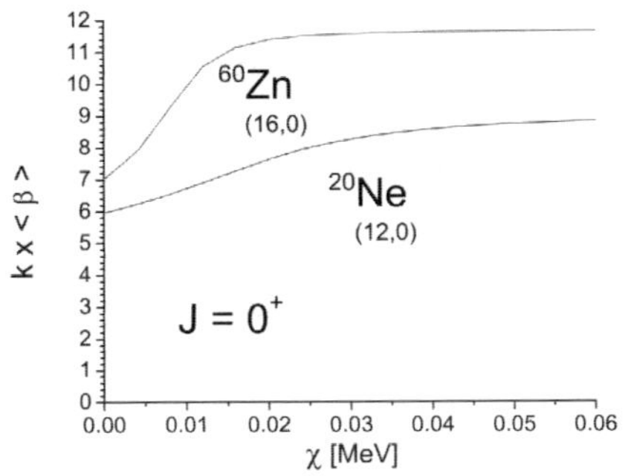

Figure 6. Deformation parameter β for ^{20}Ne and ^{60}Zn in the pairing-plus-quadrupole model in two shells.

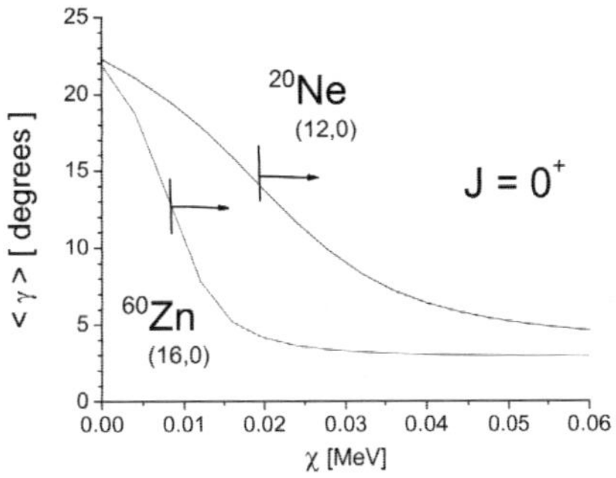

Figure 7. Triaxiality parameter γ for ^{20}Ne and ^{60}Zn in the pairing-plus-quadrupole model in two shells.

where $k = \sqrt{\frac{5}{9\pi}} A < r^2 >$ with A being the total number of nucleons, $< r^2 >$ - the mean square radius of the system, and C_2 and C_3 are the invariant Casimir operators of second and third order, respectively. The results shown in Figures 6 and 7 for the ground states of the two nuclear systems, reveal a steady rise (fall) in β (γ) for both nuclei with the increase of the parameter χ. Beyond the point marked with an arrow in Figure 7, the values of χ lead to eigenfunctions, composed primarily (at least 50 percent) of the leading irreducible representation.

In conclusion, the study presented here for pairing-plus-quadrupole Hamiltonian in two oscillator shells should serve as a base for more elaborated investigations of these and other interesting nuclear systems. More sophisticated interactions should be developed and tested in order to reveal the role of the particles from the unique-parity sector as well as the effectiveness of various truncations on the model space.

Acknowledgments

K.P.D. would like to acknowledge the financial support by the Organizing Committee for his participation at the School.

References

[1] Elliott J P 1958 *Proc. Roy. Soc. London*, Ser. A **245** 128; **A245** 562
[2] Ratna Raju R D, Draayer J P and Hecht K T 1973 *Nucl. Phys.* A **202** 433
[3] Vargas C, Hirsch J G and Draayer J P 2001 *Nucl. Phys.* A **690** 409
[4] Vargas C E, Hirsch J G and Draayer J P 2002 *Nucl. Phys.* A **697** 655
[5] Popa G, Hirsch J G and Draayer J P 2000 *Phys. Rev.* C **62** 064313; Vargas C et al. 2000, *Phys. Rev.* C **61**, 031301(R)
[6] Drumev K P 2008 Ph. D. Dissertation, Louisiana State University
[7] Bohr A Mottelson B R and Pines D 1958 *Phys. Rev.* **110** 936
[8] Belyaev S T 1959 *Mat. Fys. Medd. Dan. Vid. Selsk.* **31** No. 11
[9] Kisslinger L S and Sorensen R A 1960 *Mat. Fys. Medd. Dan. Vid. Selsk.* **32** No. 9
[10] Bes D R and Sorensen R A 1969 *Adv. Nucl. Phys.* **vol. 2** 129
[11] Kumar K and Baranger M 1965 *Nucl. Phys.* **62** 113
[12] Bahri C, Escher J and Draayer J P 1995 *Nucl. Phys.* A **592** 171
[13] Escher J, Bahri C, Troltenier D and Draayer J P 1998 *Nucl. Phys.* A **633** 662
[14] Kota V K B and Castilho Alcaras J A 2006 *Nucl. Phys.* A **764** 181
[15] Pang S C 1969 *Nucl. Phys.* A **128** 497
[16] Castaños O, Draayer J P and Leschber Y 1988 *Z. Phys.* A **329** 33; Leschber Y and Draayer J P 1987 *Phys. Lett.* B **190** 1

Journal of Physics: Conference Series **366** (2012) 012016 doi:10.1088/1742-6596/366/1/012016

Multinucleon transfer reactions: an overview of recent results

E Fioretto[1], L Corradi[1], S Szilner[2], D Bazzacco[3], E Farnea[3], A Gadea[4], D Mengoni[3], D Montanari[3], C Michelagnoli[3], T Mijatović[2], G Montagnoli[3], F Recchia[3], F Scarlassara[3], A M Stefanini[1], C A Ur[3], J J Valiente-Dobón[1], G Pollarolo[5], N Soić[2] and the PRISMA and CLARA collaborations

[1] INFN - Laboratori Nazionali di Legnaro, Viale dell'Università 2, Legnaro (PD), I-35020, Italy
[2] Ruđer Bošković Institute, Bijenička 54, Zagreb, HR-10001, Croatia
[3] Dipartimento di Fisica dell'Università di Padova and INFN, Via Marzolo 8, Padova, I-35131, Italy
[4] Instituto de Fisica Corpuscolar, CSIC-Universidad de Valencia, Valencia, E-46071, Spain
[5] Dipartimento di Fisica Teorica dell'Università di Torino and INFN, Via P. Giuria 1, Torino, I-10125, Italy

E-mail: enrico.fioretto@lnl.infn.it

Abstract. Large acceptance magnetic spectrometers, such as PRISMA installed at Laboratori Nazionali di Legnaro, gave a further boost to the renewed interest for multinucleon transfer reactions in the last decade. The large solid angles of these devices and the high resolving powers of their detection systems allowed to investigate the transfer process around and well below the Coulomb barrier and to perform nuclear structure studies in several mass regions of the nuclide chart when coupled with large γ-ray arrays such as CLARA. Selected results obtained with the PRISMA-CLARA set-up in odd argon isotopes populated by using the multinucleon transfer process and in sub-barrier transfer measurements are presented in this contribution. The status of an ancillary detector which is being developed for PRISMA in order to perform kinematical coincidence measurements is also reported.

1. Introduction

Multinucleon transfer in heavy ion reactions is a mechanism ranging from the quasi-elastic regime (i.e. few nucleon transfer and low total kinetic energy loss (TKEL)) to the deep-inelastic collisions (i.e. many nucleon transfer and large TKEL) which takes into account the largest fraction of the total reaction cross section at energies close to the Coulomb barrier [1]. The renewed interest in the last decade for multinucleon transfer reactions benefited from the construction of the new generation tracking spectrometers, based on the trajectory reconstruction, and the use of the state-of-art large area particle detectors.

PRISMA [2, 3, 4] is the large acceptance magnetic spectrometer designed to be used with heavy-ion beams accelerated at energies up to $E = 10A$ MeV by means of the Tandem-ALPI accelerator complex of LNL. It can operate as a standalone device or coupled to large γ-ray arrays such as the CLARA [5] set-up (until March 2008) or the AGATA Demonstrator (presently). Its coupling with the CLARA array allowed to make in-beam γ-spectroscopy of moderately neutron-rich nuclei populated

Journal of Physics: Conference Series **366** (2012) 012016 doi:10.1088/1742-6596/366/1/012016

by multinucleon transfer reactions through the identification of individual excited states and their population pattern. The experimental campaign of the PRISMA-CLARA set-up started in 2004 and has been completed at the end of March 2008 making use of about 50% of the total beam-time available at the Tandem/PIAVE-ALPI accelerator complex of LNL. Experiments were mainly addressed to obtain information on the shell evolution and the onset of new regions of deformation (collectivity, critical point symmetries) in medium-mass moderately neutron-rich nuclei.

2. Population of particle-vibration states in multinucleon transfer reactions

The coupling of single particle degrees of freedom to nuclear vibration quanta is very important for the understanding of the transfer strength distribution. These effects, still largely unexplored, are essential for the description of many basic states in the vicinity of closed shells. To this end we studied the population of states with a particle-phonon character in neutron transfer channels produced in the ^{40}Ar+^{208}Pb reaction [6]. The ^{40}Ar beam was extracted from an ECR ion source and accelerated by means of the superconducting Linac ALPI at $E_{\rm lab}$ = 255 MeV onto a 300 μg/cm^2 ^{208}Pb target. The yields of the projectile like fragments have been measured with PRISMA at three different angles $\theta_{\rm lab}$ = 46°, 54° (≈ $\theta_{\rm grazing}$) and 59° in order to cover most of the transfer flux in the reaction. The coincident γ-rays were detected with the CLARA array, located in the hemisphere opposite to PRISMA. The normalization for the different measured angles was ensured by a silicon SSBD monitor detector positioned at a forward angle.

The γ spectra measured in coincidence with ^{40,41,42,43}Ar corresponding to inelastic scattering, +1*n*, +2*n* and +3*n* channels are plotted in figure 1. They contain transitions from particle states as well as from states involving combinations of single-particle with a collective boson. New transitions have been identified in ^{41}Ar and ^{43}Ar ($9/2^- \rightarrow 7/2^-$ and $11/2^- \rightarrow 7/2^-$), and in ^{42}Ar ($6^+ \rightarrow 4^+$). A very strong population of the 2^+ states has been observed in ^{40}Ar and ^{42}Ar that act as cores in odd isotopes when a neutron is added. The energies, spins and parities of identified states agrees well with the results of *sd-pf* large-scale shell model (SM) calculations [7]. In ^{41}Ar and ^{43}Ar we observed, in addition to the known γ transitions of the low-lying states, strong lines at 1629.7(3) keV and at 1527.4(5) keV which we attribute to the population of the yet unknown $11/2^-$ states. These states can be understood as a coupling of collective boson to single-particle states (i.e. $|2^+, (f_{7/2})^1\rangle$ giving an $11/2^-$ stretched configuration). It is expected that the properties of these particle-phonon states are to a large extent determined by the properties of the corresponding phonon states. Figure 2 shows the comparison of the measured and SM calculated energies for the 2^+ and $11/2^-$ states of argon isotopes in the N = 20 – 28 region evidencing an excellent agreement for all argon isotopes shell. Solid circles are SM calculated energies, open squares are the adopted levels, whereas open triangles and the cross symbol correspond to the energies of $11/2^-$ in ^{41}Ar and ^{43}Ar from our experiment and in ^{45}Ar from ref. [8].

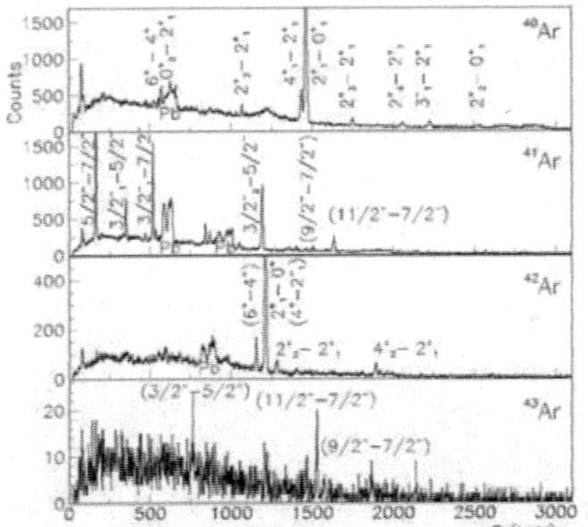

Figure 1. Doppler corrected γ-ray spectra for ^{40,41,42,43}Ar.

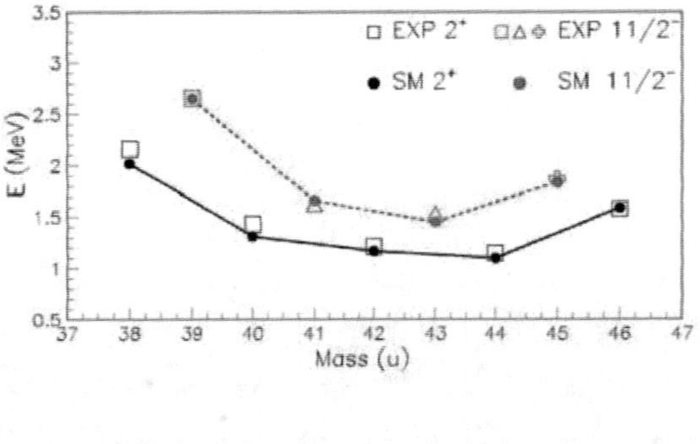

Figure 2. Energies of the 2+ and 11/2− states of argon isotopes with N = 20–28.

Journal of Physics: Conference Series **366** (2012) 012016 doi:10.1088/1742-6596/366/1/012016

The behavior of their energies displays that the evolution of the collectivity, in the even isotopes (2^+ energies) and in the odd isotopes ($11/2^-$ energies), is very similar. This further corroborates the particle-phonon character of these $11/2^-$ states. We expect that heavy ion induced transfer reactions populate states of similar character in more neutron rich isotopes. Argon isotopes with N≥28 have been populated in ^{238}U+^{48}Ca reaction [9], and the populated states in the ^{47}Ar behave similarly to odd-argon isotopes discussed here.

Experimental transfer yields have been interpreted within a reaction model [10] that explicitly treats the internal degrees of freedom of the two ions in terms of elementary modes, surface vibration and single particles. The significant population of particle-vibration states, reached via neutron transfer, demonstrates the importance of excitation of the states whose structure can be explained with the same degrees of freedom which are needed in the reaction model, i.e. coupling of the valence neutron to the vibration quanta.

3. Sub-barrier transfer measurements

In recent years there has been a growing interest in studying nuclear dynamics processes at energies well below the Coulomb barrier, in particular sub-barrier fusion reactions. This same energy range is also ideal to investigate transfer processes, which are strongly connected with fusion in an overlapping range of energies and angular momenta. In transfer reactions at sub-barrier energies nuclei follow almost pure Coulomb trajectories, excitation energies are restricted to few MeV and uncertainties in calculations associated with optical potentials can be minimized. These peculiar conditions should allow to extract more quantitative information on the mechanism of multiple transfer processes, in particular on the relative contribution of single particle and more complex degrees of freedom which include nucleon-nucleon correlations. However, available data for heavy ion transfer reactions in the sub-barrier region are extremely scarce or almost not existing due to the significant experimental difficulties of this kind of measurements (such as the strongly backward peaked angular distributions, the low kinetic energy for the backscattered projectile like fragments and the low cross sections). A suitable way to overcome these limitations is to make use of inverse kinematics detecting the lighter target like fragments with magnetic spectrometers at very forward angles [11, 12].

We recently measured with the spectrometer PRISMA the excitation functions for the neutron transfer channels populated in the inverse kinematics ^{96}Zr+^{40}Ca reaction [13] from the Coulomb barrier (330 MeV) to ~ 25% below (275 MeV). Projectile and target are closed or near-closed shell nuclei for both neutrons and protons, thus representing a good reference for a quantitative comparison with theoretical calculations. The ^{96}Zr beam was accelerated by the Tandem+ALPI accelerator complex of LNL onto a 50 μg/cm^2 $^{40}CaF_2$ target supported on a 15 μg/cm^2 C backing. Mass spectra of the target like fragments, measured with magnetic spectrometer PRISMA placed at 20°, evidenced the population of more than four neutron pick-up channels at energies close to the Coulomb barrier while at sub-barrier energies only one or at most two neutron transfers survive.

Making use of semiclassical conditions, one can extract the transfer probability P_{tr} as function of the distance of closest approach D, with P_{tr} defined as the ratio of transfer cross sections to the Rutherford one. This representation is significant only if semiclassical conditions are fulfilled and one deals with (almost) pure Coulomb trajectories. The case studied here well fulfills these requirements, with the further advantage that the Q-value distributions at the measured sub-barrier energies are quite narrow and corresponding to few MeV of excitation energy. At large ion-ion separation the radial behaviour of the form factor is governed by the exponential form of the bound-state wave function and the transfer probability is approximated by $P_{tr}(\theta) \cong e^{-2\alpha D(\theta)}$, where the parameter α is related to the binding energy E_b of the transferred nucleon, $\alpha = (2mE_b)^{1/2}/\hbar$, and $D(\theta)$ is the distance of closest approach. The excitation functions of transfer processes vs the distance of closest approach D are thus represented (in a semi-logarithmic plot) by straight lines with a slope $-\alpha$. Such behaviour is independent on the way in which transfer proceeds, as a successive process or as a simultaneous transfer. Contradictory results have been obtained in previous experiments around the Coulomb barrier, where slopes smaller than predicted were found [12], and at lower energies where no anomaly in the slope behavior has been clearly identified [14].

The transfer probabilities extracted from the yields of the $+1n$, $+2n$ and $+3n$ transfer channels as a function of the distance of closest approach D are shown in figure 3, together with the solid lines which are the results of the fitting procedure. Data for the $+4n$ channel are available only at the highest energies, therefore a reliable fit could not be performed. The extracted experimental slopes agree well with those expected by the binding energies. Given the correct behaviour of P_{tr} and keeping in mind the simplified assumptions mentioned before, we can make a phenomenological analysis which compares the probabilities for transfer channels with those expected from an independent particle transfer mechanism. It turns out that $P_{2n} = 3(P_{1n})^2$ and $P_{3n} = 3(P_{1n})^3$.

The two-neutron transfer channel has been analyzed with a semiclassical model that calculated, in the successive approximation, transitions to 0^+ states. Figure 4 shows the results of these calculations where the full line represents the inclusive transfer probability for one neutron transfer, the dotted line the ground state to ground state transition for the two-neutron transfer and the dashed line the transition to the first 0^+ excited state at 5.76 MeV in ^{42}Ca. It appears that the transfer probability for the transition to the excited 0^+ state in ^{42}Ca is much larger than the ground state one. But, by considering only 0^+ transitions the experimental cross section is underestimated by a factor of ~3. This enhancement is ascribed to transitions to states with large angular momentum and to transitions of non-natural character, indicating that more complex two-particle correlations have to be considered in the transfer process.

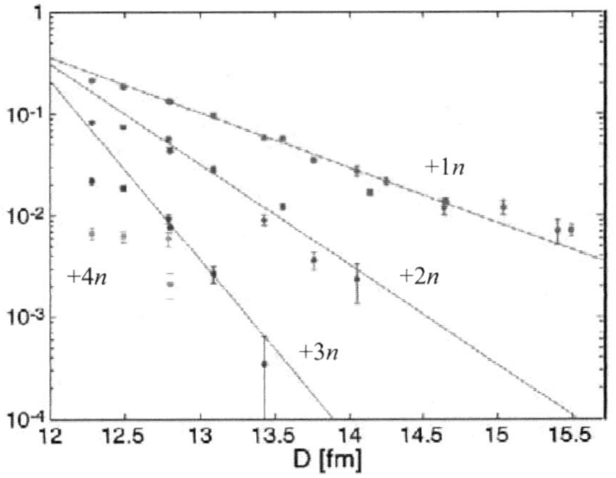

Figure 3. Extracted transfer probabilities P_{tr} as function of the distance of closest approach D for neutron transfer channels.

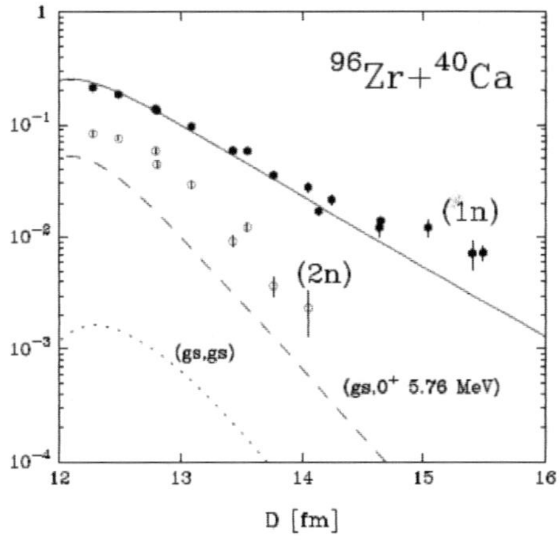

Figure 4. Theoretical transfer probabilities for one- and two-particle transfer (lines) in comparison with the experimental data (points).

4. A new ancillary detector for the PRISMA spectometer

A relevant aspect to be further investigated in transfer reactions that involves heavy ions is the influence of secondary processes, evaporation and fission that is important for the heavy partner. The determination of the survival probability against fission of heavy targetlike fragments (TLF) would help to understand how effectively multinucleon transfer reactions may be used to populate heavy nuclei [15]. We remark that data on the transfer induced fission are very scarce.

In order to check the relevance of the fission process in the population of the heavy fragments, we are planning to perform kinematical coincidence measurements where light fragments identified at the focal plane of PRISMA will be used to tag heavy partners entering into position sensitive device located at the correlation angle in the scattering chamber. To this end the magnetic spectrometer

Journal of Physics: Conference Series **366** (2012) 012016 doi:10.1088/1742-6596/366/1/012016

PRISMA is being equipped with a position sensitive detector composed of a single-sided silicon strip detector (SSSD) with a thickness of 300 μm and an active area of 5x5 cm^2. The detector is segmented in 16 resistive providing *X* and *Y* position information, timing and energy signals. In order to minimize the number of electronic channels, one end of each strip is grounded through a 100 Ω resistor and the position along the strip is obtained from the amplitude of the signal collected on the other end. The readout scheme of the detector is shown in figure 5. Energy and position resolutions of about 80 keV and 1 mm along the strip, respectively, were obtained in laboratory tests performed with 5.486 MeV α particles. A preliminary in-beam test has been performed by using the ^{40}Ca+^{90}Zr reaction at E_{lab} = 120 MeV. Figure 6 shows the *X-Y* scatter-plot (Zr ions) obtained with the entrance detector of PRISMA tagged by elastically scattered Ca ions entering into the SSSD. It was placed at 10 cm of distance from the target and covered with an aluminum mask composed of 8 by 10 holes, 1 mm diameter spaced 1.5 mm. Correlated Zr events in PRISMA cover only about half detector.

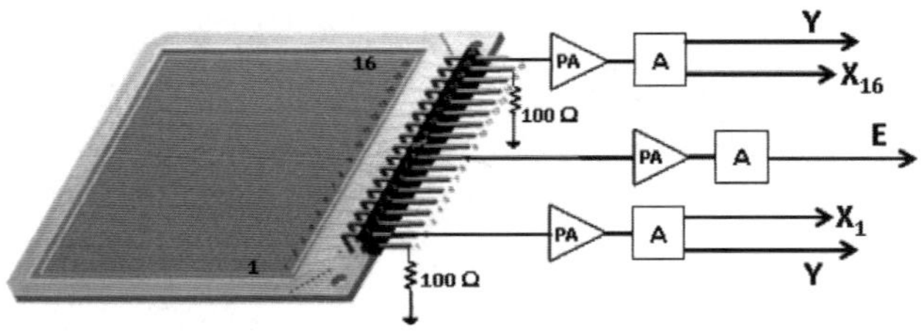

Figure 5. Si strip detector for kinematical coincidence measurements.

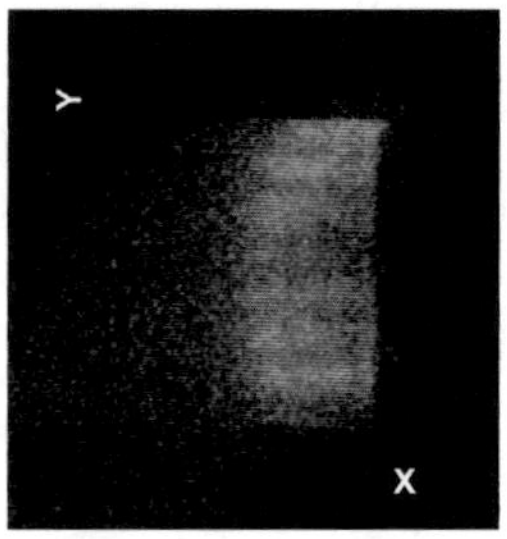

Figure 6. *X-Y* plot measured with the start detector of PRISMA.

5. Summary

In odd Ar isotopes populated via neutron transfer in the ^{40}Ar+^{208}Pb reaction, we observed a significant population of proposed $11/2^-$ states which well match a stretched configuration of the valence neutron coupled to the vibration quanta. The properties of such states are closely connected with the properties of the vibration quanta, allowing to follow the development of collectivity in odd argon isotopes. An enhancement by a factor 3 has been evidenced in the transfer probability extracted from the excitation function of the +2*n* transfer channel populated in the inverse kinematics reaction ^{96}Zr+^{40}Ca. A new ancillary detector for the PRISMA spectrometer will allow to study the effect of secondary processes such as the fission in the population of the TLF yields.

6. References

[1] Corradi L *et al* 2009 *J. Phys.* G **36** 113101
[2] Stefanini A M *et al* 2002 *Nucl. Phys.* A **701** 217c
[3] Montagnoli G *et al* 2005 *Nucl. Instr. and Meth.* A **547** 455
[4] Beghini S *et al* 2005 *Nucl. Instr. and Meth.* A **551** 364
[5] Gadea A *et al* 2004 *Eur. Phys. J.* A **20** 193
[6] Szilner S *et al* 2011 *Phys. Rev.* C **84** 014325
[7] Nowacki F and Poves A 2009 *Phys. Rev.* C **79** 014310
[8] Mengoni D *et al* 2010 *Phys. Rev* C **82** 024308
[9] Bhattacharyya S *et al.* 2008 *Phys. Rev. Lett.* **101** 032501
[10] Winther A 1994 *Nucl. Phys.* A **572** 191; Winther A 1995 *Nucl. Phys.* A **594** 203
[11] Jiang C L *et al* 1994 *Phys. Lett.* B **337** 59
[12] Jiang C L *et al* 1998 *Phys. Rev.* C **57** 2393
[13] Corradi L *et al* 2011 *Phys. Rev.* C **84** 034603
[14] Rehm K E *et al* 1993 *Phys. Rev.* C **47** 2731; Roberts R B *et al* 1993 *Phys. Rev.* C **47** R1831
[15] Corradi L *et al* 2002 *Phys. Rev.* C **66** 024606

Fixing the moment of inertia in the Bohr Hamiltonian through Supersymmetric Quantum Mechanics

D Bonatsos[1], P E Georgoudis[1], D Lenis[1], N Minkov[2] and C Quesne[3]

[1] Institute of Nuclear Physics, N.C.S.R. "Demokritos", GR-15310 Aghia Paraskevi, Attiki, Greece

[2] Institute of Nuclear Research and Nuclear Energy, Bulgarian Academy of Sciences, 72 Tzarigrad Road, 1784 Sofia, Bulgaria

[3] Physique Nucléaire Théorique et Physique Mathématique, Université Libre de Bruxelles, Campus de la Plaine CP229, Boulevard du Triomphe, B-1050 Brussels, Belgium

E-mail: georgoudis@inp.demokritos.gr

Abstract. A well known problem of the Bohr-Mottelson model is the behavior of nuclear moments of inertia. The coefficient of the angular part (γ, θ_i) of the kinetic energy is proportional to β^2. It is shown how a SUSYQM treatment of the β vibrations, emerges moments of inertia compatible with the experimental picture. This emergence is the result of the SUSYQM solution of the radial equation. Fitting results for the ^{162}Dy and ^{238}U in ground state, β_1 and γ_1 bands are also presented.

1. Introduction

Quesne and Tkachuk began to study SUSYQM methods for non-pointlike quantum oscillators [1, 2], that is harmonic oscillators with non-pointlike excitations. In principle, in such an oscillator the Heisenberg uncertainty relations are modified and this guides the modification of the canonical commutation relations. In [3] the equivalence of such an oscillator with a Schroedinger equation of a position dependent mass problem was established. In [4] a Schroedinger equation of the Bohr-Mottelson type was presented for the case of a mass dependent on the β degree of freedom. In this equation the coefficient of the angular part of the kinetic energy is modified as,

$$\beta^2 \sin^2\left(\gamma - \frac{2}{3}\pi k\right) \rightarrow \frac{\beta^2}{f^2(\beta)} \sin^2\left(\gamma - \frac{2}{3}\pi k\right). \quad (1)$$

$f(\beta)$ is called the deformation function which shapes the moments of inertia and its form will be determined by SUSYQM mathematical techniques.

2. SUSYQM for the β vibrations

In order to present a SUSYQM discussion of the β vibrations, a phase space of the β degree of freedom with commutation relations

$$[\beta, p_\beta] = i\hbar f(\beta), \quad (2)$$

Journal of Physics: Conference Series **366** (2012) 012017 doi:10.1088/1742-6596/366/1/012017

can be formed, based on [3, 5]. Because of the presence of $f(\beta)$, in the second quantization procedure the ladder operators will not be as usual but modified in general as,

$$A^{\pm} \to A^{\pm}(a,\mu,\nu) = \mp\sqrt{f(a;\beta)}\frac{d}{d\beta}\sqrt{f(a;\beta)} + W(\mu,\nu;\beta). \tag{3}$$

Here a deformed momentum operator is introduced through the deformation function $f(a;\beta)$ and the superpotential $W(\mu,\nu;\beta)$ which signals the SUSYQM method. From the parameters (a,μ,ν), only a will remain free. The Hamiltonian corresponding to these ladder operators will of course give good quantum numbers for the stationary states of β-vibrations, characterized by a function $R(\beta)$.

The principal SUSYQM demand states that the action of the Hamiltonian operator to the ground state shall give zero. The parameter ε_0 is introduced, which is assumed to be the energy of the ground state, and therefore the SUSYQM method is valid for the Hamiltonian,

$$A^{+}(a,\mu,\nu)A^{-}(a,\mu,\nu) = H - \varepsilon_0, \tag{4}$$

which gives zero eigenvalue for the vacuum. This Hamiltonian shall correspond to the radial equation which is [5],

$$HR = -\left(\sqrt{f}\frac{d}{d\beta}\sqrt{f}\right)^2 R + 2uR = 2\epsilon R. \tag{5}$$

This correspondence emerges the equation,

$$W^2(\mu,\nu;\beta) - f(\beta)W'(\mu,\nu;\beta) + \varepsilon_0 = 2u(\beta). \tag{6}$$

Now, the main result of [3] is that the Schroedinger equation of Eq. (5) is also obtained for the case of a position dependent mass problem, as discussed in [4, 6] with a change in the potential,

$$u \to u_{eff} = u + \frac{1}{4}ff'' + \frac{1}{6}(f')^2. \tag{7}$$

Therefore the energy of the ground state ε_0 can be determined from u_{eff} . This can be done if the specific potential $u_{eff}(\beta)$ has known superpotential and deformation function.

2.1. Shape invariance and the Davidson potential

The Schroedinger equation is known to be exactly solvable for the Davidson potential [7] . Shape invariance states that a potential gives exact solutions if and only if retains the same functional dependence under the change of its parameters. In Figure 1, shape invariance is shown for the Davidson's parameter β_0 which is fitted to a specific nucleus, namely the minimum reflects the ground state of the β vibrations.

We extend the shape invariance condition for the effective potential. This means that u_{eff} should retain the Davidson behavior for every change in the parameters, with

$$W^2(\mu,\nu;\beta) - f(\beta)W'(\mu,\nu;\beta) + \varepsilon_0 = 2u_{eff}(\beta) = k_1\beta^2 + k_0 + \frac{k_{-1}}{\beta^2}. \tag{8}$$

In [8], classes of shape invariant potentials have been studied with the identification of their corresponding superpotentials and deformation functions. The superpotential and deformation function for the Davidson case are,

$$W(\beta) = \frac{\mu}{\beta} + \nu\beta \quad , \quad f(\beta) = 1 + a\beta^2. \tag{9}$$

With this deformation function, moments of inertia receive a nice correction as can be seen in Figure 2. Paramater a controls in a certain nucleus the mass dependence on deformation. A theoretical expectation of this parameter is lacking, and a first, numerical picture can be seen in [5] for the γ unstable and axially symmetric prolate nuclei.

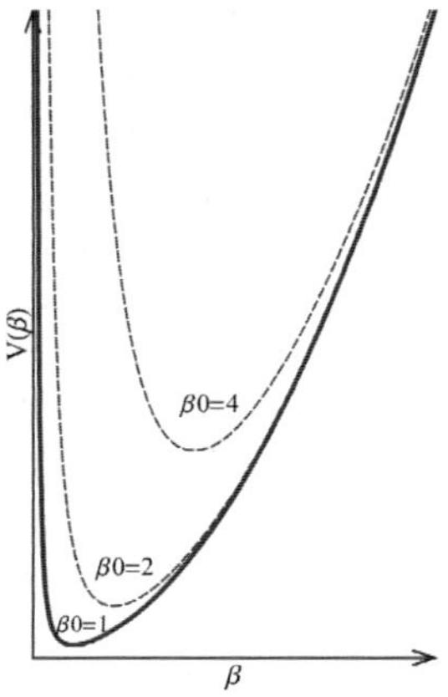

Figure 1. The Davidson Potential $V(\beta) = \beta^2 + \frac{\beta_0^4}{\beta^2}$ and its shape invariant behavior under a parameter shift.

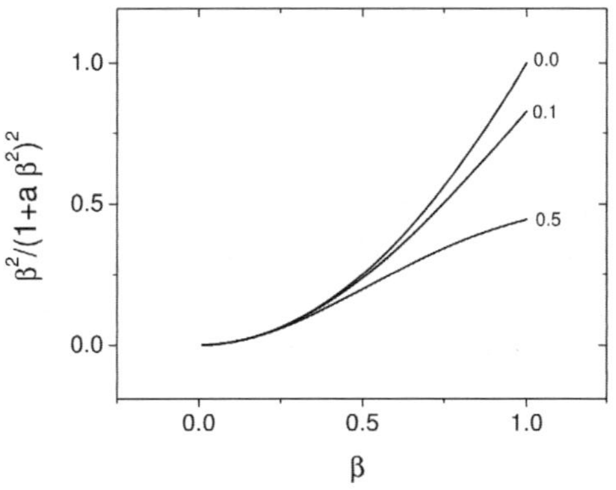

Figure 2. Moments of inertia with respect to β, for various values of a. $a = 0.0$ reflects the original Bohr Hamiltonian.

3. Energy spectrum

From Ref. [4] the coefficients of the effective potential are,

$$k_1 = 2 + a^2(12 + \Lambda), \qquad k_0 = a(13 + 2\Lambda), \qquad k_{-1} = 2 + \Lambda + 2\beta_0^4, \tag{10}$$

with $\Lambda = \tau(\tau + 3)$ for γ unstable and $\Lambda = \frac{L(L+1)-K^2}{3} + (6c)(n_\gamma + 1)$ for the axially symmetric behavior. The parameter c controls γ stiffness and n_γ is the quantum number for γ vibrations.

With these equations the energy of the ground state is determined from the expressions,

$$\mu(\mu + 1) = k_{-1}, \qquad \nu(\nu - a) = k_1, \qquad 2\mu\nu + \mu a - \nu + \varepsilon_0 = k_0, \tag{11}$$

$$\mu = -\frac{1}{2}\left(1 + \sqrt{9 + \Lambda + 8\beta_0^4}\right), \qquad \nu = \frac{a}{2}\left(1 + \sqrt{1 + \frac{8 + 4a^2(12 + \Lambda)}{a^2}}\right). \tag{12}$$

From these equations the energy of the ground state ϵ_0 is found to be

$$\epsilon_0 = \frac{19}{4}a + \frac{5}{2}a + \frac{1}{2}\sqrt{a^2 + 4k_1} + \frac{a}{2}\sqrt{1 + 4k_{-1}} + \frac{1}{4}\sqrt{(a^2 + 4k_1)(1 + 4k_{-1})} + a\Lambda. \tag{13}$$

Table 1. Normalized [to the energy of the first excited state, $E(2_1^+)$] energy levels of the ground state band (gsb) and the β_1 and γ_1 bands of ^{162}Dy and ^{238}U, obtained from the energy spectrum of Eq. (14) using the parameters given in [5], compared to available experimental data [9].

L	^{162}Dy exp	^{162}Dy th	^{238}U exp	^{238}U th	L	^{162}Dy exp	^{162}Dy th	^{238}U exp	^{238}U th
	gsb	gsb	gsb	gsb		γ_1	γ_1	γ_1	γ_1
0	0.00	0.00	0.00	0.00	2	11.0	11.2	23.6	24.7
2	1.00	1.00	1.00	1.00	3	11.9	12.1	24.6	25.5
4	3.29	3.30	3.30	3.31	4	13.2	13.3	25.9	26.7
6	6.80	6.80	6.84	6.86	5	14.7	14.7	27.4	28.1
8	11.41	11.41	11.54	11.57	6	16.4	16.5	29.2	29.8
10	17.04	17.01	17.27	17.33	7	18.5	18.5	31.2	31.7
12	23.57	23.49	23.97	24.06	8	20.7	20.8	33.5	33.9
14	30.90	30.74	31.51	31.63	9	23.3	23.3	36.0	36.3
16	38.90	38.70	39.82	39.97	10	25.9	26.0	38.8	39.0
18	47.58	47.28	48.78	48.98	11	29.0	28.9	41.7	41.9
20			58.31	58.61	12	31.4	32.1	44.9	45.0
22			68.31	68.77	13	35.5	35.5	48.3	48.3
24			78.71	79.44	14	39.4	39.9	51.9	51.8
26			89.46	90.55	15			55.7	55.5
28			100.57	102.08	16			59.7	59.4
30			112.10	113.99	17			63.9	63.4
					18			68.2	67.7
	β_1	β_1	β_1	β_1	19			72.7	72.0
0	17.3	15.7	20.6	20.6	20			77.3	76.6
2	18.0	16.7	21.5	21.6	21			82.1	81.3
4	19.5	19.0	23.5	24.0	22			87.0	86.1
6	21.9	22.6			23			91.9	91.0
8	24.6	27.4			24			97.0	96.1
					25			102.1	101.3
					26			107.4	106.6
					27			112.7	112.0

Actually this is the energy of the ground state band for the β vibration, because of the Λ dependence in each case. The energy for all the bands, for the β vibrations is found to be

$$\epsilon_n = \frac{1}{2}[k_0 + \frac{1}{2}a(3 + 2\Delta_1 + 2\Delta_2 + \Delta_1\Delta_2) + 2a(2 + \Delta_1 + \Delta_2)n + 4an^2], \qquad n = 0, 1, 2, \ldots, \quad (14)$$

with $\Delta_1 \equiv \sqrt{1 + 4k_{-1}}$, $\Delta_2 \equiv \sqrt{1 + 4\frac{k_1}{a^2}}$. The ground state band is obtained from $n = 0$, while the quasi-β_1 band is obtained from $n = 1$, and the quasi-β_2 band is obtained from $n = 2$.

4. Fitting

In [5] the above energy spectrum was fitted for the γ-unstable and axially symmetric prolate nuclei. The fitting measure was the Gaussian error,

$$\sigma = \sqrt{\frac{\sum_{i=1}^{n}(E_i(exp) - E_i(th))^2}{(n-1)E(2_1^+)^2}}. \quad (15)$$

Journal of Physics: Conference Series **366** (2012) 012017 doi:10.1088/1742-6596/366/1/012017

In the axially symmetric prolate case the free parameters are (β_0, a, c). A comparison to experimental data [9] is presented in Table 1, which shows fitting results for the ground state, β_1 and γ_1 bands in the cases of ^{238}U and ^{162}Dy. Both the bandheads and the spacings within bands are in general well reproduced. This is particularly true for the ground state and the γ_1 bands. The deviation in the gsb of ^{162}Dy reaches 0.6% at $L = 18$, while in the gsb of ^{238}U it reaches 1.7% at $L = 30$. The experimental levels of the γ_1 band of ^{162}Dy (up to $L = 14$) extend over 28.4 energy units, while the corresponding theoretical predictions spread over 28.7 units, the difference being of the order of 1%. Similarly in ^{238}U the experimental spread of the γ_1 band (up to $L = 27$) is 89.1 energy units, while the theoretical one is 87.3 units, the difference being of the order of 2%.

On the other hand, the theoretical level spacings within the β_1 bands are larger than the experimental ones. It is worth mentioning that results for $B(E2)$ transition rates in axially symmetric prolate nuclei [5], reveal a similar behavior. $B(E2)$s are in overall good agreement with the experimental data, apart from the $\beta_1 \to$ gsb transitions. The theoretical predictions are overestimating these transitions. This behavior along with the β_1 bands could be seen as an effect of the Davidson potential. It rises too fast in its right part as seen in Fig. 1, thus producing a large gap between the ground state and the β_1 bands and in addition increasing their interlevel spacing. However, as it has been pointed out in [10], the form of the quadrupole operator should be tested by changing the mass coefficients of the Bohr Hamiltonian.

References

[1] Quesne C, Tkachuk V M, *J.Phys.A* A36 (2003) 10373.
[2] Quesne C, Tkachuk V M, *J.Phys.A* A37 (2004) 10095.
[3] Quesne C, Tkachuk V M, *J.Phys.A Math. Gen.* 37 (2004) 4267.
[4] Bonatsos D, Georgoudis P, Lenis D, Minkov N, Quesne C, *Phys.Let.B* 683 (2010) 264.
[5] Bonatsos D, Georgoudis P E, Lenis D, Minkov N, Quesne C, *Phys. Rev. C* 83 (2011) 044321.
[6] Bonatsos D, these proceedings.
[7] Davidson P M, *Proc. R. Soc. London Ser. A* 135 (1932) 459.
[8] Bagchi B, Banerjee A, Quesne C, Tkachuk V M, *J. Phys. A: Math. Gen.* 38 (2005) 2929.
[9] Nuclear Data Sheets, as of December 2005.
[10] Jolos R V, von Brentano P, *Phys. Rev. C* 77 (2008) 064317.

Journal of Physics: Conference Series **366** (2012) 012018 doi:10.1088/1742-6596/366/1/012018

Measurements of Dynamical Dipole in isospin asymmetric fusion reactions

A. Giaz[1], A. Corsi[1,2], F. Camera[1], A. Bracco[1], F.C.L. Crespi[1], S. Leoni[1], R. Nicolini[1], V. Vandone[1], G. Benzoni[3], N. Blasi[3], S. Brambilla[3], B. Million[3], O. Wieland[3], M. Cinausero[4], M. Degelier[4], F. Gramegna[4], V.L. Kravchuk[4], T. Marchi[4], V. Rizzi[4], L. Bardelli[5], S. Barlini[5], M. Bini[5], S. Carboni[5], G. Casini[5], M. Chiari[5], A. Nannini[5], G. Pasquali[5], S. Piantelli[5], G. Poggi[5], G. Baiocco[6], M. Bruno[6], M. D'agostino[6], L. Morelli[6], V. Vannini[6], M. Colonna[7], M. Di Toro[7], C. Rizzo[7], P. Bednarcyk[8], M. Ciemala[8], M. Kmiecik[8], A Maj[8], K. Mazurek[8], W. Menczynski[8], R. Alba[9], C. Maiolino[9], D. Santonocito[9], D. Montanari[10,2], A. Ordine[11]

[1]Univerità degli studi di Milano - INFN sezione di Milano, Milano, Italy
[2]CEA, Centre de Saclay, IRFU/Service de Physique Nuclaire, F-91191 Gif-sur-Yvette, France
[3]INFN sezione di Milano, Milano, Italy
[4]INFN Laboratori Nazionali di Legnaro, Legnaro (PD), Italy
[5]Universitá di Firenze and INFN sezione di Firenze, Firenze, Italy
[6]Universitá di Bologna and INFN sezione di Bologna, Bologna, Italy
[7]Universitá di Catania and INFN Laboratori Nazionali del Sud, Catania, Italy
[8]The Henryk Niewodniczanski Institute of Nuclear Physics, PAN, Kracow, Poland
[9]INFN Laboratori Nazionali del Sud, Catania, Italy
[10]Universitá di Padova and INFN sezione di Podova, Padova, Italy
[11]INFN sezione di Napoli, Napoli, Italy

E-mail: agnese.giaz@mi.infn.it

Abstract. In heavy ion nuclear reactions the process leading to complete fusion is expected to produce pre-equilibrium γ-ray emission, if particular conditions are met. Indeed, when there is an N/Z asymmetry between projectile and target, charge equilibration takes place with a collective dipole oscillation, called Dynamical Dipole (DD), associated to a γ-ray emission. The existing experimental data concerning this pre-equilibrium γ-ray emission are still rather scarce and manly concentrated in the A$\simeq$132 mass region. The very preliminary results concerning the measurement of the DD γ-ray emission in the fusion reaction ^{16}O (E$_{lab}$=192 MeV) + ^{116}Sn at 12 MeV/u will be presented and compared with the γ yield measured for the same reaction at 8.1 and 15.6 MeV/u. The present experiment aims at the measurement of the total emission yield of the DD at 12 MeV/u where the predicted theoretical yield does not completely reproduce the experimental data. The experiment has been performed at the INFN Legnaro Laboratories using the GARFIELD-HECTOR array.

1. Introduction

The study of the collective properties of a nuclear system is a powerful tool to understand the structure which lies inside the nucleus. A successful technique which has been used in this field

Journal of Physics: Conference Series **366** (2012) 012018 doi:10.1088/1742-6596/366/1/012018

is the measurement of the decay of the highly collective Giant Dipole Resonance (GDR). In fact, GDR can be used as a probe for the internal structure of hot nuclei and, in addition, constitutes a clock for the thermalization process. Using the fusion-evaporation reaction, it has been recently possible to study i) the yield of the high-energy gamma-ray emission of the Dynamical Dipole which takes place during the fusion process if there an N/Z asymmetry and ii) the degree of isospin mixing at T=0 in the decay of ^{80}Zr [1].

In heavy ion nuclear reactions the process leading to complete fusion is expected to produce pre-equilibrium gamma-ray emission. If there is isospin asymmetry between the projectile and the target, it is possible to form a time dependent dipole moment which is associated with gamma radiation emission, called dynamical dipole (DD). This phenomenon is expected to depend on the energy of the projectile and on the size of the dipole moment calculated as

$$D(0) = \frac{NZ}{A}|X_Z(0) - X_N(0)| = \frac{r_0(A_p^{1/3} + A_t^{1/3})}{A} Z_p Z_t \left| \frac{N_t}{Z_t} - \frac{N_p}{Z_p} \right|. \tag{1}$$

The existing experimental data are still rather scarce and, although some systematic exists in the A$\simeq$132 mass region, the data do not seem to follow the theoretical predictions concerning the dependence of the DD yield emission with beam energy. Experimetally, it was obseved that Dynamical Dipole emission increases with beam energy up to $\approx$10 MeV/u, then the yield significantly decreases [2], [3], [4]. Qualitatively, an increase of the DD yield with beam energy is expected as the dynamics in the neck region between projectile and target, where the DD oscillation develops, become faster. A decreasing yield for higher beam energies is similarly expected due to the damping related to fast processes like pre-equilibrium neutron emission and proton-neutron direct collisions that reduces the isospin asymmetry and damps the isovector oscillation [5]. The DD yield dependence with beam energy is the result of the interplay between these two phenomena.

The model which better describes the Dynamical Dipole uses the dynamical evolution of the dipole in the framework of BNV prescription and employs the bremsstrahlung expression for the calculation of the photon yield. It is important to point out that the model critically depends on the nuclear equation of state with its symmetry term and on its density dependence.In ^{16}O + ^{116}Sn the dependence on the nuclear EOS with its symmetry term is not so strong, instead this dependence become significantly in neutron rich nuclei such as ^{132}Sn. The theoretical model predict that the dependence on the symmetry term of EOS is clear visible for ^{132}Sn + ^{58}Ni [6]; so it is necessary to study this phenomena in neutron rich nuclei. Consequently, the dynamical dipole emission, being related to the isospin asymmetry in the entrance channel, is also affected by the value of the symmetry energy and in the future could be used to estimate a value for it.

The understanding of this mechanism is relevant in the study of exotic nuclear matter especially in view of the future availability of intense radioactive beams that are expected to give a boost to this kind of measurements expecially if ^{132}Sn beams will be available with high intensity.

An experimental campaign on the DD yield has been performed at LNL [14] (Laboratori Nazionali di Legnaro), using GARFILED-HECTOR arrays [7]. During this campaign the DD emission in the fusion reaction ^{16}O (E_{lab}=192 MeV) + ^{116}Sn has been measured in function of beam energy (in particular at 8.1MeV/u, at 12MeV/u and at 15.6MeV/u); data at 8.0 and 15.6 MeV/u have been already discussed in [8]. Some preliminary results of an experiment which has measured the DD emission in the fusion reaction ^{16}O (E_{lab}=192 MeV) + ^{116}Sn will be presented. The experiment aims at the measurement of the total emission yield of the DD at an intermediate value of beam energy (12 MeV/u). The experiment set up is composed by the GARFIELD array (for the measurement of light charged particles), by an array of phoswich detectors (for the measurement of fusion residues) and by the HECTOR+HELENA arrays of BaF_2 detectors

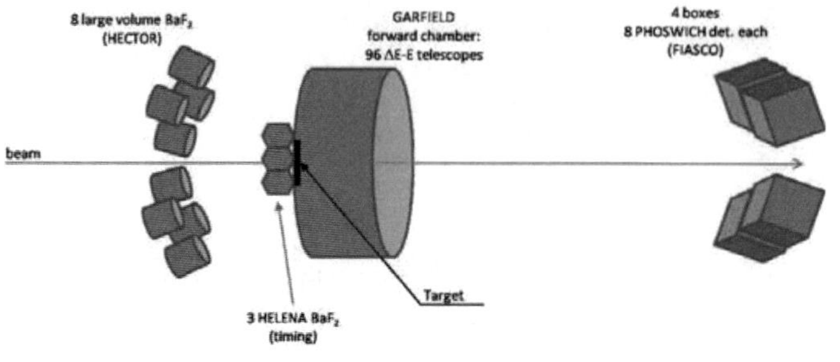

Figure 1. A sketch view of the experimental setup.

(the first for the measurement of high-energy gamma rays, the second to discriminate neutron and as time reference).

2. Experimental set up

The experimental setup was composed by different detection systems in order to provide correlated information on γ rays and particle emission in coincidence with the evaporation residue. The setup is illustrated in figure 1. In particular, the forward chamber of the GARFIELD apparatus is dedicated to the measurement of light charged particles (LCP) emitted during the different steps of the fusion-evaporation process.

The FIASCO [11] array consists of four boxes placed in the very forward direction (5°- 13°) to be used as an evaporation residue trigger and as a proton/alpha multiplicity counter. Each box contains eight phoswich detectors made of three layer scintillators. GARFIELD and FIASCO detectors use fully digitized electronics [12], [13] that allows on-line pulse shape analysis of both GARFIELD CsI(Tl) and FIASCO multilayer scintillators. This allows to perform LCP pulse shape identification in GARFIELD and evaporation residue accurate selection in FIASCO.

Eight large volume BaF_2 scintillators from HECTOR array are used for the measurement of the high-energy γ rays [7] [14].

Two additional groups of small (3"x3") BaF_2 scintillators from the HELENA array have been used. The first group was placed close to the target to provide a time reference to be compared with the one obtained from the pulsed beam radiofrequency (RF), while the second one was mounted in the forward direction for the measurement of pre-equilibrium neutrons.

3. Preliminary results

The analysis of the data has been focused to produce the high energy γ ray spectrum emitted in the fusion evaporation process (see the data in figure 3). From the data analysis we have estimated $\approx$50000 coincidence events between high energy γ rays (10-20 MeV) and one evaporation residue. A preliminary comparison with Statistical Model shows evidence of an excess yield between 10-20MeV (see figure 3) which is attributed to Dynamical Dipole emission. The extra yield has been compared with the previous results obtained at 8 MeV/u and 15.6 MeV/u (see [8]). The measured data shows an increasing behavior (see figure 3) similar to that of [3].

The measured pre-equilibrium contribution has been compared to the theoretical predictions from BNV-model calculations [5]. The BNV model has as input of the simulation i) the nuclei and parameters of the reaction; ii) the nuclear Equation of State (EOS) as function of his density (The model uses two different parametrization of EOS: Asy-Stiff and Asy-Soft) [5]; iii) the Nucleon-Nucleon (N-N) in medium cross section. The output of the model is the nucleon dynamics from BNV equation during fusion reaction for an isovector oscillation on isolated

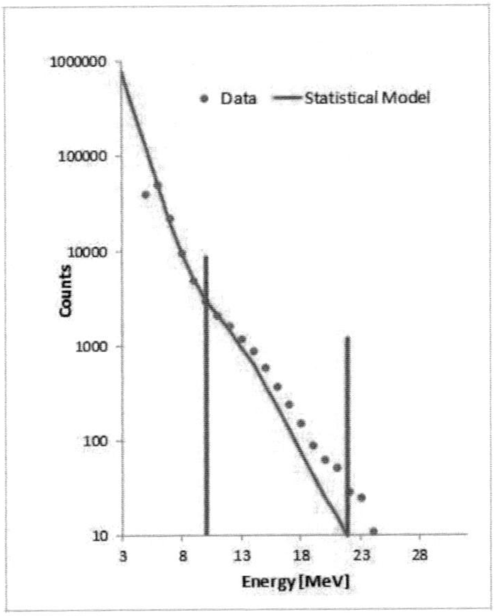

Figure 2. The comparison between the BaF_2 energy spectrum (blue points) with prediction by the statistical model (straight line). An excess of counts between 10-20MeV is visible. The two spectra have been normalized at 7 MeV.

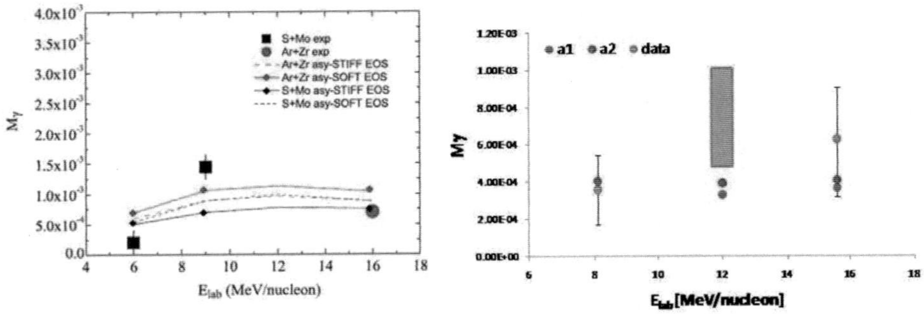

Figure 3. The DD γ-emission yield measured in mass region A=132 for beam energy ranging from 6 to 15 MeV/u. Left pannel: data reported in reference [3]. Right pannel: data points at 8 and 15.6 MeV/u are from [8], in green the experimental poit (indicate with data), in red the point from the BNV model of asy-soft parametrization (indicated with a2), in blue the poit of asy-stiff parametrization (indicated with a1). For the point at 12 MeV/u, as the analysis is not completed yet, we report the region where we expect to have the measured total gamma yield.

nucleus (GDR-like). This model has been used to extimate the DD γ multiplicity for differnt impact parametres at the three beam energies (8.1 - 12 - 15.6 MeV/u). In the reaction ^{16}O (Elab=192 MeV) + ^{116}Sn there is not enough sensitivity to different EOS parametrizations. The DD γ multiplicity has been calculate with the BNV model, for the three different beam energies and for different impact parametrer (b=0, 2, 4, 5, 5.5, 6 and 6,5 fm). From this calculation we obtain that the DD contribution decreases when the impact parameter increase; for impact parameter $b = 7$ there is not fusion and at high impact parameter (bigger than 4) the DD contribute became smaller and seems to be costant untill $b = 6.5$. For this reason it is very important to choose the right weight of each impact parameter. The DD γ multiplicity increases, with the beam energy, but this increasemnt is rather small, namely of the order of 8%. Furthemore, the different nuclear equation of state parametrizations used in calculation do not seem to produce a good agreement with measured data [8], [2], [3]. This difference between the expected and measured DD total γ-ray yield calls for further investigation, e.g. performing new theoretical calculations with different parameterization of N-N cross section and a more detailed description of the pre-equilibrium particle emission. For this reason, it is necessary to well understand the DD emission yield in function of the beam energy to be prepared for radioactive beams with extreme N/Z Ratios.

The Angular distribution of γ-rays measured in the backward hemisphere (covered by HECTOR detectors) will be extracted from data at 12 MeV/u and the comparison with the one obtained within theoretical model will provide a much deeper detailed check of the model.

4. Conclusion

The measured γ-ray spectra has been compared with the statistical γ decay of $^{132}Ce^*$ and the measured pre-equilibrium Dynamical Dipole emission yield has been compared with that given by the BNV-model. The theoretical model displays an almost constant trend in function of the beam energy for the DD γ multiplicity instead for the experimental data there is a differnt trend. More detailed theoretical BNV calculation (in collaboration with Catania) are in progress. The general trend of the calculations shows a small increase of DD multiplicity with the beam energy (at a fixed impact parameter) and a decreases of the DD contribute as the of impact parametre increases (at a fixed beam energy). For the future the angular distribution (at 12MeV/u) will be extrated from the experimental data and the pre-equilibrium neutron multiplicity (in collaboration with Bologna and Firenze) will be calculeted.

References

[1] Corsi A. et al., *Phys. Rev. C.*, **84** (2011) 041304
[2] Pierroutsakou D. et al., *Phys. Rev. C*, **71** (2005) 056405
[3] Pierroutsakou D. et *Phys. Rev. C*, **80** (2009) 024612
[4] Martin B. et al., *Phys. Lett. B*, **664** (2008) 47
[5] Baran V. et al., *Phys. Rev. Lett.*, **87** (2001) 18
[6] Baran V. et al., *Phys. Rev. Lett.*, **79** (2009) 021603
[7] Kmiecik M. et al., *Phys. Rev. C.*, **70** (2004) 064317
[8] Corsi A. et al., *Phys. Rev. Lett.*, **679** (2009) 197
[9] Gramegna F. et al., *Proceedings of 2004 IEEE Nuclear Science Symposium*,(2004)
[10] Gramegna F. et al., *Nucl. Inst. Meth. A*, **389** (1997) 474
[11] Bini M. et al., *Nucl. Inst. Meth. A*, **515** (2003) 497
[12] Bardelli L. et al., *Nucl. Inst. Meth. A*, **572** (2007) 882
[13] Bardelli L. et al., *Nucl. Inst. Meth. A*, **570** (2007) 127
[14] Wieland O. et al., *Phys. Rev. Lett.*, **97** (2006) 012501

Journal of Physics: Conference Series **366** (2012) 012019 doi:10.1088/1742-6596/366/1/012019

Electron-induced proton knockout from neutron rich nuclei

C. Giusti[1], A. Meucci[1], F. D. Pacati[1], G. Co'[2], V. De Donno[2],
[1]Dipartimento di Fisica Nucleare e Teorica, Università degli Studi di Pavia and INFN, Sezione di Pavia, via Bassi 6 I-27100 Pavia, Italy
[2]Dipartimento di Fisica, Università del Salento, Lecce, and INFN, Sezione di Lecce, via Arnesano, I-73100 Lecce, Italy
E-mail: `Carlotta.Giusti@pv.infn.it`

Abstract. We study the evolution of the $(e, e'p)$ cross section on nuclei with increasing asymmetry between the number of neutrons and protons. The calculations are done within the framework of the nonrelativistic and relativistic distorted-wave impulse approximation. In the nonrelativistic model phenomenological Woods-Saxon and Hartree-Fock wave functions are used for the proton bound-state wave functions, in the relativistic model the wave functions are solutions of Dirac-Hartree equations. The models are first tested against experimental data on ^{40}Ca and ^{48}Ca nuclei, and then they are applied to a set of spherical calcium isotopes.

1. Introduction

The understanding of the evolution of nuclear properties with respect to the proton to neutron asymmetry is one of the major topics of interest in modern nuclear physics. It is going to extend our knowledge about the effects of isospin asymmetry on the nuclear structure and is also relevant to the study of the origin and the limits of stability of matter in the universe.

Nuclear reactions represent our main source of information on the properties of atomic nuclei. Direct reactions, where the external probe interacts with only one, or a few, nucleons of the target nucleus, give insight into the single-particle (s.p.) properties of the many-body nuclear system. In particular, the $(e, e'p)$ reaction, where a proton is emitted with a direct knockout mechanism, represents a clean probe to explore the proton-hole states structure of the nucleus [1, 2, 3]. With respect to similar reactions with hadronic probes, such as, e.g., the $(p, 2p)$ reaction, the $(e, e'p)$ reaction can exploit the advantages of the electromagnetic interaction, that is well known and relatively weak, if compared with the nuclear interaction.

Several decades of experimental and theoretical work on electron scattering have provided a wealth of information on nuclear structure and dynamics [2, 3]. In the last thirty years many high-resolution exclusive $(e, e'p)$ experiments [1, 2, 3, 4, 5] have provided accurate information on the s.p. structure of stable closed-shell nuclei. The separation energy and the momentum distribution of the removed proton, which allows to determine the associated quantum numbers, have been obtained. From the comparison between the experimental and theoretical cross sections it has been possible to extract the spectroscopic factors, which give a measurement of the occupation of the different shells and, as a consequence, of the effects of nuclear correlations, which go beyond a mean field (MF) description of nuclear structure.

These studies can be extended to exotic nuclei. In upcoming years the advent of radioactive ion beams facilities will provide a large amount of data on unstable nuclei. A new generation of electron colliders that use storage rings, under construction at RIKEN [6] and GSI [7], will offer unprecedented opportunities to study the structure of exotic unstable nuclei through electron scattering in the ELISe experiment at FAIR [8] and the SCRIT project at RIKEN [9].

In this work models developed and successfully applied to the analysis of the available experimental data for the exclusive $(e, e'p)$ knockout reactions are used to make predictions for the $(e, e'p)$ cross sections on exotic nuclei. Our models are based on the distorted-wave impulse approximation (DWIA), within a nonrelativistic and a relativistic framework, and are here applied to a set of calcium isotopes. We have chosen calcium isotopes since data are available from NIKHEF experiments for the doubly magic nuclei ^{40}Ca and ^{48}Ca [10, 11]. We first compare our models with these data, then we apply them to some even-even isotopes, *i.e.*, ^{40,48,52,60}Ca nuclei, where the s.p. levels below the Fermi surface are fully occupied. In this manner we can work with spherical systems and minimize pairing effects. Some results are presented and discussed in the next section. More results can be found in [12]

2. Results

Our DWIA calculations have been carried out with the same program `DWEEPY` [13, 14] which was used for the analyses of the NIKHEF data. The program includes the effects of the final-state interactions (FSI) between the emitted proton and the remaining nucleus, that are described in the calculations by the phenomenological optical potential of [15], as well as of the Coulomb distortion of the electron wave functions. Although based on a nonrelativistic DWIA model, `DWEEPY` contains relativistic effects in the kinematics and in the nuclear current operator. In the comparison with the ^{40}Ca$(e, e'p)$ and ^{48}Ca$(e, e'p)$ data we have repeated the original analyses of [10, 11], where the bound-state wave functions are calculated with a phenomenological Woods-Saxon (WS) well whose radius was determined to fit the experimental momentum distribution and whose depth was adjusted to give the experimentally observed separation energy of the bound final state. This choice was able to describe, with a high degree of accuracy, the shape of the experimental momentum distributions at missing-energy values corresponding to specific peaks in the energy spectrum. In order to reproduce the size of the experimental cross sections, reduction factors must be applied to the calculated cross sections. These factors are identified with the spectroscopic factors and their deviation from the predictions of the MF approximation is interpreted as the effect of nucleon-nucleon correlations.

The results obtained with WS wave functions are here compared with those obtained by solving Hartree-Fock (HF) equations with the Gogny-like finite-range D1M interaction [16].

In a third approach we have performed $(e, e'p)$ calculations with the relativistic DWIA (RDWIA) model of [17, 18, 19, 20, 21, 22], where the s.p. bound-state wave function is the Dirac-Hartree solution from a relativistic Lagrangian written in the context of the relativistic MF theory (RMF) [23] and the scattering wave function is solution of the Dirac equation with the relativistic energy-dependent and A-dependent EDAD1 optical potential [24].

In Fig. 1 our DWIA and RDWIA results are compared with the ^{40}Ca$(e, e'p)$ and ^{48}Ca$(e, e'p)$ NIKHEF data [10] for the knockout of a proton from the $1d_{3/2}$ s.p. level. Data for the ^{40}Ca$(e, e'p)$ reaction were taken in the so-called parallel and $(\boldsymbol{q}, \omega)$ constant kinematics. For the ^{48}Ca$(e, e'p)$ reaction data were taken only in parallel kinematics. In parallel kinematics the momentum of the outgoing proton $\boldsymbol{p}'$ is kept fixed and is taken parallel to the momentum transfer $\boldsymbol{q}$. Different values of the missing momentum $p_{\rm m}$, which is the recoil momentum of the residual nucleus, are obtained by varying the electron scattering angle and, as a consequence, q. In $(\boldsymbol{q}, \omega)$ constant kinematics $\boldsymbol{q}$ and the outgoing proton energy are kept constant and different values of $p_{\rm m}$ are obtained by varying the angle of the outgoing proton. Experimental data are usually presented as a function of $p_{\rm m}$ in terms of the reduced cross section [3], *i.e.*, the cross section divided by a

Journal of Physics: Conference Series **366** (2012) 012019 doi:10.1088/1742-6596/366/1/012019

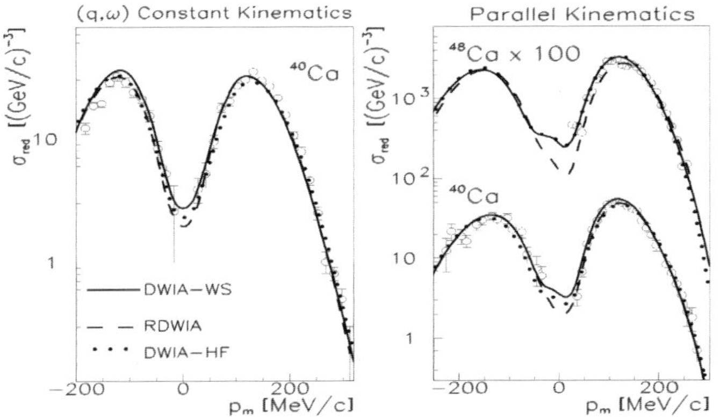

Figure 1. Reduced cross sections of the ^{40}Ca$(e,e'p)$ and ^{48}Ca$(e,e'p)$ reactions as a function of $p_{\rm m}$ for the transition to the $3/2^+$ ground state of ^{39}K and to the $3/2^+$ excited state at 0.36 MeV of ^{47}K in $(\boldsymbol{q},\omega)$ constant kinematics (left panel), with incident electron energy $E_0 = 483.2$ MeV, electron scattering angle $\vartheta = 61.52°$, and $q = 450$ MeV$/c$, and in parallel kinematics (right panel), with $E_0 = 483.2$ MeV. The outgoing proton energy is $T' = 100$ MeV in both kinematics. Line convention: DWIA-WS (solid lines), DWIA-HF (dotted lines), RDWIA (dashed line). The experimental data are taken from [10].

kinematical factor and by the elementary off-shell electron-proton scattering cross section of [25]. In the plane-wave impulse approximation (PWIA), where FSI are neglected and the cross section is factorized into the product of a kinematical factor, the elementary electron-proton scattering cross section, and the hole spectral function, the reduced cross section is the squared Fourier transform of the hole wave function, and can be interpreted as the momentum distribution of the emitted proton when it was inside the nucleus. In DWIA this factorization is destroyed by FSI, but the reduced cross section remains an interesting quantity that can be regarded as the nucleon momentum distribution modified by FSI.

All the theoretical results provide a good description of the shape of the experimental data. A reduction factor has been applied to reproduce the size. This factor has been determined by a fit of the calculated reduced cross sections to the data over the whole missing-momentum range considered in the experiment. The reduction factors applied to the DWIA-WS results (0.49 in $(\boldsymbol{q},\omega)$ constant kinematics and 0.65 in parallel kinematics for ^{40}Ca, 0.55 for ^{48}Ca) are the same as in the data analysis of [10]. The corresponding factors for the DWIA-HF results are 0.51, 0.64, 0.55, and for the RDWIA results are 0.49, 0.69, 0.52. In all the calculations the reduction factors obtained for ^{40}Ca$(e,e'p)$ reaction in $(\boldsymbol{q},\omega)$ constant kinematics are about 20-25% lower than those obtained in parallel kinematics. The source of this difference is not clear, but it reflects the uncertainties in the identification of the spectroscopic factor as a simple reduction factor of the theoretical results with respect to the experimental data. The reduction factors obtained in all our model calculations for the ^{48}Ca$(e,e'p)$ reaction are consistently lower than those obtained for the ^{40}Ca$(e,e'p)$ reaction in the same parallel kinematics.

The predictions of our models for $1d_{3/2}$ knockout from ^{40,48,52,60}Ca nuclei in $(\boldsymbol{q},\omega)$ constant and parallel kinematics are shown in Fig. 2. We remark that the evolution of the cross section with respect to the change of the neutron number is the same in all the panels: the ^{40}Ca lines are always above the other ones, and the size of the curves decreases with the increasing number of neutrons. This behavior is clearer in the DWIA-WS results and becomes less evident in the other cases, especially in the RDWIA ones. The behavior of the s.p. hole wave functions with

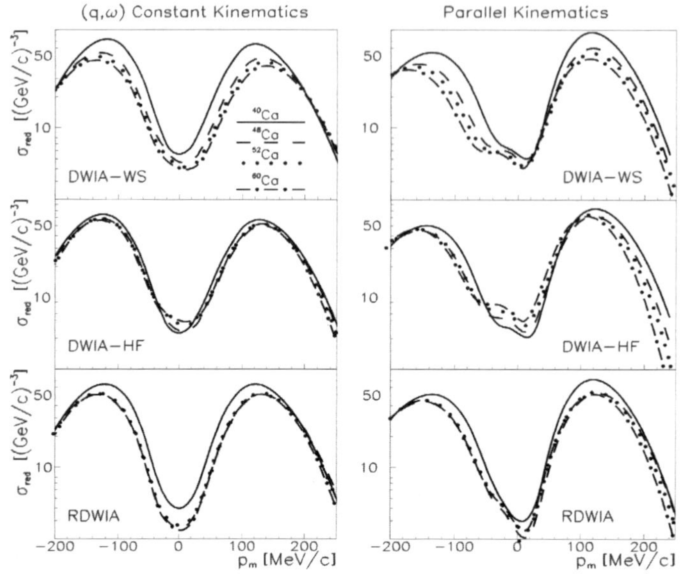

Figure 2. Reduced cross section of the $(e, e'p)$ reaction for $1d_{3/2}$ knockout from ^{40}Ca (solid lines), ^{48}Ca (dashed lines), ^{52}Ca (dotted lines), and ^{60}Ca (dot-dashed lines), as a function of $p_{\rm m}$, calculated in $(\boldsymbol{q}, \omega)$ constant (left panels) and parallel kinematics (right panels) with DWIA-WS (top panels), DWIA-HF (middle panels), and RDWIA (bottom panels).

increasing neutron number shows a different trend for the different models [12]. The dependence of the wave functions on the proton to neutron asymmetry is responsible for only a part of the differences in the reduced cross sections. While in PWIA the reduced cross section contains only information on the bound-state wave function, in DWIA this information is modified by the contribution of the other ingredients of the model, such as FSI and the electron-nucleon interaction. All these contributions are intertwined in the calculated cross section and, in general, they cannot be easily disentangled. The difference between the results of the two kinematics in Fig. 2 is basically due to the different effects of the distortion produced by the optical potential. These effects strongly depend on kinematics and are larger in parallel than in $(\boldsymbol{q}, \omega)$ constant kinematics. Moreover, the trend of the calculated cross sections with the increasing neutron number is significanlty affected by the A-dependence of the optical potential [12].

No reduction factor has been applied to the results shown in Fig. 2. The comparison with data in Fig. 1 gives a significant quenching of the measured cross sections with respect to the predictions of the MF model. The quenching is different for the ^{40}Ca$(e, e'p)$ and ^{48}Ca$(e, e'p)$ reactions and increases with the neutron number. A quenching depending on the neutron number can be expected for all the isotopes and would give further differences on the reduced cross sections than those shown in Fig. 2.

3. Summary and conclusions

We have presented and discussed $(e, e'p)$ cross sections for a set of calcium isotopes with the aim of studying their evolution with respect to the change of the neutron number.

The nonrelativistic DWIA and relativistic RDWIA models used for the calculations were widely and successfully applied to the analysis of the available $(e, e'p)$ data over a wide range of stable nuclei. The results obtained with three different MF descriptions of the hole wave function of the knocked out proton have been compared.

All the three models give a good and similar description of the experimental data on ^{40}Ca and ^{48}Ca. The general behavior of the cross sections with respect to the increasing number of neutrons is analogous for all the three models: the reduced cross sections are larger and narrower for the lighter isotopes, and evolve by lowering and widening with increasing neutron number. Although the evolution of the s.p. bound-state wave functions is different for the three models, the dependence of the wave functions on the proton to neutron asymmetry is responsible for only a part of the differences in the calculated cross sections. An important and crucial contribution is also given by FSI, which are described in our models by phenomenological optical potentials. The optical potential affects both the size and the shape of the cross section in a way that strongly depends on kinematics. In particular, its imaginary part, that gives a reduction of the calculated cross section, can affect the values of the spectroscopic factors obtained from the comparison between data and theoretical results. The dependence of the optical potential on the proton to neutron asymmetry is an interesting problem that deserves careful investigation.

From recent studies there are indications that the spectroscopic factors and the effects of correlations depend on the proton to neutron asymmetry. In general, the quenching of quasiparticle orbits, and hence correlations, become stronger with increasing separation energy.

Measurements of the exclusive quasifree $(e, e'p)$ cross section on nuclei with neutron excess would offer a unique opportunity for studying the dependence of the properties of bound protons on the neutron to proton asymmetry. The present results can serve as a useful reference for future experiments. The comparison with data can confirm or invalidate the predictions of the model and will test the ability of the established nuclear theory in the domain of exotic nuclei.

This work was partially supported by the MIUR through the PRIN 2009 research project.

References

[1] Frullani S, Mougey J 1984 *Adv. Nucl. Phys.* **14** 1
[2] Boffi S, Giusti C and Pacati F D 1993 *Phys. Rep.* **226** 1
[3] Boffi S, Giusti C, Pacati F D and Radici M 1996 *Electromagnetic Response of Atomic Nuclei* Oxford Studies in Nuclear Physics, Vol. 20
[4] Bernheim M *et al.* 1982 *Nucl. Phys.* A **375** 381
[5] de Witt Huberts P K A *et al.* 1990 *J. Phys.* G **16** 507
[6] Suda T, Maruyama K, Tanhihata I 2001 *RIKEN Accel. Prog. Rep.* **34** 49
[7] GSI report 2006, `http://www.gsi.de/GSI-Future/cdr/`
[8] Antonov A N *et al.* 2011 *Nucl. Instr. and Methods in Physical Research* A **637** 70
[9] Suda T 2010 *Proposal for Nuclear Physics Experiment at RIBF NP1006-SCRIT01*
[10] Kramer G J 1990 *Ph.D. thesis, Universiteit Amsterdam*
[11] Kramer G J, Blok H P, Lapikás L 2001 *Nucl. Phys.* A **679** 267
[12] Giusti C, Meucci A, Pacati F D, Co' G, De Donno V 2011 *Phys. Rev.* C **84** 014604
[13] Giusti C, Pacati F D 1987 *Nucl. Phys.* A **473** 717
[14] Giusti C, Pacati F D 1988 *Nucl. Phys.* A **485** 461
[15] Schwandt P *et al.* 1982 *Phys. Rev.* C **26** 55
[16] Goriely S, Hilaire S, Girod M, Péru S 2009 *Phys. Rev. Lett.* **102** 242501
[17] Meucci A, Giusti C, Pacati F D 2001 *Phys. Rev.* C **64** 014604
[18] Meucci A, Giusti C, Pacati F D 2001 *Phys. Rev.* C **64** 064615
[19] Meucci A 2002 *Phys. Rev.* C **65** 464601
[20] Meucci A, Giusti C, Pacati F D 2002 *Phys. Rev.* C **66** 034610
[21] Radici M, Meucci A, Dickhoff W H 2003 *Eur.Phys. Jou.* A **17** 65
[22] Tamae T *et al.* 2009 *Phys. Rev.* C **80** 064601
[23] Horowitz C J, Murdoch D P, Serot B D 1991 *Computational Nuclear Physics I: Nuclear Structure* Springer-Verlag, Berlin
[24] Cooper E D *et al.* 1993 *Phys. Rev.* C **47** 297
[25] de Forest Jr. T 1983 *Nucl. Phys.* A **392** 232

Journal of Physics: Conference Series **366** (2012) 012020 doi:10.1088/1742-6596/366/1/012020

Search for one-phonon mixed-symmetry states in the radioactive nucleus ^{140}Nd

K A Gladnishki[1], G Rainovski[1,2], P Petkov[3,4], J Jolie[4], N Pietralla[2], A Blazhev[2], A Damyanova[1], M Danchev[1], A Dewald[4], C Fransen[4], M Hackstein[4], O Möller[2], T Pissulla[4], M Reese[2] and W Rother[4]

[1]Faculty of Physics, St. Kliment Ohridski University of Sofia, 1164 Sofia, Bulgaria
[2]Institut für Kernphysik, Technische Universität Darmstadt, D-64289 Darmstadt, Germany
[3]Bulgarian Academy of Sciences, Institute for Nuclear Research and Nuclear Energy, 1784 Sofia, Bulgaria
[4]Institut für Kernphysik, Universität zu Köln, D-50937 Köln, Germany

E-mail: kag@phys.uni-sofia.bg

Abstract. Low-spin excited states of ^{140}Nd have been studied via the ^{140}Ce(^{3}He,3n)^{140}Nd reaction. The results from the data analysis show that one of the candidates for the one-phonon mixed symmetry state of ^{140}Nd, namely the 2^+_3 state at 2140 keV with an effective lifetime of 220(90) fs, decay with a fast $M1$ transition to the 2^+_1 state. Therefore consequently this state can be treated as, at least a fragment of the one-phonon MSS of ^{140}Nd. This is the first example where mixed symmetry character is tentatively assigned to a state of an unstable nucleus from the mass $A \approx 140$ region based on the data on absolute $M1$ transition rates.

1. Introduction

The isovector states with mixed proton-neutron character, the so called mixed symmetry states (MSSs) are currently of interst because their properties allow some parts of the proton-neutron interaction to be studied[1]. Available information on MSSs of vibrational nuclei is summarized in a recent review article [2]. The best examples of MSSs in stable nuclei are found in the mass $A \approx 90$ region [2, 3, 4, 5]. However, there are only a few MSSs identified in the mass $A \approx 130$ region [6, 7, 8]. The main reason for the small number of studied cases originates from the fact that the stable open-shell even-even isotopes in this mass region have relatively low abundance. This comprises an experimental problem since an unambiguous identification of MSSs can only be done on the basis of measured large absolute B($M1$) values. This experimental information can be obtained through series of experiments [2] which in the case of low-abundant or unstable isotopes is not always possible. Projectile Coulomb excitation reactions in combination with a large γ-ray array detectors were suggested as a solution to this methodological problem [9]. By using projectile Coulomb excitation reactions and the Gammasphere array at the Argonne National Laboratory, the one-phonon MSSs in several low-abundant stable nuclei were identified, namely ^{134}Xe [10], ^{138}Ce [9], ^{136}Ce [11], and ^{130,132}Xe [12]. This experimental technique can also be applied to radioactive ion beams (RIBs). The measured $E2$ and $M1$ strength distributions between the lowest 2^+ states allow for direct and unambiguous identification of the $2^+_{1,\mathrm{ms}}$ state.

These extensive data demonstrate not only the experimental accessibility of MSSs by inverse kinematics Coulomb excitation reactions on a light target, but also reveals novel interesting physics phenomena. Examples are given by the data on the $N = 80$ isotones ^{138}Ce [9] and ^{134}Xe [10]. In contrast to the isotone ^{136}Ba [4], the $2^+_{1,\mathrm{ms}}$ state of ^{138}Ce is strongly mixed with a nearby 2^+ state, probably full symmetry states (FSS), suggesting that the microscopic structure can have a dramatic influence on the properties of the MSSs. The observed mixing in ^{138}Ce is attributed to the lack of *shell stabilization* [9] at the $\pi g_{7/2}$ sub-shell closure. This hypothesis has been partially confirmed by the observation of a single, well pronounced one-phonon MSS in ^{134}Xe [10]. The data on MSSs of stable $N = 80$ isotones and the suggested mechanism of shell stabilization have also initiated theoretical investigations. The properties of MSSs of stable $N = 80$ isotones were studied with the quasiparticle-phonon model (QPM) [13] and the large scale shell model (SM) [14]. Both models have demonstrated that the splitting of the $M1$ strength in ^{138}Ce is a genuine shell effect caused by the specific shell structure and the pairing correlations [13, 14]. The results from the shell model calculations [14] also show that the experimental information on MSSs provides a tool to determine the pairing matrix elements of realistic interactions. To fully demonstrate the shell stabilization mechanism, the investigation of MSSs in the $N = 80$ isotonic chain have to be extended beyond the proton number $Z = 58$, *i.e.*, to the unstable nuclei $^{140}_{60}\mathrm{Nd}_{80}$ and $^{142}_{62}\mathrm{Sm}_{80}$. The theoretical models differ in their predictions for the properties of MSSs of ^{140}Nd. SM calculations predict a single isolated MSS for ^{140}Nd [14], while the QPM predicts a fragmentation [13]. This situation prompts for an experimental identification of MSSs in ^{140}Nd. An experiment [15] was performed recently in an attempt to identify the candidates for the one-phonon $2^+_{1,\mathrm{ms}}$ in ^{140}Nd. The measurement showed that the 2^+_3 and the 2^+_4 states of ^{140}Nd at 2140 keV and 2332 keV, respectively, decay predominantly by $M1$ transitions to the 2^+_1 state. The measured $E2/M1$ multipole mixing ratios for the 1366-keV and the 1559-keV transitions are $\delta = -0.08(8)$ and $\delta = -0.19(9)$, respectively. Both states have to be considered as candidates for the one-phonon $2^+_{1,\mathrm{ms}}$ of ^{140}Nd [15]. As

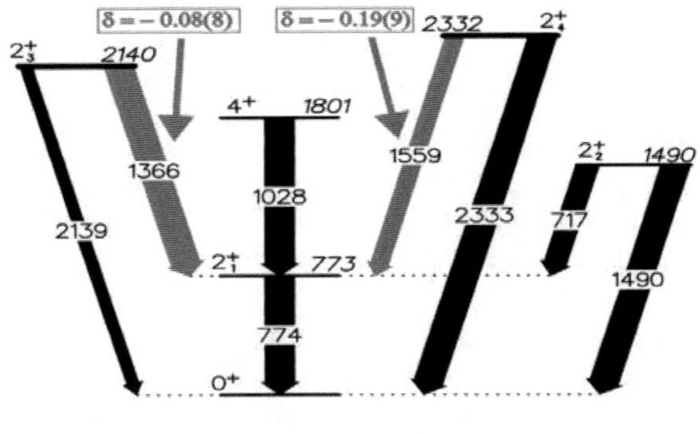

Figure 1. Partial level scheme of ^{140}Nd. The candidates for the one-phonon MSS [15] and the transition which were analyzed in order to extract the lifetimes are indicated.

a result, neither the available experimental information, nor the theoretical predictions allow for any decisive conclusion on whether and how the effect of shell stabilization is present in ^{140}Nd. Apparently, this can be resolved only by measuring the absolute $M1$ strengths of the 1366.2-keV and 1558.6-keV transitions (see figure 1). Such measurements are possible by using projectile Coulomb excitation of the radioactive ion beam ^{140}Nd. However, taking into account the high demand for beam time at RIB facilities and the fact that ^{140}Nd is not a very exotic nucleus, it first should be examined to what extent the present physics problem can be solved by using experimental techniques which are based on stable beams. The purpose of this study is to determine the lifetimes of the states, that are candidates for one-phonon $2^+_{1,\mathrm{ms}}$ of ^{140}Nd (see figure 1) in a light-ion fusion-evaporation reaction, induced by an accelerated stable beam.

2. Experimental details

The experiment was performed at the FN Tandem Facility in Cologne (Germany). The excited states of ^{140}Nd were populated using the ^{140}Ce(^{3}He, $3n$)^{140}Nd reaction. The beam of ^{3}He was accelerated to energy of 19.8 MeV and then delivered to the target consisting of 0.8 mg/cm^2 thick ^{140}Ce layer deposited on 2 mg/cm^2 thick Ta foil (the stopper). The ^{3}He beam and the beam energy were chosen as a compromise between the requirements to populate the low-spin non-yrast states and to provide sufficient recoil velocity for DSAM measurements of lifetimes in the range of hundred femtoseconds or shorter. The emitted γ-rays were detected by five coaxial high purity Germanium (HPGe) detectors arranged in a ring at angle $\theta = 143°$ with respect to the beam axis and one EUROBALL HPGe cluster detector build up of seven segments and positioned at forward angle so that its central segment laid on the beam axis and the other six formed a ring at angle $\theta = 19°$ (see figure 2).

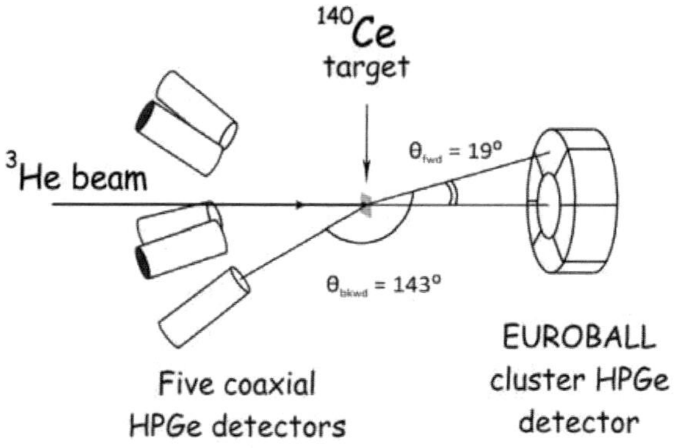

Figure 2. Schematic drawing of the experimental setup.

The three polar angles at which the detectors were positioned formed three rings. R_0 is the ring to which the center segment of the EUROBALL cluster detector belongs ($\theta_0 = 0°$), R_1 contains the six outer segments of the EUROBALL cluster detector ($\theta_1 = 19°$) and R_2 is formed from the five coaxial HPGe detectors at angle $\theta_2 = 143°$. For this arrangement of the detectors, eight types of $\gamma\gamma$-coincidence matrices were sorted. The full (or the total) matrix was used in the analysis for the identification of the observed transitions. Gates were set on the 774 keV and 1028 keV transitions (see figure 1 and figure 3) and so the excited states populated in the reaction were determined (see figure 1). Most of the observed γ-rays connect energy levels of ^{140}Nd which are known from previous studies [15, 16]. However, there are several γ-lines which are in strong coincidences with the γ-rays from low-spin states of ^{140}Nd but cannot be placed unambiguously in the level scheme. For example, the 1491-keV transition is in coincidence with the 774-keV transition and with a new 1888-keV transition. The 1888-keV transition is in coincidence only with the 1491-keV transition. This suggests that the 1888-keV transition represents a decay to the ground state of a new level which is fed by the 1491-keV transition. Unfortunately, no transitions were observed with energies that add up to 1114 keV which is the energy difference between the new level at 1888 keV and the first 2^+ state at 774 keV. Besides these ambiguities, the spectra in figure 3 clearly demonstrate that the previously known 1366-keV and the 1559-keV transitions (which represent the $M1$ decay of the candidates for the one-phonon MSS of ^{140}Nd [15]) directly feed the 2_1^+ state.

The lifetime analysis was performed in "singles"-like regime because of the small yield of the states of interest (2140 keV and 2332 keV), and the coincidence matrices R_1R_2 and R_2R_1 were

Journal of Physics: Conference Series **366** (2012) 012020 doi:10.1088/1742-6596/366/1/012020

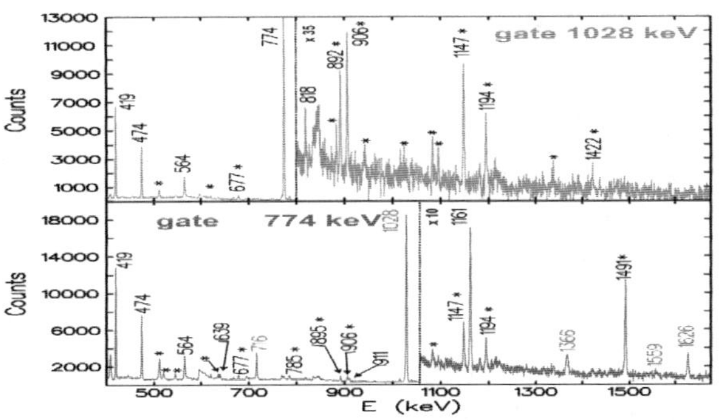

Figure 3. Gated γ-ray spectra from the total $\gamma\gamma$ matrix. The known transitions of ^{140}Nd are marked with their energies, new transitions which most likely belong to ^{140}Nd are labelled with their energies and stars while possible contaminants are labelled with stars only. The transitions directly feeding the 2_1^+ state are indicated.

used for purification of the spectra. A gate was set on the transition de-exciting the 2_1^+ state, namely the 774-keV transition, for which no Doppler-shifted component was observed.

From the obtained spectra the line-shape of the 1366-keV transition ($2_3^+ \rightarrow 2_1^+$) was analyzed. The slowing-down process of the recoiling nuclei was simulated with a modified version [17, 18] of the program code DESASTOP [19]. According to the calculations performed, the recoils needed in average 430 fs to come to rest. This time is longer than the expected lifetime of the state of interest meaning that the Doppler-shift method is applicable for the determination of the line-shape. The results from the fitting at forward and backward angles are presented in figure 4. The evaporation of neutrons was also taken into account in the MC simulation. The database of about 10000 velocity histories was additionally randomized with respect to the experimental setup by taking into account the positions of the detectors and their finite size. More details of our approach for the Monte Carlo simulation can be found in Refs.[17, 18].

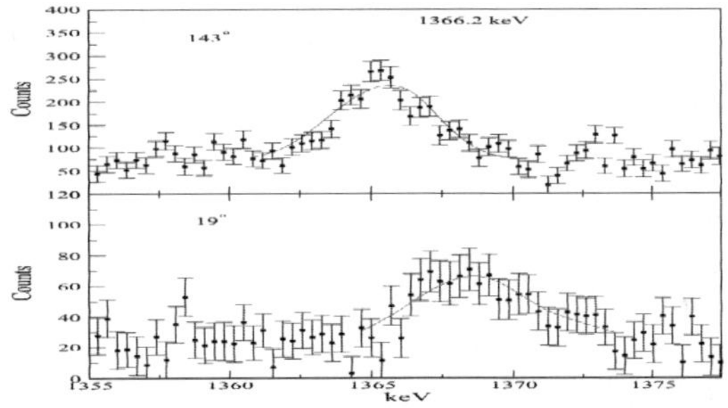

Figure 4. Line-shape analysis of the 1366.2-keV transition at backward (top panel) and forward (bottom panel) angles with a gate set on the fully stopped 774-keV transition.

The used gating procedure introduces uncertainties related to the unobserved feeding of the level of interest and for this reason the deduced lifetime is an effective lifetime.

$$\tau_{\text{eff}}(2140keV) = 220(90)\, fs \tag{1}$$

Taking into account the previously measured multipole mixing ratio of the 1366 keV transition and the estimated branching ratio for the decay of the 2140 keV transition [15] the B($M1$) value is determined:

$$B(M1; 2_3^+ \rightarrow 2_1^+) > 0.07^{+0.05}_{-0.02}\, \mu_N^2 \tag{2}$$

This lower limit for the $M1$ strength clearly identifies the 2_3^+ state of ^{140}Nd as a rapidly decaying fragment of the one-phonon MSS. For comparison, the $M1$ strength from the weaker fragment of the one-phonon MSS of ^{138}Ce, the 2_3^+ state at 2143 keV, is 0.058(6) μ_N^2 [9]. However, the question whether the decay of the 2_3^+ state of ^{140}Nd accounts for the total $M1$ strength remains unclear. Nuclear structure models predict a total $M1$ strength of about 0.3 μ_N^2 [14] for the decay of the $2^+_{1,\mathrm{ms}}$ state of ^{140}Nd. This value can be in agreement with the present observation if a finite feeding time from above is included in the fitting procedure for the lifetime. The data did not allow for the extraction of the lifetime of the 2^+ state at 2332 keV excitation energy. Its 1559-keV decay transition was barely observed at forward angles. However, a small centroid shift of about -0.9(4) keV was estimated for this transition at backward angles. This is an indication that 2332-keV state also undergoes a fast $M1$ decay.

In summary, we have attempted to identify the one-phonon MSS of ^{140}Nd using the DSAM in the reaction ^{140}Ce(^{3}He,3n)^{140}Nd. An effective lifetime of 220(90) fs was measured for one of the candidates for the one-phonon MSS of ^{140}Nd. This fast $M1$ decay identifies the 2_3^+ state at 2140 keV, at least, as a fragment of the one-phonon MSS of ^{140}Nd. However, the data are not conclusive on whether this decay exhausts the total $M1$ strength and whether the one-phonon MSS of ^{140}Nd is fragmented or not.

2.1. Acknowledgments
We thank Ch. Stoyanov and D. Tarpanov for discussion on the QPM calculations for ^{140}Nd. G. R. is a research fellow of the Alexander von Humboldt foundation. This work is supported by the partnership agreement between the University of Cologne and University of Sofia, by the DFG under grants Nos. Pi 393/2-2, SFB 634, Jo 391/3-2, by the German-Bulgarian exchange programme under grant Nos. D/08/02055, by the BgNSF under contract DO 02-219, and by the Helmholtz International Center (HIC) for FAIR.

References
[1] Iachello F 1984 *Phys. Rev. Lett.* **53** 1427
[2] Pietralla N, von Brentano P and Lisetskiy A F 2008 *Prog. Part. Nucl. Phys* **60** 225 and references therein.
[3] Pietralla N *et al.* 1999 *Phys. Rev. Lett.* **83** 1303
[4] Pietralla N *et al.* 2001 *Phys. Rev.* C **64** 031301
[5] Werner V *et al.* 2001 *Phys. Lett.* B **550** 140
[6] Molnár G *et al.* 1988 *Phys. Rev.* C **37** 898; Fazekas B *et al.* 1992 *Nucl. Phys.* A **548** 249
[7] Wiedenhöver I *et al.* 1997 *Phys. Rev.* C **56** R2354
[8] Pietralla N *et al.* 1998 *Phys. Rev.* C **58** 796
[9] Rainovski G *et al.* 2006 *Phys. Rev. Lett.* **96** 122501
[10] Ahn T *et al.* 2009 *Phys. Lett.* B **679** 19
[11] Ahn T *et al.* 2011 *Phys. Rev.* C submitted
[12] Coquard L *et al.* 2010 *Phys. Rev.* C **82** 024317
[13] Lo Iudice N, Stoyanov Ch and Tarpanov D 2008 *Phys. Rev.* C **77** 044310 and references therein.
[14] Sieja K, Martinez-Pinedo G, Coquard L and Pietralla N 2009 *Phys. Rev.* C **80** 054311
[15] Williams E *et al.* 2009 *Phys. Rev.* C **80** 054309
[16] Ponomarev V Yu *et al.* 1996 *Nucl. Phys.* A **601** 1
[17] Petkov P *et al.* 1998 *Nucl. Phys.* A **640** 293
[18] Petkov P *et al.* 1999 *Nucl. Instrum. Methods Phys. Res.* A **431** 208
[19] Winter G, 1983 *ZfK Rossendorf Report ZfK* 497; Winter G 1983 *Nucl. Instrum. Methods* **214** 537

Level scheme investigation of ^{102}Rh

N. Goutev[1], M.S. Yavahchova[1], D. Tonev[1,2], G. de Angelis[2], P. Petkov[1], R.K. Bhowmik[3], R.P. Singh[3], S. Muralithar[3], N. Madhavan[3], R. Kumar[3], M. Kumar Raju[4], J. Kaur[5], G. Mahanto[3], A. Singh[5], N. Kaur[5], R. Garg[6], A. Sukla[7], Ts. K. Marinov[1], S. Brant[8]

[1] Bulgarian Academy of Sciences, Institute for Nuclear Research and Nuclear Energy, Sofia, Bulgaria
[2] INFN, Laboratori Nazionali di Legnaro, Legnaro, Italy
[3] Inter-University Accelerator Center, New Delhi, India
[4] Nuclear Physics Department, Andhra University, Visakhapatnam, India
[5] Department of Physics, Punjab University, Chandigarh, India
[6] Department of Physics and Astrophysics, Delhi University, New Delhi, India
[7] Department of Physics, Banaras Hindu University, Varanasi, India
[8] Department of Physics, Faculty of Science, Zagreb University, Zagreb, Croatia

Abstract. Excited states in ^{102}Rh were populated in the fusion-evaporation reaction ^{96}Zr(^{11}B, 3n)^{102}Rh. The angular correlations and the electromagnetic character of some of the gamma-ray transitions observed were investigated in detail. A new chiral candidate sister band was found in the level-scheme of ^{102}Rh.

1. Introduction

Chirality is a novel feature of rotating nuclei which is among the most studied phenomena in nuclear physics during the last decade. A spontaneous breaking of the chiral symmetry can take place for configurations where the angular momenta of the valence protons, valence neutrons, and the core are mutually perpendicular [1]. Under such conditions the angular momenta of the valence particles are aligned along the short and long axes of the triaxial core, while the angular momentum of the core is aligned along the intermediate axis. In such a case the projections of the angular momentum vector on the three principal axes can form either a left- or a right- handed system and therefore the system expresses chirality. Since the chiral symmetry is dichotomic, its spontaneous breaking by the axial momentum vector leads to a pair of degenerate $\Delta I = 1$ rotational bands, called chiral doublet bands. Pairs of bands, possibly due to the breaking of the chiral symmetry in triaxial nuclei, have been recently found in the mass regions $A \sim 130$ [2, 3, 4], $A \sim 105$ [5, 6, 7, 8, 9] and $A \sim 195$ [10].

Due to the underlying symmetry, the pair of chiral twin bands should exhibit systematic properties [11]. The yrast and the side bands should be nearly degenerate. In the region where chiral symmetry sets up the $B(E2)$ values of the electromagnetic transitions de-exciting analogous states of the chiral twin bands should be almost equal. Correspondingly, the $B(M1)$ values should exhibit odd-even staggering. The $B(M1)$ values for $\Delta I = 1$ transitions connecting the side to the yrast band should have odd-even staggering which is out of phase with respect to the $B(M1)$ staggering for transitions de-exciting states in the yrast band. The last condition means that the $B(M1)_{In}/B(M1)_{Out}$ staggering in the side band should be in phase with the

Journal of Physics: Conference Series **366** (2012) 012021 doi:10.1088/1742-6596/366/1/012021

$B(M1)$ staggering in the yrast band. In many cases the energy degeneracy of the twin chiral bands is observed but the transition probabilities are different, like in the case of ^{134}Pr [12, 13].

The goal of the present work is to check for existence of a chiral sister band of the yrast band in the nucleus of ^{102}Rh. In previous works [5, 6, 7], an island of chiral candidates has been proposed around ^{104}Rh. Next to identification of the twin chiral bands some lifetimes have been reported for the yrast band of ^{104}Rh [7]. Theoretical studies of the chiral phenomenon within the framework of the adiabatic and configuration-fixed constrained triaxial Relativistic Mean Field (RMF) approaches have been recently performed in order to investigate the triaxial shape coexistence and possible chiral doublet bands [14]. A new phenomenon, the existence of multiple chiral doublets, i.e. more than one pair of chiral doublet bands in one single nucleus has been predicted for ^{106}Rh and neighboring nuclei. The ^{102}Rh nucleus, according to the work of Meng et al. [14], is one of the candidates to present such a phenomenon. These results stimulated our work to check for existence of chirality in ^{102}Rh.

2. Experiment

Excited states in ^{102}Rh were populated using the reaction ^{96}Zr(^{11}B,3n)^{102}Rh at a beam energy of 36 MeV. The beam was delivered by the 15-UD Pelletron accelerator at the Inter University Accelerator Center in New Delhi. The target consisted of 0.9 mg/cm^2 ^{94}Zr, enriched to 96.5%, which was evaporated onto a 8 mg/cm^2 gold backing. The recoils were leaving the target with a mean velocity v of 0.9(1)% of the velocity of light, c. The type of the target and backing were chosen such in order to ensure lifetime determination in the sub-picosecond region using the Doppler-Shift Attenuation Method (DSAM).

The de-exciting gamma-rays were registered by the Indian National Gamma Array (INGA), whose 15 Clover detectors were accommodated in a 4π geometry [15]. For the purposes of the DSAM analysis the detectors of INGA were grouped into rings with approximately the same position with respect to the beam axis. The rings where appreciable Doppler-shifts are observed are these at angles of 32, 57, 123 and 148 degrees. Gain matching and efficiency calibration of the Ge detectors were performed using ^{152}Eu and ^{133}Ba radioactive sources.

The events were unfolded and sorted into different γ-γ coincidence matrices. For the lifetime analysis, matrices were constructed where one of the gamma-rays was detected at an angle with a significant Doppler-shift whereas on the other axis detection by all detectors of INGA was allowed. To measure intensities, a matrix was constructed where the gamma-rays were detected on both axes by all detectors of INGA.

3. Data analysis and results

3.1. Angular correlation analysis

The angular correlation function $W(\theta_1, \theta_2, \phi)$ for a cascade of two consecutive transitions describes the probability to register, in coincidence, two γ rays emitted from an excited nucleus at angles θ_1 and θ_2 with respect to the beam axis and angle ϕ between the two scattering planes. The angular correlation function is given by the expression:

$$W(\theta_1, \theta_2, \phi) = \sum_{\lambda_1,\lambda,\lambda_2} B_{\lambda_1}(I_1) A_{\lambda}^{\lambda_1\lambda_2}(\gamma_1) A_{\lambda_2}(\gamma_2) H_{\lambda_1,\lambda,\lambda_2}(\theta_1, \theta_2, \phi) \tag{1}$$

This function mainly depends on the spins of the initial, intermediate, and final states, on the multipolarities of the transitions and their multipole mixing ratios. The term $B_{\lambda_1}(I_1)$ describes the orientation of the upper nuclear state, the term $A_{\lambda}^{\lambda_1\lambda_2}(\gamma_1)$ the orientation of the intermediate state due to emission of γ_1, and $A_{\lambda_2}(\gamma_2)$ the emission of γ_2. With $H_{\lambda_1,\lambda,\lambda_2}(\theta_1, \theta_2, \phi)$ is denoted the angular function. It is reduced to an ordinary Legendre polynomial if any of the λ's vanishes. Detailed information about the angular correlation function and the theoretical formalism behind it can be found in [16].

The angular correlation function is subject to certain symmetries. Accordingly, the pairs of detectors can be sorted in unique geometry groups, which are characteristic of the multidetector spectrometer used. In the present work the angular correlation analysis was carried out with the computer code CORLEONE [17]. This code allows to determine the geometry groups of the spectrometer as well. Twenty five unique geometry groups were found for the INGA spectrometer. The large number of correlation groups ensures precise determination of the multipole mixing coefficients.

In theory, the γ-γ coincidence matrices obtained for all pairs of detectors from one geometry group should be the same and therefore can be summed-up to improve the signal-to-noise ratio. In practice though this is far from being true because each detector has a different efficiency curve. Furthermore, the efficiency of the detectors under high load (during the experiment) may differ [18] from the efficiency found with standard sources. Therefore, besides determining the shape of the efficiency curves with the sources, one has to adjust their height as appropriate to the experimental conditions by using a set of lines of the sample itself as an internal intensity standard. The quality of the efficiency calibration should then be evaluated by applying angular correlation analysis to a known cascade of pure $E2$ transitions. Two cascades, $4_1^+\to2_1^+\to0_1^+$ and $6_1^+\to4_1^+\to2_1^+$, in the even-even nucleus of ^{102}Ru were chosen by us as the most appropriate for the evaluation.

The quality of the efficiency calibration of the detectors is demonstrated in Fig. 1, where the good fit between the experimental results for the angular correlation function and the theoretical predictions are clearly seen. The same approach was applied to investigate data from a GASP experiment and it provided excellent results [19]. We need to stress that this is the first experiment where angular correlation analyses are performed using the approach [17]. The results obviously are very successful.

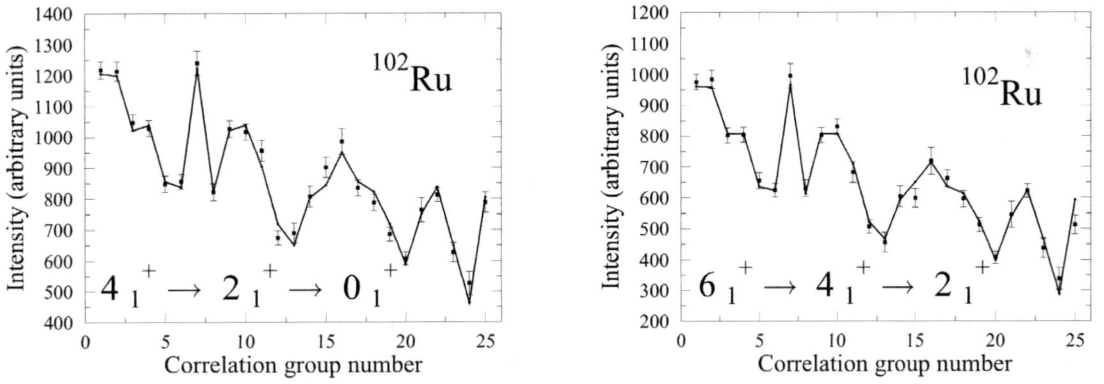

Figure 1. Angular correlation patterns for the cascades $4_1^+\to2_1^+\to0_1^+$ (on the left hand side) and $6_1^+\to4_1^+\to2_1^+$ (on the right hand side) involving transitions of ^{102}Ru. For these spin hypotheses, the best fit confirms the pure $E2$ character of the transitions.

3.2. Linear polarization measurements

Each clover detector is made up of four Ge segments that are packed together side by side. Photons that are Compton scattered in one segment can be detected efficiently in an adjacent segment. The scattering is polarization sensitive. The gamma-rays are more efficiently scattered perpendicularly to their polarization. The mutually perpendicular pairs of segments of one Clover detector therefore form a four-fold Compton polarimeter. The efficiency with which the pairs of segments of one Clover detector register the Compton scattering events was determined with the ^{152}Eu and ^{133}Ba sources that emit gamma rays of natural polarization isotropically.

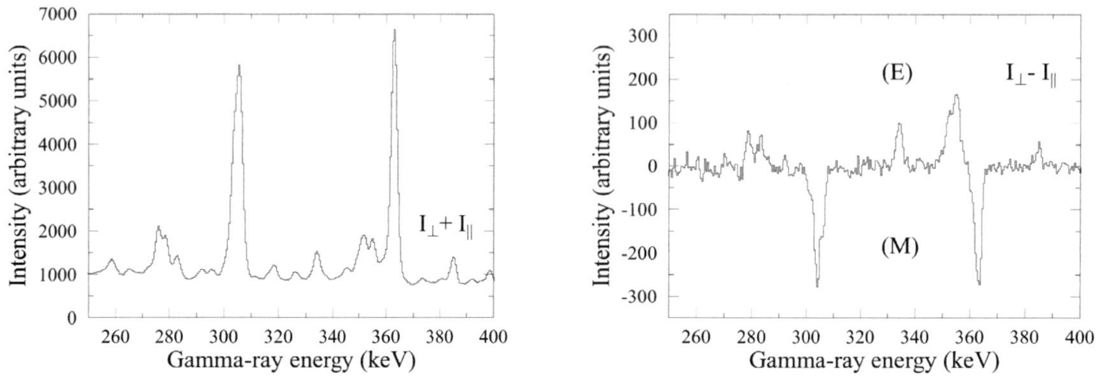

Figure 2. Sum (on the left hand side) and difference (on the right hand side) of the coincidence spectra registered by the perpendicular and parallel arms of the composite Compton polarimeter.

In the present experiment we used all the four Clover detectors from the ring at 90 degrees with respect to the beam as one composite Compton polarimeter. The coincidence signals from the respective arms of the polarimeter, perpendicular and parallel to the reaction plane, were summed-up to improve the statistics. The sum and difference of the coincidence spectra, registered by the perpendicular and parallel arms of the composite polarimeter, are shown in Fig. 2. The difference spectrum in Fig. 2 reflects the linear polarization of the transitions observed. The negative lines correspond to transitions of predominantly magnetic character while the positive lines correspond to transitions of predominantly electric character. For example, the negative lines that appear around 305.8 and 363.0 keV in Fig. 2 are due to $M1/E2$ transitions in ^{102}Rh [20]. According to DCO ratios measurements [20], the lines around 283.3, 334.4, 352.1 and 385.4 keV are due to transitions of dipolar character in ^{102}Rh. The positive sign of these lines in the linear polarization spectrum, Fig. 2, proves that the corresponding transitions are of predominantly electric ($E1/M2$) character.

The results obtained from linear polarization measurements were essential for the spin and parity assignments of the new candidate for chiral sister band in ^{102}Rh. These results are published at the present work for the first time. The newly established level-scheme of the ^{102}Rh will be published in a forthcoming paper [21].

4. Conclusions

In order to investigate the level-scheme and to determine lifetimes of the excited states in the chiral candidate nucleus of ^{102}Rh we have performed an experiment at the IUAC in New Delhi using the INGA spectrometer. The usual way of ordering of transitions in a level-scheme is according to the γ-ray relative intensities, the γ-γ coincidence relationships, and the γ-ray energy sum. Next to these approaches we have used for the first time for INGA spectrometer the powerful techniques of angular correlation analysis [17] and linear polarization measurements. Based on the known transitions of the level-scheme published in [20] we have succeeded to extend the level-scheme with a new $\Delta I = 1$ band with a negative parity. The comparison of the excitation energies of the analogous states of the two sister bands in ^{102}Rh together with their electromagnetic properties will give an answer whether the chiral phenomenon is present in this nucleus [21].

Acknowledgments

D.T. express his gratitude to Ivanka Necheva for her outstanding support. M.Y. and D.T. acknowledge support from the European Operational Programme HRD, contract BGO 051 P O 001/07/3.3 - 02/53 with the Bulgarian Ministry of Education.

References

[1] Frauendorf S 2001 *Rev. Mod. Phys.* **73** 463
[2] Starosta K*et al.* 2001 *Phys. Rev. Lett.* **86** 971
[3] Grodner E *et al.* 2006 *Phys. Rev. Lett.* **97** 172501
[4] Mukhopadhyay S *et al.* 2006 *Phys. Rev. Lett.* **97** 172501
[5] Timár J *et al.* 2006 *Phys. Rev. C* **73** 011301
[6] Vaman C*et al.* 2004 *Phys. Rev. Lett.* **92** 032501
[7] Suzuki T*et al.* 2008 *Phys. Rev. C* **78** 031302
[8] Timár J*et al.* 2004 *Phys.Lett. B* **598** 178
[9] Joshi P *et al.* 2004 *Phys.Lett. B* **595** 135
[10] Lawrie E A *et al.* 2008 *Phys. Rev. C* **78** 021305
[11] Brant S, Tonev D, de Angelis G and Ventura A 2008 *Phys. Rev. C* **78** 034301
[12] Tonev D *et al.* 2006 *Phys. Rev. Lett.* **96** 052501
[13] Tonev D *et al.* 2007 *Phys. Rev. C* **76** 044313
[14] Meng J *et al.* 2006 *Phys. Rev. C* **73** 037303
[15] Muralithar *et al.* 2008 *Proc. DAE-BRNS Symp. on Nucl. Phys.* **53** 221
[16] Krane K S, Steffen R M and Wheeler R M 1973 *Nucl. Data Tables* **11** 351
[17] Wiedenhöver I *et al* 1998 *Phys. Rev.* C **58** 721
[18] Nordlund A 1996 *Nucl. Instrum. Meth. A* **368** 731
[19] Tonev D *et al.* 2011 *Joural Phys.Conf. Ser.* **267** 012048
[20] Gizon J *et al.* 1999 *Nucl. Phys. A* **658** 97
[21] Yavahchova M S *et al.to be published.*

Journal of Physics: Conference Series **366** (2012) 012022 doi:10.1088/1742-6596/366/1/012022

DSA lifetime measurements in ^{132}La in the context of spontaneous chiral symmetry breaking accompanied by the new S-symmetry.

Ernest Grodner

Faculty of Physics, University of Warsaw, Institute of Experimental Physics, ul. Hoża 69, PL-00681 Warsaw, Poland.

E-mail: `ernest.grodner@fuw.edu.pl`

Abstract. The lifetimes of the excited states belonging to three rotational bands of ^{132}La have been measured with help of the DSA technique. The obtained results are interpreted in frame of the chiral symmetry breaking and recently postulated S-symmetry. Explanation of the absence of the characteristic B(M1) staggering in the chiral partner bands of ^{132}La is suggested.

1. Introduction

Symmetry considerations dominate modern quantum physics from theoretical as well as experimental standpoint. Time-reversal is one of the fundamental symmetry incorporated in the description of the atomic nucleus. The possibility of time reversal symmetry breaking appears in the Cabibbo-Kobayashi-Maskawa matrix of the standard model, however, it is not associated with low energy nuclear excitations where time-reversal invariance rules the quantum description of the strong interaction. Even if the time-reversal is fundamentally conserved, there is a possibility of its spontaneous breakdown. Observation of spontaneous time-reversal symmetry breaking via the eigenstate properties of the nucleus has a general difficulty since there are no eigenstates of the time-reversal operator. A solution of that problem came with the hypothesis of the chiral symmetry breaking [1] where the symmetry operator is a combination of the π-rotation and the time-reversal. Apparently, spontaneous chiral symmetry breaking has been observed in Cs isotopes [2, 3] as presence of characteristic gamma selection rules in the chiral rotational bands. Recently, those selection rules have been explained in frame of new S-symmetry related to the specific structure of the chiral nuclei [4]. In the present work the chirality phenomenon is investigated on the example of ^{132}La DSA lifetime measurements.

2. Experimental method

The excited states of ^{132}La were populated in the $^{122}Sn(^{14}N, 4n)^{132}La$ reaction. The ^{14}N beam of 70 MeV energy was provided by the U-200P cyclotron at the Heavy Ion Laboratory of the University of Warsaw. The ^{122}Sn target, 10mg/cm^2 thick, acted also as a stopper. About 10^8 $\gamma - \gamma$ coincidences were collected by the OSIRIS II multidetector array consisting of 10 ACS germanium detectors placed at angles 25°, ±38°, 63°, ±90°, 117°, ±142° and 155° with respect to the beam axis.

The analysis of the Doppler broadened lineshapes of the γ transitions was carried out by using the programs COMPA, GAMMA and SHAPE [5]. The software includes Monte-Carlo simulation of the recoils production (COMPA) as well as the slowing down process, γ emission and registration (GAMMA). The lifetime analysis was performed starting from highest observed excited levels that decay to the levels of interest. In this way the influence of all observed transitions on the lifetime value has been taken into account. The unobserved feeding – side feeding – was described by the model presented in detail in Ref.[5]. The parameters of the side-feeding model were the same for all studied states and were experimentally determined from the analysis of the feeding intensities and lineshapes of high-energy levels of ^{131}La. The determined side-feeding parameters have been verified by comparing the lifetime of the $I^{\pi}=31/2^{-}$ yrast level in ^{131}La obtained with the method described above and with the method less sensitive to the side-feeding distribution (gating from above and narrow gate techniques) [5]. The stopping power parameters being input data for the GAMMA code were measured in an additional experiment with the use of semi-thick target method [6]. The uncertainties of the stopping-power parameters were accounted for the final lifetime uncertainties.

The $\gamma - \gamma$ coincidence events collected by the OSIRIS II detection setup were sorted into 10 coincidence matrices. Each matrix contained energies of gamma quanta registered by a selected detector being in coincidence with gammas registered by any of the remaining detectors. In this way a coincidence spectrum from a selected HPGe spectrometer was obtained whereas the others played the role of the trigger. The $\gamma - \gamma$ angular correlation coefficients were the input parameters for the GAMMA code. This is particularly important when a sum of coincidence spectra is used for the DSA analysis.

3. Results of the DSA study

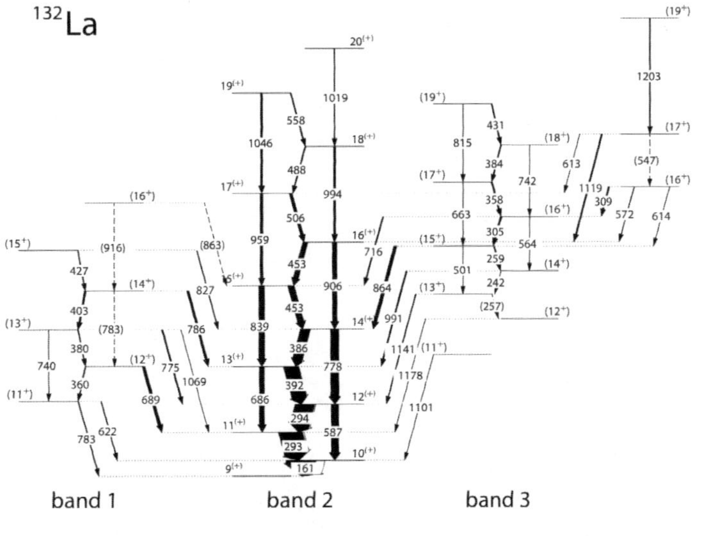

Figure 1. Relevant part of the level scheme of ^{132}La. Band 1 and band 2 built on the $\pi h_{11/2} \otimes \nu h_{11/2}^{-1}$ configuration as observed in the present experiment. Band 3 - newly found rotational band. The arrow widths are proportional to the γ-intensities observed during the measurement.

Relevant part of the level scheme of ^{132}La observed in our experiment is shown in Fig.1. Additionally to the previously known band 1 and band 2 [7, 8] a new rotational band (band 3

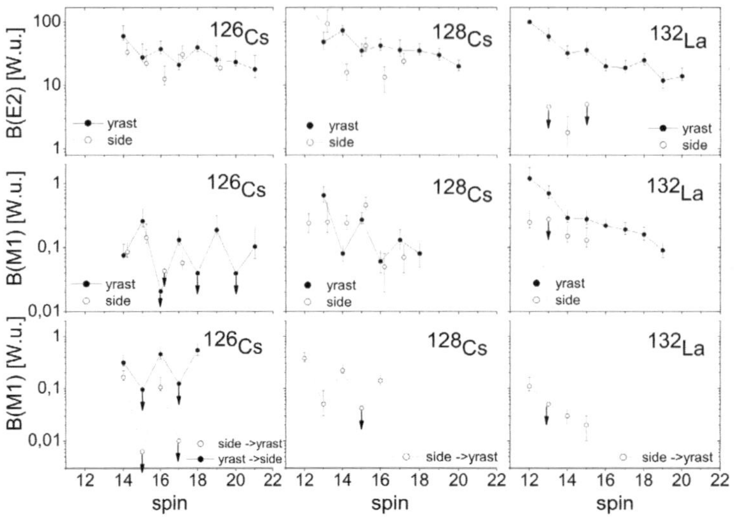

Figure 2. Comparison of γ transition probabilities in the partner bands of ^{126}Cs (left column), ^{128}Cs (middle), and ^{132}La (right column). Upper and middle rows – B(E2) and B(M1) values for in-band transitions, bottom row – B(M1) values for inter-band transitions.

in Fig.1) connected with the yrast one with six strong linking transitions has been found [9]. The spin and parity assignment of the levels belonging to band 1 and 2 follows Ref.[7, 8]. The assignment of band 3 has not been uniquely determined. For the band 3 $\rightarrow$ band 2 linking transitions the R_{DCO} and lifetime values allow $\Delta I = \pm 1, \pm 2$ without parity change and $\Delta I = 0$ where parity change is not excluded. The $\pi h_{11/2} \otimes \nu h_{11/2}$ configuration has been assigned for band 1 and band 2 which were proposed as the candidates for the chiral partner bands [7]. The lifetimes of 13 levels belonging to those bands were extracted from our DSA experimental data. We also observed the Doppler broadening of the transitions coming from the newly found band 3 where lifetimes of 7 excited states have been determined. The results of the DSA analysis are listed in Table 1 and shown in Fig.2 together with similar data obtained for the ^{126,128}Cs nuclei [2, 3].

4. Discussion

The obtained gamma transition probabilities show a striking difference in the electromagnetic behavior of ^{132}La and ^{126,128}Cs observed in the partner bands. The B(M1) and B(E2) values in the yrast band are systematically different then the corresponding side-band values in ^{132}La while in ^{126,128}Cs both bands present similar electromagnetic properties. According to Ref.[10] the ^{132}La lifetime results show that the chiral symmetry cannot be broken strongly. The energy splitting between the partner bands in ^{132}La supports the above conclusion. The average splitting is around 350 keV in ^{132}La while in Cs isotopes it is twice smaller, around 150-200 keV. This unquestionably shows that in the chiral scenario of ^{132}La the energy barrier between left- and right-handed configurations has to be smaller than in odd-odd Cs nuclei. A deviation from the full $\gamma = 30°$ triaxial deformation can be one of the reasons for the lowering of the barrier. Another feature distinguishing ^{132}La form ^{128,126}Cs are gamma selections rules along the partner bands observed as specific B(M1) staggering.

As it has been discussed in Refs.[10, 3] a characteristic feature results from the breaking of the chiral symmetry, namely, if the inband transition is strong then the corresponding interband one should be weak and vice-versa. This is consistent with the observed selection rules. In Cs isotopes, strong inband E2 transitions are observed while the interband ones are not seen. Also, the inband B(M1) staggering has opposite phase to the interband one. The described feature tells us how the inband and interband B(M1) staggering should be related to each other but does not explain the origin of the staggering itself. In contrast to Cs nuclei, no B(M1) staggering is observed in ^{132}La.

Recently, new explanation of the B(M1) staggering in the partner bands have been reported [4] as appearance of new additional symmetry called the S-symmetry. The new symmetry is a combination of the parity operation in the five dimensional space of deformation of the core and an exchange of states of the unpaired particles. The B(M1) staggering in the context of the S-symmetry should appear when the expectation value of the triaxial deformation is $\gamma = 30°$ (independently on the gamma-softness) and the partner bands are built on two quasiparticle states where odd proton and odd neutron are in the same j-shell. It was shown in Ref.[11] that even little deviation from $\gamma = 30°$ triaxial deformation causes vanishing of the B(M1) staggering.

5. summary

In the present paper lifetimes of the excited states belonging to rotational bands of 132La are analyzed in the context of the spontaneous chiral symmetry breaking. In contrast to Cs isotopes, the probabilities of the gamma transitions observed in the chiral partner bands (band 1 and band 2 in Fig. 1) of 132La do not show characteristic B(M1) staggering. Recently a new symmetry, has been introduced in order to explain the appearance of the chiral gamma selection rules. In frame of the new symmetry the $\gamma \neq 30°$ nonaxial deformation is the possible mechanism responsible for absence of the electromagnetic selection rules in ^{132}La.

5.1. Acknowledgments

This work was supported in part by the Polish Ministry of Science under contract No. N N202 169736.

References

[1] S. Frauendorf, Jie Meng, Nucl. Phys. **A617**, 131 (1997).
[2] E. Grodner, J. Srebrny, A. A. Pasternak, I. Zalewska, T. Morek, Ch. Droste, J. Mierzejewski, M. Kowalczyk, J. Kownacki, M. Kisieliński, S. G. Rohoziński, T. Koike, K. Starosta, A. Kordyasz, P. J. Napiorkowski, M. Wolińska-Cichocka, E. Ruchowska, W. Płóciennik, J. Perkowski, Phys. Rev. Lett. **97**, 172501 (2006).
[3] E. Grodner, I. Sankowska, T. Morek, S.G. Rohoziński, Ch. Droste, J. Srebrny, A.A. Pasternak, M. Kisieliński, M. Kowalczyk, J. Kownacki, J. Mierzejewski, A. Król, K. Wrzosek Phys. Lett. **B703**, 46 (2011).
[4] L. Próchniak, S.G. Rohoziński, Ch. Droste, K. Starosta, Acta Phys. Pol. **B42**, 465 (2011).
[5] E. Grodner, A.A. Pasternak, Ch. Droste, T. Morek, J. Srebrny, J. Kownacki, W. Płóciennik, A.A. Wasilewski, M. Kowalczyk, M. Kisieliński, R. Kaczarowski, E. Ruchowska, A. Kordyasz,M. Wolińska, Eur. Phys. J **A27**, 325 (2006).
[6] E. Grodner, Ch. Droste, T. Morek, J. Srebrny, A.A. Pasternak, A. Wasilewski, W. Płóciennik, E. Ruchowska, J. Kownacki, Acta Phys. Pol. **B34**, 2447 (2003).
[7] K. Starosta, C. J. Chiara, D. B. Fossan, T. Koike, T. T. S. Kuo, D. R. LaFosse, S. G. Rohoziński, Ch. Droste, T. Morek, J. Srebrny, Phys. Rev. **C65**, 044328 (2002).
[8] J. Timár, D. Sohler, B. M. Nyako, L. Zolnai, Zs. Dombradi, E. S. Paul, A. J. Boston, C. Fox, P. J. Nolan, A. J. Sampson, H. C. Scraggs, A. Walker, J. Gizon, A. Gizon, D. Bazacco, S. Lunardi, C. M. Petrache, A. Astier, N. Buforn, P. Bednarczyk, N. Kintz, Eur. Phys. J. **A16**, 1 (2003).
[9] J. Srebrny, E. Grodner, T. Morek, I. Zalewska Ch. Droste, J. Mierzejewski, A.A. Pasternak, J. Kownacki, J. Perkowski, Acta Phys. Pol. **B36**, 1063 (2005).
[10] E. Grodner, Int. J. Mod. Phys. **E20**, 380 (2011).
[11] Ch.Droste, S.G.Rohoziński, K.Starosta, L.Próchniak, E.Grodner, Eur.Phys.J. **A42**, 79 (2009).

Table 1. ^{132}La results of the lifetime study.

band	I^π	τ [ps]	E_γ [keV]	$\sigma\lambda$	I_γ [a.u.]	$B(\sigma\lambda)$ [W.u.]
1	12^+	$0.61^{+0.19}_{-0.14}$	360	M1	4	$0.25^{+0.13}_{-0.05}$
			689	M1	10	$0.11^{+0.05}_{-0.02}$
1	13^+	> 0.8	740	E2	0.3	< 1.5
			380	M1	4	< 0.14
			775	M1	5	< 0.02
1	14^+	$1.15^{+0.40}_{-0.26}$	783	E2	0.5	$1.18^{+1.4}_{-0.8}$
			403	M1	6	$0.15^{+0.07}_{-0.03}$
			786	M1	8	$0.03^{+0.01}_{-0.01}$
1	15^+	$1.43^{+0.52}_{-0.38}$	427	M1	3	$0.13^{+0.07}_{-0.03}$
			827	M1	13	$0.02^{+0.01}_{-0.01}$
2	12^+	$0.68^{+0.28}_{-0.18}$	587	E2	24	200^{+50}_{-20}
			294	M1	66	$1.2^{+0.6}_{-0.2}$
2	13^+	$0.53^{+0.12}_{-0.10}$	686	E2	17	59^{+19}_{-9}
			392	M1	50	$0.70^{+0.22}_{-0.11}$
2	14^+	$0.97^{+0.23}_{-0.19}$	778	E2	27	32^{+10}_{-5}
			385	M1	33	$0.29^{+0.09}_{-0.04}$
2	15^+	$0.60^{+0.16}_{-0.13}$	839	E2	18	36^{+13}_{-6}
			454	M1	21	$0.28^{+0.10}_{-0.05}$
2	16^+	$0.77^{+0.14}_{-0.15}$	906	E2	16	20^{+6}_{-3}
			452	M1	17	$0.22^{+0.07}_{-0.02}$
2	17^+	$0.62^{+0.13}_{-0.12}$	959	E2	9	19^{+6}_{-3}
			507	M1	10	$0.19^{+0.06}_{-0.03}$
2	18^+	$0.54^{+0.12}_{-0.11}$	994	E2	7	25^{+7}_{-4}
			487	M1	4	$0.16^{+0.05}_{-0.03}$
2	19^+	$0.75^{+0.18}_{-0.14}$	1046	E2	7	12^{+4}_{-3}
			559	M1	5	$0.09^{+0.03}_{-0.02}$
2	20^+	$1.27^{+0.40}_{-0.28}$	1019	E2	3	14^{+5}_{-2}
3	12^+	$0.69^{+0.21}_{-0.14}$	257	M1	0.8	$0.47^{+0.28}_{-0.19}$
			1142	M1	3	$0.02^{+0.08}_{-0.05}$
3	13^+	$0.72^{+0.25}_{-0.20}$	242	M1	1	$0.5^{+0.3}_{-0.1}$
			991	M1	5	$0.03^{+0.17}_{-0.06}$
3	14^+	$0.65^{+0.22}_{-0.17}$	501	E2	0.4	$23^{+0.14}_{-0.07}$
			259	M1	4	$0.63^{+0.30}_{-0.12}$
			865	M1	11	$0.05^{+0.02}_{-0.01}$
3	15^+	$1.06^{+0.47}_{-0.31}$	564	E2	0.6	15^{+14}_{-5}
			305	M1	6	$0.53^{+0.34}_{-0.11}$
			718	M1	4	$0.03^{+0.02}_{-0.01}$
3	16^+	$1.15^{+0.37}_{-0.25}$	663	E2	0.6	11^{+8}_{-5}
			358	M1	5	$0.5^{+0.2}_{-0.1}$
3	17^+	$1.07^{+0.29}_{-0.24}$	742	E2	0.5	$8.1^{+5.4}_{-3.5}$
			384	M1	4	$0.44^{+0.16}_{-0.07}$
3	18^+	$0.81^{+0.18}_{-0.16}$	815	E2	0.5	$7.2^{+4.3}_{-3.2}$
			432	M1	4	$0.41^{+0.12}_{-0.06}$
3	19^+	0.83 ± 0.07	431	M1	4	$0.41^{+0.10}_{-0.05}$
			815	E2	1	$7.1^{+3.6}_{-2.6}$

Recent advances in the application of dynamical supersymmetry to describe atomic nuclei.

J. Jolie, Ch. Bernards, S. Heinze, J.M. Régis, T. Thomas

Institut fuer Kernphysik, Universitaet zu Koeln, Zuelpicher Str. 77, D-50937 Cologne, Germany

E-mail: jolie@ikp.uni-koeln.de

Abstract. In recent years we have investigated further the use of supersymmetry in the Au-Pt regio. In the extended supersymmetry which takes account of the neutron-proton degree of freedom, the even-even Hg isotopes ^{196,198}Hg could be described as two proton fermions coupled to a neutron boson core. We have performed experiments on ^{193}Au at the 10 MV Tandem accelerator in Cologne using the HORUS and Double ORANGE spectrometers and fast LaBr3(Ce) scintillators to perform $\gamma-\gamma$, $\gamma-e^-$ and $e^- - e^-$ coincidences and on ^{196}Hg at the Yale accelerator using YRAST Ball. Both level schemes could be considerably extended and corrected. This allowed a detailed comparison with the predictions of the U(6/4) supersymmetry for the positive parity states in ^{193}Au and ^{196}Hg.

1. Introduction

In 1980, Iachello introduced dynamical supersymmetries to describe bosonic and fermionic systems [1]. In the following year the U(6/4) supersymmetry was compared to the positive parity-states of Au and Pt isotopes [2, 3], showing for the first time that U(6/4) is able to describe odd gold isotopes. In 1984, Van Isacker *et al.* [4] introduced the so-called extended supersymmetry by including the proton-neutron degree of freedom and thus were able to describe sets of four neighboring nuclei: even-even ^{196}Pt, odd-neutron ^{197}Pt, odd-proton ^{197}Au, and odd-odd ^{198}Au. Such a supermultiplet is also called *magical quartet* or *magical square*. About 15 years later, experimental evidence was found for a new neighboring magical quartet consisting of ^{194,195}Pt and ^{195,196}Au [5, 6, 7].

These magical quartets contain odd Au isotopes on which the experimental knowledge is rather scarse as many were studied in the early 1970's [8, 9]. It is of interest whether other Au isotopes can be described using Bose-Fermi symmetry and/or supersymmetry and to improve the experimental knowledge on odd Au isotopes. Therefore we started recently a campaign to extend the knowledge on odd Au isotopes. Here, we present results for ^{193}Au [10, 11].

In 1986 the magic square was further extended to include the even-even Hg isotopes. They are described as two proton fermions coupled to $N_{B\nu}$ neutron boson[12]. Only recently these calculations were tested in ^{198}Hg[13] and ^{196}Hg[14].

In the next section the Bose-Fermi symmetry and supersymmetry are shortly reviewed. Section 3 gives an overview of the new results for the ^{196}Hg isotope obtained from a beta decay experiment performed in Yale. Section 4 describes the experiment on ^{193}Au performed at the Cologne Tandem accelerator. Then, a common description of the isotones ^{193}Au and ^{192}Pt in the supersymmetric U(6/4) scheme is investigated.

2. The $U^B(6) \times U^F(4)$ Bose-Fermi symmetry and the U(6/4) supersymmetry

The model used to describe odd proton Au isotopes, is the Interacting Boson Fermion Model 1 (IBFM) [15] in the $U^B(6) \times U^F(4)$ limit[16]. Hereby, a fermion in the $\pi 2d_{\frac{3}{2}}$ orbital is coupled to a system of N_B bosons using the $\mathrm{U}^B(6) \otimes \mathrm{U}^F(4)$ algebra. Isomorphisms in the sub-algebra structure of $\mathrm{U}^B(6) \otimes \mathrm{U}^F(4)$ can be found between the boson and fermion algebras. Such an isomorphism exists between $\mathrm{U}^B(6) \supset \mathrm{SO}^B(6)$ and $\mathrm{U}^F(4) \supset \mathrm{SU}^F(4) \simeq \mathrm{SO}^F(6)$ groups. The generators g_k of these subgroups $\mathrm{SO}^B(6)$ and $\mathrm{SO}^F(6)$ commute and a linear combination of these generators closes under commutation and thus forms a boson-fermion algebra $\mathrm{O}^{BF}(6)$=Spin(6). The group chain of the Hamiltonian with the quantum numbers generated by the groups is:

$$\begin{array}{ccccccccc} \mathrm{U}^B(6) \otimes \mathrm{U}^F(4) & \supset & \mathrm{SO}^B(6) \otimes \mathrm{SU}^F(4) & \supset & \mathrm{O}^{BF}(6) & \supset & \mathrm{O}^{BF}(5) & \supset & \mathrm{SU}^{BF}(2) \\ [N_B] \quad [1^{N_F}] & & \langle\sigma\rangle\langle\alpha_1,\alpha_2,\alpha_3\rangle & & \langle\sigma_1,\sigma_2,\sigma_3\rangle & & (\tau_1,\tau_2) & & J \end{array} \tag{1}$$

with N_F=1. The Hamiltonian written in form of a linear combination of Casimir operators, neglecting constant terms that only contribute to the binding energy, is:

$$H = DC_2(\mathrm{SO}^B(6)) + AC_2(\mathrm{SO}^{BF}(6)) + BC_2(\mathrm{O}^{BF}(5)) + CC_2(\mathrm{SU}^{BF}(2)). \tag{2}$$

The corresponding energy eigenfunction of the Hamiltonian can be derived from the eigenfunction of the Casimir operators of the subgroups:

$$E = D\sigma(\sigma+4) + A(\sigma_1(\sigma_1+4)+\sigma_2(\sigma_2+2)+\sigma_3^2) + B(\tau_1(\tau_1+3)+\tau_2(\tau_2+1)) + C(J(J+1)). \tag{3}$$

For this Hamiltonian, it is possible to find an embedding superalgebra to the Bose-Fermi symmetry [1, 16]. The generators of this graded Lie algebra now also consist of mixed boson-fermion creation and annihilation operators. While the Bose-Fermi symmetry preserves the boson and fermion numbers separately, the supersymmetry only preserves the total number of particles $N = N_B + N_F$. The embedding algebra of $U^B(6) \otimes U^F(4)$ is:

$$\begin{array}{ccc} \mathrm{U}(6/4) & \supset & \mathrm{U}^B(6) \otimes \mathrm{U}^F(4). \\ \downarrow & & \downarrow \\ [N] & & [N_B][1^{N_F}] \end{array} \tag{4}$$

In case that a fermion is annihilated and a boson created, the number of fermions is $N_F = 0$ and the problem can be described within the Interacting Boson Model (IBM - 1). The eigenvalues of the IBM in the O(6) limit are given by the formula:

$$E = \overline{A}\sigma(\sigma+4) + B\tau(\tau+3) + CJ(J+1), \tag{5}$$

with $\overline{A} = A + D$. In the opposite case a boson is converted into a fermion and we have a system with two fermions and $N-2$ bosons. While this situation generally would describe so-called broken pair states a different interpretation follows when considering an neutron-proton version of (4):

$$\begin{array}{ccc} \mathrm{U}(6/4) & \supset & \mathrm{U}^B_\nu(6) \otimes \mathrm{U}^F_\pi(4). \\ \downarrow & & \downarrow \\ [N] & & [N_{B\nu}][1^{N_{F\pi}}] \end{array} \tag{6}$$

.

and atomic nuclei having two proton holes in the Z=82 shell and neutron boson holes below N=126. This is exactly the case for the even-even Hg isotopes next to the magic squares [12, 13].

Journal of Physics: Conference Series **366** (2012) 012023 doi:10.1088/1742-6596/366/1/012023

3. Beta-decay study of ^{196}Hg

The beta-decay experiment was performed using the ESTU Tandem Van de Graaff Accelerator of the WNSL of Yale University using a ^{198}HgS target. With a 35 MeV proton beam the (p,3n) reaction produced ^{196}Tl which decays to ^{196}Hg with a half life of 1.5 hours. Gamma-gamma coincidences and angular correlations were measured using the YRAST spectrometer. The main results not found in the in-beam experiment of [14] are the non-existence of a $1,2^+$ state at 958 and the new determination of the spin 2^+ of the 1450 keV state, which before was assigned 0^+ [17].

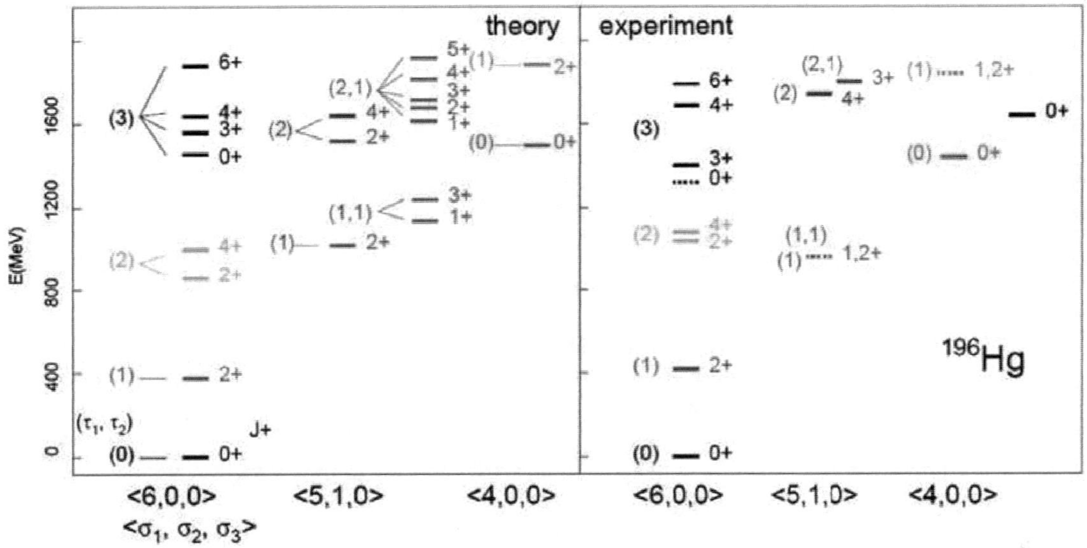

Figure 1. (Color online) Comparison between experiment and theory for ^{196}Hg. Also given are the values for the quantum numbers. The dashed experimental states were not confirmed by this experiment and the one described in [14].

In Figure 1 a fit of the experimental positive parity states using eq. (2) is shown. The quantum numbers and reduction rules used are given in [13]. The parameter D was artificially set to a high negative value. The other parameters are (all in keV): A= 78.3, B= 81.4 and C= 7.9. The obtained agreement is good and of similar quality as the one in ^{198}Hg [13]. However, the parameters differ from those obtained for a common description of the Pt, Au nuclei.

4. In-beam study of ^{193}Au.

The experiments on ^{193}Au were performed at the Cologne Tandem accelerator by impinging a 14 MeV proton beam onto a 1.3 and $0.2\frac{mg}{cm^2}$ ^{194}Pt target. The (p,2n) reaction was used which at this energy yields an angular momentum transfer of about 3.5 $\hbar$. The HORUS spectrometer [19], the Orange spectrometers[10] and LaBr3(Ce) scintillators[20] were used. During the experiments HORUS was equipped with 12 high-purity germanium detectors on the edges and the faces of a cube to detect the γ transitions of excited states. The setup of the spectrometer allows the analysis of $\gamma\gamma$ angular correlations. The data were sorted in so-called correlation group matrices, which consist of detector pairs defined by specific angles. The method using the HORUS spectrometer is described in more detail in Refs. [19, 21]. By fitting theoretical angular distributions described in Refs. [23, 22] to the data, spins, multipolarities, multipole mixing

Journal of Physics: Conference Series **366** (2012) 012023 doi:10.1088/1742-6596/366/1/012023

ratios, and eventually parities are obtained. The fit was performed with the computer code CORLEONE [24]. An example of a correlation fit for the 861 keV-258 keV cascade from the $3/2^+$ level at 1119 keV in ^{193}Au is shown in Fig. 1. Both, the spin and the mixing ratios, could be clearly determined.

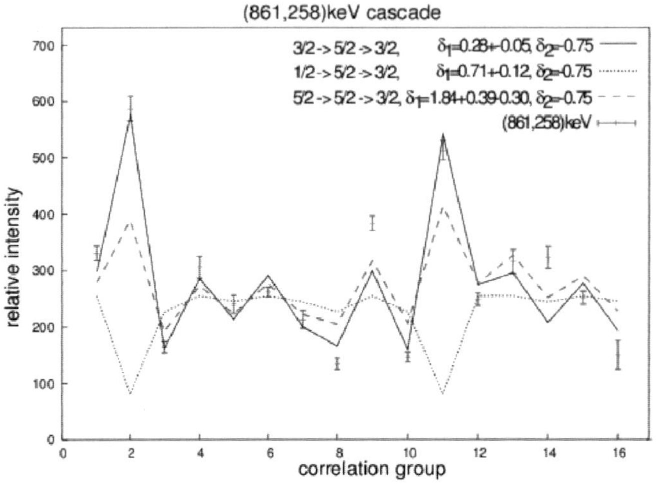

Figure 2. (color online) Example of a $\gamma\gamma$ correlation fit.

By analyzing the $\gamma\gamma$ coincidence matrices, numerous new transitions and states in ^{193}Au could be identified. As an example, The spins of all low-lying positive-parity states in ^{193}Au were determined except for a state at 828 keV, where the fits cannot distinguish between spin $\frac{1}{2}$ and $\frac{3}{2}$.
With the ORANGE electron spectrometers and fast LaBr3(Ce) scintillators $\gamma - \gamma$, $\gamma - e^-$ and $e^- - e^-$ coincidences were performed using the fast timing technique. Using the deconvolution method, the centroid shift technique and the new mirror symmetric centroid difference method [20] several new lifetimes could be measured. Unfortunatelly, all are of states with negative parity. The results are described in detail in Refs. [10,11] to which we refer the interested reader.

For ^{193}Au it was investigated whether a U(6/4) supersymmetry with ^{192}Pt exists. A common fit of both nuclei yielded $\overline{A}$= -0.0432, B= 0.0403 and C= 0.0173 (all in MeV)[11]. It is clear that the fit of ^{193}Au shown in Figure 3 gives only a reasonable description.

5. Summary

We have studied ^{196}Hg after beta decay using the YRAST Ball at Yale university and ^{193}Au using the (p,2n) reaction at the Cologne Tandem accelerator. Using angular correlation and lifetime measurements the experimental knowledge on these nuclei could be considerably extended and corrected. A good fit of ^{196}Hg can be obtained considering coupling two proton fermions to a neutron boson core. A less good but still reasonable description is found using the U(6/4) supersymmetry for ^{193}Au-^{192}Pt.

Journal of Physics: Conference Series **366** (2012) 012023 doi:10.1088/1742-6596/366/1/012023

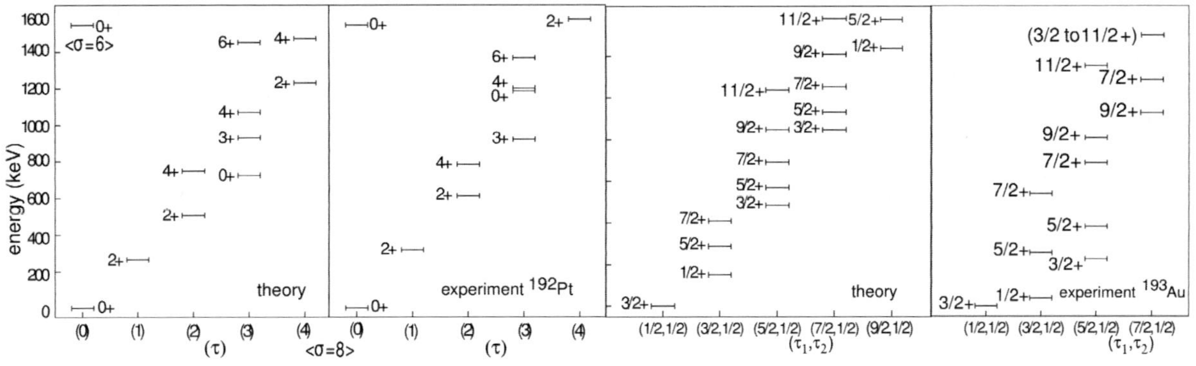

Figure 3. Fit of ^{193}Au-^{192}Pt using the U(6/4) supersymmetry.

References

[1] F. Iachello, Phys. Rev. Lett. **44**, 772 (1980).
[2] J.L. Wood, Phys. Rev. C **24**, 1788 (1981).
[3] J. Vervier, Phys. Lett. B **100**, 383 (1981).
[4] P. Van Isacker, J. Jolie, K. Heyde, and A. Frank, Phys. Rev. Lett. **54**, 653 (1985).
[5] A. Metz, *et al.*, Phys. Rev. Lett. **83**, 1542 (1999).
[6] J. Gröger *et al.*, Phys. Rev. C **62**, 064304 (2000).
[7] H.-F. Wirth *et al.*, Phys. Rev. C **70**, 014610 (2004).
[8] B. Fogelberg, A. Baecklin, V. Berg, and S. G. Malmskog, Nuclear Physics **A153**, 301 (1970).
[9] Y. Gono, R. M. Lieder, M. Müller-Veggian, A. Neskas, and C. Mayer-Böricke, Nucl. Phys. **A327**, 269 (1979).
[10] J.-M. Régis, Th. Materna, G. Pascovici, S. Christen, A. Dewald, C. Fransen, J. Jolie, P. Petkov, K.O. Zell, Rev. Sci. Instr. **81**, 113505 (2010).
[11] T. Thomas, C. Bernards, J.-M. Régis, S. Heinze, C. Fransen, J. Jolie, D. Radeck, M. Albers, K.O. Zell, subm. to Phys. Rev.C (2011)to be publ.
[12] J. Jolie PhD-thesis, Rijksuniversiteit Gent (1986).
[13] C. Bernards, S. Heinze, J. Jolie, C. Fransen, A. Linnemann, and D. Radeck, Phys Rev C 79 (2009) 054307.
[14] C. Bernards, S. Heinze, J. Jolie, M. Albers, C. Fransen, and D. Radeck, Phys. Rev. C 81 (2010) 024312.
[15] F. Iachello and P. Van Isacker, *The Interacting Boson Fermion Model* (Cambridge University Press, Cambridge, England, 1991).
[16] A.B. Balantekin, I. Bars and F. Iachello Nucl. Phys. **A370**, 284 (1981).
[17] Ch. Bernards, PhD thesis, University of Cologne (2011).
[18] F. Pühlhofer, Nucl. Phys. **A280**, 267 (1977).
[19] A. Linnemann, Ph.D. thesis, University of Cologne, 2005, unpublished.
[20] J.-M. Régis, G. Pascovici, J. Jolie, M. Rudigier, Nucl. Instr. Meth. A **622**, 83 (2010).
[21] D. Radeck, M. Albers, C. Bernards, L. Bettermann, A. Blazhev, C. Fransen, S. Heinze, J. Jolie, and D. Mücher, Nucl. Phys. **A821**, 1 (2009).
[22] K.S. Krane, R.M. Steffen and R.M. Wheeler, At. Nucl. Data tables 11, 351 (1973).
[23] K.S. Krane and R.M. Steffen, Phys. Rev. C **2**, 724 (1970).
[24] I. Wiedenhöver, code CORLEONE, University of Cologne, 1995, unpublished.
[25] http://www.nndc.bnl.gov
[26] S. Heinze, Ph.D. thesis, University of Cologne, (2008).

Journal of Physics: Conference Series **366** (2012) 012024 doi:10.1088/1742-6596/366/1/012024

Experiments with massive uranium targets - on the way to technologies for Relativistic Nuclear Energy

M.Kadykov[1], A.Baldin[1], E.Belov[2], V.Chilap[2], A.Chinenov[2], V.Furman[1], M.Galanin[2], N.Gundorin[1], V.Kolesnikov[2], Yu.Kopach[1], A.Rogov[1], N.Ryazansky[2], S.Solodchenkova[2], S.Tyutyunnikov[1]

[1]*Joint Institute for Nuclear Research, Dubna, Russia*
[2]*Center of Physical&Technical Projects «Atomenergomash», Moscow, Russia*

Abstract. It is considered an essentially new scheme of the electro-nuclear method - nuclear relativistic technologies (RNT). This is based on the formation and use an extremely hard neutron spectrum inside deep subcritical active core. It is shown that the development and application of RNT may be promising for solving the problem of utilization of spent nuclear fuel and the global challenges of energy. The results of the first experiments carried out at JINR, indicate a validity of the basic principles of RNT, in particular, doubling the gain of the power of the deuteron beam, irradiating the massive (315 kg) uranium target, with increasing the beam energy from 1 to 4 GeV.

1. Introduction

There are two main reasons that hinder the wide dissemination of nuclear power in the world:
- unsettled in the modern concept of nuclear energy utilization problem of spent nuclear fuel (SNF);
- lack of inventories of raw materials (uranium-235) for hundreds of years.

Modern fast and thermal reactors operate at a controlled fission chain reaction with mean neutron energy about her, or substantially below 0.2 MeV. Subcritical multiplying systems, initiated by accelerators (electronuclear system or Accelerator Driven Systems - ADS) can, in principle, to work with much more hard neutron spectrum. However, the vast majority of ADS scheme is proposed use the same "reactor" neutron spectrum, implemented in sub-critical systems with a $keff \sim 0.94 \div 0.98$.

In the fission neutron spectrum, the threshold minor actinide burning is ineffective because of their high threshold (~ 1 MeV). Transmutation of long-lived radioactive waste from the spent fuel is very bad closed due to multistep reactions that lead to the emergence of new long-lived radioactive isotopes.

Thus it becomes clear that the only real prospect of radical solutions to the problems of modern nuclear power is the use of more hard than fission's, the neutron spectrum.

For practical realization of this path has been worked out principally new scheme electronuclear method, based on the relativistic nuclear technologies (RNT)[1].

This scheme is aimed at creation extremely hard neutron spectrum inside of the multiplying system. It is expected that such a spectrum would permit to "burn" for energy production natural (depleted) uranium or thorium, and simultaneously utilize the long-lived components of spent nuclear fuel of nuclear power plants.

RNT scheme is based on the implementation of the following basic principles[1].

1. Using the deep subcritical active core (AC) of natural (depleted) uranium or thorium the size of which provides minimal leakage of neutrons. (Below this core is called the quasi-infinite).

2. An increase in energy of incident particles up to ~ 10 GeV instead of 1 GeV as in the traditional ADS schemes.

3. Using as a target for incident beam the material of AC.

4. Using a scanning divergent incident beam to reduce by several orders of power release density in the central region of AC serving as neutron productive target

5. Creation of compact powerful linear accelerator based on Russian original scheme BWLAP.

6. Application as a load of AC encapsulated fuel elements from uranium or thorium, as well as spent nuclear fuel, without its preliminary radiochemical reprocessing.

7. Using the technology of high temperature helium coolant for primary circuit.

In difference of traditional reactor and ADS schemes the neutron spectrum in the RNT AC volume is determined apart (n,f), (n,γ) and (n,n' γ)-reactions by large set of competing inelastic processes, in particular, by multi-step cascade reactions as well as by threshold (n,xn)-reactions. The hardest part of the neutron spectrum is formed by high energy neutrons generated at first stages of intra-nuclear cascades. The obtained neutron spectrum allows "burning" out the AC material and the minor actinides placed in this system.

A significant increase in the energy of incident particles up to 10 GeV allows an order to reduce the required current of the accelerator at the same beam power and greatly increase a fraction of the energy beam, which goes on generation of hard neutron field in the AC. This is determined, in particular by the increasing role of meson production in growth of neutron multiplicity and the hardness of the neutron spectrum with increasing beam energy in quasi-infinite multiplying system.

2. The results of the first experiments on the basic physics of RNT

In June 2009 a series of experiments with the target assembly «Quinta» irradiated by deuteron beam from JINR Nuclotron with energies of 1 and 4 GeV were carried out [2]. This assembly shown in Fig.1 consists of the uranium target placed in a lead blanket thickness of 10 cm with the input beam window size of the 150x150 mm. The target consists of three sections of hexagonal aluminum containers with an inscribed diameter of 284 mm, each of which is placed for 61 cylindrical uranium block. In front of the target and between its sections as well as behind it there are 4 detector probes. In these experiments, the first time in the study of accelerator driven systems the integral characteristics of fission in AC were examined by measuring the time spectra of delayed neutrons (DN). They were recorded with a detector assembly "Isomer-M" and stilbene detector.

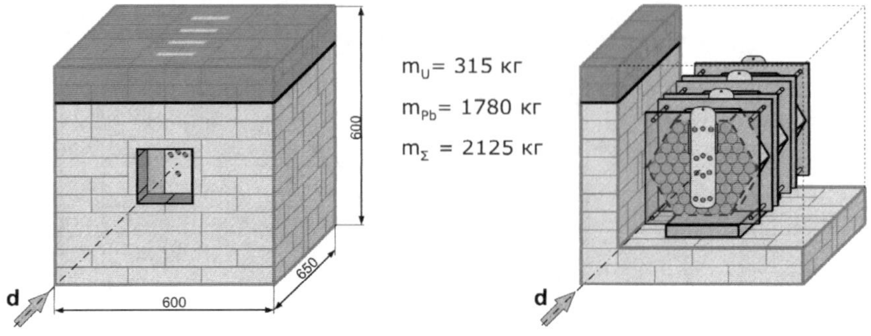

Fig.1. Target assembly "Quinta".

Along with the measurements of the DN yield during experiments the methods of measuring spatial-energy characteristics of neutron fields inside and on the surface of the target assembly were tested using sets of activation detectors.

Fig.2 shows the time dependence of neutron yield from a uranium target irradiated by deuterons with energies of *Ed* = 1 and 4 GeV (indicated by 2 and 3 respectively), as well as from geometrically identical lead target for *Ed* = 4 GeV (labeled by 1). The incident deuteron beam (duration of pulse ~ 500 ms, repletion rate ~ (8 ÷ 9) 1/s) had fine temporary structure defined by features of the beam extraction from the Nuclotron.

Analysis of the time spectra of DN presented in Fig.2 shows that with increasing deuteron energy from 1 to 4 GeV, the number of fissions, and hence the total energy release in the uranium target increases (8.7 ± 1.2) and (10.3 ± 1.5) times from the data obtained by the "Isomer-M" and stilbene detector respectively. So the beam power gain has to grow at least two times. Note that the error values given are determined mainly by accuracy of monitoring the deuteron beam current.

In November 2009, at Nuclotron new experiment was carried out with the target set-up "Energy + Transmutation "("E+T") [3] irradiated by 4 GeV deuterons. The "E + T" set-up consists of a central lead target surrounded by 200 kg blanket from metallic natural uranium (see Fig.3). Beside that the lead-uranium assembly was placed in thick (~300 mm) and dense (ρ=0.7 g/cm3) polyethylene box serving as a reflector and a moderator. In parallel with measurements made on the program of the collaboration "Energy plus Transmutation" it was performed measurements of the time dependence of neutron yields. In fig.4 it is shown the time dependence of neutron yields from the target set-ups «Quinta» and «E+T» obtained by detector assembly «Isomer-M» at incident deuteron energy 4 GeV.

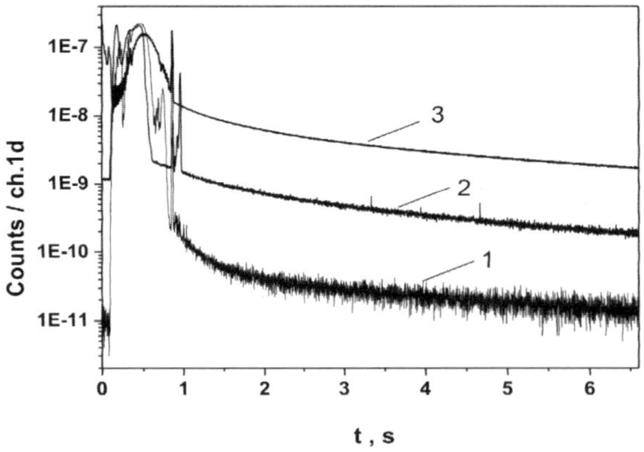

Fig.2. The time dependence of the neutron yield from the geometrically identical lead and uranium targets. 1 - (Pb+d) for *Ed* = 4 GeV; 2 and 3 (U+d) for *Ed* = 1 and 4 GeV.

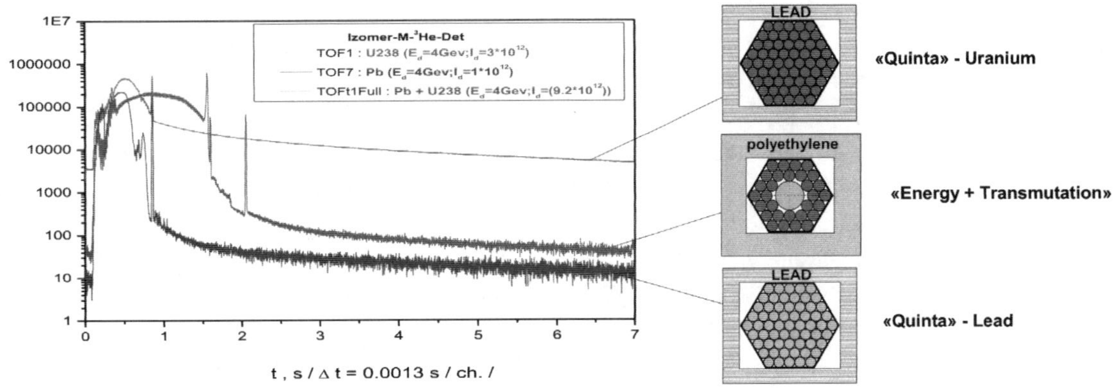

Fig.3. The time dependence of neutron yields from different target assemblies for *Ed* = 4 GeV.

In Fig.3 it is shown the time dependence of neutron yields from the target set-ups «Quinta» and «E+T» obtained by detector assembly «Isomer-M».

It is seen from the above picture that at the same beam energy the DN yield and respectively the number of fissions in the «E+T» target assembly is approximately by 2 orders of magnitude smaller than for the «Quinta» one. This could be related with the usage of the intermediate lead target in the set-up «E+T» and also with the small thickness of the uranium blanket. Beside a presence of the thick layer of polyethylene surrounding the target assembly has to make a resulting neutron spectrum softer in comparison with the same for «Quinta» set-up. All these factors could lead to decreasing of the number of fissions in the «E+T» set-up.

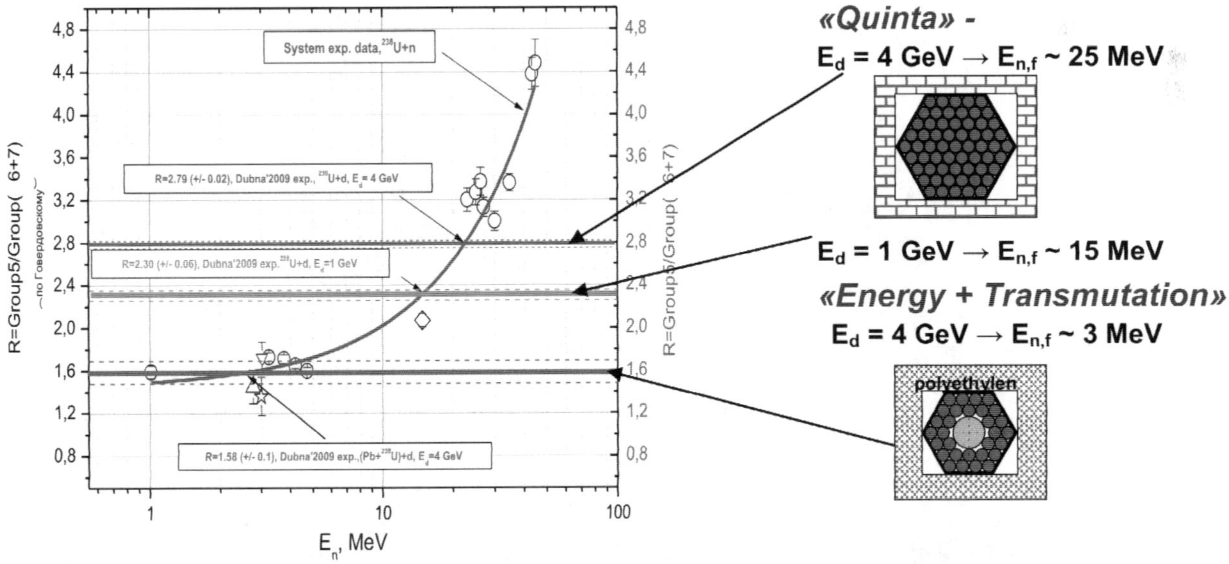

Fig. 4. Comparison of neutron energy dependence of the weight ratios of 5-th to (6+7-th) DN groups from 238U(n,f)-reaction and similar values extracted from DN time spectra measured in present work.

In Fig.4 the systematic of weight ratios of the abovementioned groups in dependence on neutron energy for 238U(n,f)- reaction is presented together with the respective ratios (horizontal lines with error corridors) extracted from analysis of the DN time spectra measured for the uranium target assembly «Quinta» at the deuteron energy of 1 and 4 GeV as well as for the (Pb+238U) assembly «E+T» at *Ed* = 4 GeV.

As follows from Fig.4 for the uranium target assembly «Quinta» the values of the "mean neutron energy" *<En>* inducing of 238U fission are about 15 and 25 MeV for $Ed = 1$ and 4 GeV correspondingly. But for the «E+T» target assembly *<En>* is much lower and is only ~ 3 MeV at $Ed = 4$ GeV.

The experimental results show promising application of RNT. We note in particular a twofold increase power gain of the deuteron beam, irradiating a massive (315 kg) uranium target when incident energy is increased from 1 to 4 GeV. This is in qualitative contradiction with the results of many current simulations, performed by various authors.

However, the amount currently available experimental data, as well as the level of accuracy of the results of computational and theoretical work in this area are insufficient for appropriate cost-informed policy decision to create full-scale prototype of RNT installation aimed at generation electricity and processing spent nuclear fuel.

3. Towards to the creation of technology for relativistic nuclear energy

June 22, 2010 JINR PACs in particle physics has approved Project "Study of deep subcritical accelerator driven systems and possibilities of their use for energy production and transmutation of radioactive waste"[4] with the first priority of execution.

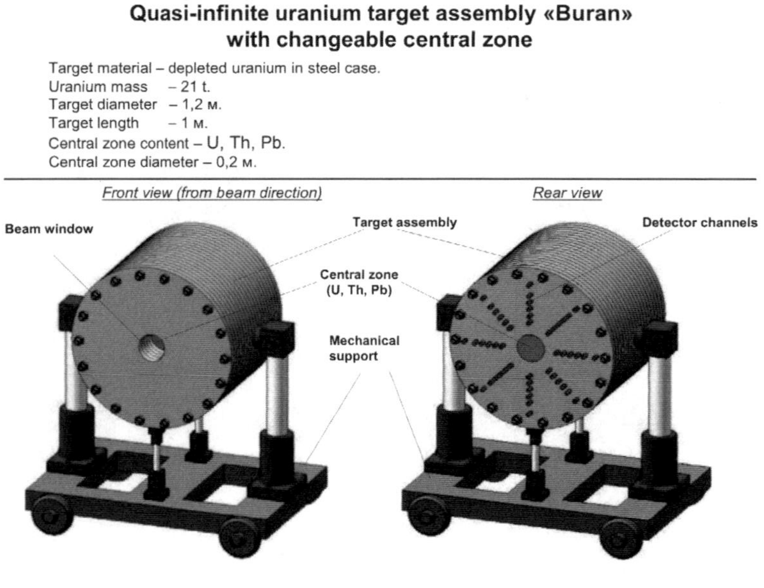

Fig.5.

Project "E&T - RAW" is aimed at experimental demonstration of the effectiveness and feasibility of the RNT scheme for processing of spent nuclear fuel and energy production.

In the framework of the project should be used two target setups:

1. Quasi-infinite uranium target assembly "Buran" a mass of ~ 21 tons, which will be a full-scale nuclear-physical model of the active core of RNT reactor (Fig.5).

2. The upgraded uranium target setting "Quinta"(Fig.6) weighing about 500kg, which simulates the central zone the target assembly "Buran".

The main physical problems have to be studied in experiments with setup "Buran":

1) the determination of the optimal energy and the type of incident particle (proton or deuteron) aimed at achieving of maximal beam power gain;

2) to study the neutron production processes and spatial distribution of neutron spectra;

3) investigation of the incident energy dependence of energy and power gain of the beam;

4) to study the dynamics of production and consumption (due to fission and (n,γ)-process) of 239Pu isotope, depending on its concentration inside of target aimed at determination of its equilibrium concentration;

5) determining the reaction rates of processing the most relevant isotopes from the spent fuel.

The main objectives of the experimental program with target assembly "Quinta" are:

1) determination of the dependence of the beam power gain on energy of incident protons and deuterons;

2) to study the incident energy dependence of the spatial and energy distributions of neutrons, spatial distributions of numbers of fission and plutonium production, as well as the anisotropy of the spectra and multiplicity of leakage neutron;

3) obtaining a set of experimental data to proceed with the modification of existing models and transport codes to improve the reliability of predicting outcomes of future experiments under the "E & T – RAW" project.

In March 2011 the first series of experiments with the upgraded uranium target assembly "Quinta" have been carried out. The target was bombarded by deuterons with energies of 2, 4 and 6 GeV. Layout of the upgraded target setup with measuring equipment is shown in Fig.7. In the near future the results will be published.

Layout of target assembly "Quinta" at the irradiation position

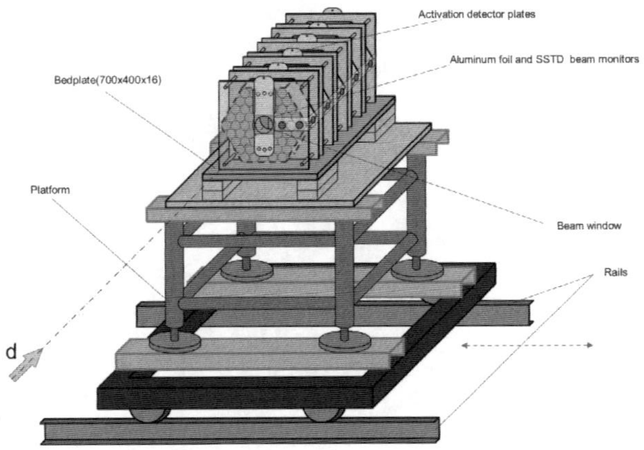

● - Displacement of SSTD and threshold activation detectors on surface of "Quinta" setup

Fig.6.

4. Conclusion

This article presents the basic physical and technical principles of the original electronuclear scheme RNT. It is shown that using RNT can be achieved by a significant increase in the power of relativistic particle beam initiating a subcritical active core. There are considered optimal conditions for the energy gain in the RNT. First results of experiments conducted on JINR Nuclotron with model AC indicate the validity of the basic physical principles of RNT. However, requires further experiments with the greater mass of AC at higher energies of incident particles to convincingly justify the application of innovative perspectives RNT for energy production and utilization of spent nuclear fuel of modern nuclear power plants.

References

1. A.Baldin et al., 2011, *Preprint JINR* **E1-2011-24**, Dubna

2. W. Furman et al., 2011, Time-dependent spectra of neutrons emitted by interaction of 1 and 4 GeV deuterons with massive natural uranium and lead targets., *Proc. Int. Conf. "Nuclear Data for Science and Technology"*, Jeju island, Korea, p.914-917.
3. M.I. Krivopustov et al., 2000, *Preprint JINR* **P1-2000-168**, Dubna, 2003 *Kerntechnik* **68**, p.48-55.
4. J.Adam et al. («E&T – RAW» Collaboration), 2010, *Preprint JINR* **E1-2010-61**, Dubna

Analytic derivation of the Alhassid–Whelan arc of regularity

D Bonatsos[1], S Karampagia[1] and R F Casten[2]

[1] Institute of Nuclear Physics, National Center of Scientific Research "Demokritos", GR–15310 Aghia Paraskevi, Attiki, Greece

[2] Wright Nuclear Structure Laboratory, Yale University, New Haven, CT 06520, USA

E-mail: bonat@inp.demokritos.gr, skaram@inp.demokritos.gr, rick@riviera.physics.yale.edu

Abstract. The Alhassid–Whelan arc of regularity is a unique valley of regularity connecting, amidst chaotic regions, the SU(3) and U(5) vertices of the symmetry triangle of the Interacting Boson Model (IBM). Using the contraction of the SU(3) algebra to the algebra of the rigid rotator in the large boson number limit of the IBM, we construct for the first time a symmetry line in the interior of the triangle, along which the SU(3) symmetry is preserved. The line extends from the SU(3) vertex to near the critical line of the first order shape/phase transition separating the spherical and prolate deformed phases. It lies within the Alhassid–Whelan arc of regularity, thus providing an explanation for its existence.

1. Introduction

The study of chaotic properties of the IBM [1], classically and quantum mechanically, led to the discovery [2, 3, 4] of a narrow strip of nearly regular behavior, connecting the U(5) and SU(3) vertices of the symmetry triangle [5] of the IBM, which was called the Alhassid–Whelan arc of regularity, depicted in Fig. 1. The presence of near regularity presupposes the existence of some underlying approximate symmetry. A recent study [6], reveals the existence of a line, which lies very close to the arc and is based on SU(3) symmetry. This line was derived using characteristic signatures of SU(3), like the degeneracies between levels of the β band and those (with the same L) of the γ band.

In this work, we will show the analytical derivation of a line of approximate SU(3) symmetry, by studying Hamiltonians that approximately commute with the SU(3) generators. The derivation takes advantage of the well–known contraction [7] of the SU(3) algebra to the $[R^5]SO(3)$ algebra [8, 9] of the rigid rotator [10].

The IBM Hamiltonian used is described is described in Section 2. In Section 3, the equation of the line corresponding to the SU(3) symmetry is derived. The conclusions are presented in Section 4. Necessary values of the commutation relations and matrix elements are given in the Appendices of [11].

2. The IBM Hamiltonian and symmetry triangle

The IBM possesses an overall U(6) symmetry, with three dynamical symmetries, labeled by subgroups of U(6), namely, U(5), SU(3) and O(6), which correspond to vibrational, rotational and $\gamma-$unstable nuclei respectively. The IBM Hamiltonian used by Alhassid and Whelan is

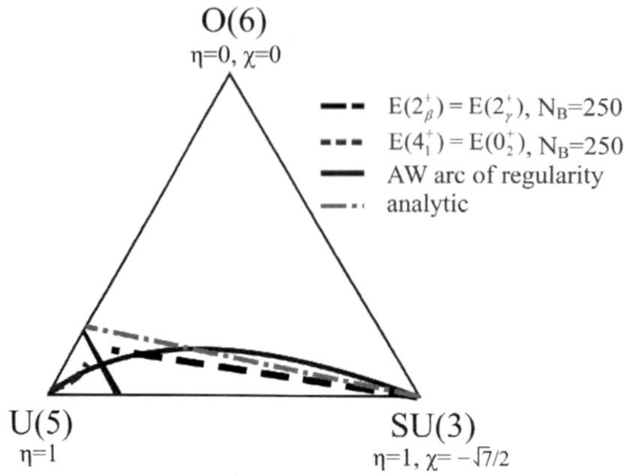

Figure 1. IBM symmetry triangle with the three dynamical symmetries, the Alhassid–Whelan arc of regularity [Eq. (9)], the present line of Eq. (8) (labelled as analytic), the loci of the degeneracies $E(2^+_\beta)$=$E(2^+_\gamma)$ (dashed line on the right for N_B=250 [6]) and $E(4^+_1)$=$E(0^+_2)$ (dotted line on the left for N=250 [6]).

$$\hat{H}(\eta,\chi) = \hat{H}_1 + \hat{H}_2 = c\left[\eta\hat{n}_d + \frac{\eta-1}{N}\hat{Q}^{(2)}_\chi \cdot \hat{Q}^{(2)}_\chi\right], \tag{1}$$

where N is the number of valence bosons, c is a scaling factor, $\hat{H}_1$ and $\hat{H}_2$ denote the first and second terms of the Hamiltonian respectively, $\hat{n}_d = d^\dagger \cdot \tilde{d} = \sqrt{5}(d^\dagger\tilde{d})^{(0)}$ is the d boson number operator, $\hat{Q}_{\chi,\xi} = (s^\dagger\tilde{d} + d^\dagger s)^{(2)}_\xi + \chi(d^\dagger\tilde{d})^{(2)}_\xi$ is the quadrupole operator.

The above Hamiltonian contains two parameters, η and χ, with the parameter η ranging from 0 to 1 and the parameter χ ranging from 0 to $-\sqrt{7}/2$. The U(5) limit corresponds to $\eta = 1$, the SU(3) limit to $\eta = 0, \chi = -\sqrt{7}/2$ and the O(6) limit to $\eta = 0, \chi = 0$.

3. The SU(3) symmetry

3.1. Commutation Relations

In order to see whether the Hamiltonian has a certain symmetry, it has to commute with the generators of that symmetry. The generators of the SU(3) algebra [1], are the angular momentum operators

$$\hat{L}_\xi = \sqrt{10}(d^\dagger\tilde{d})^1_\xi, \tag{2}$$

and the quadrupole operators

$$\hat{Q}^{(2)}_{SU(3),\xi} = (s^\dagger\tilde{d} + d^\dagger s)^{(2)}_\xi - \frac{\sqrt{7}}{2}(d^\dagger\tilde{d})^{(2)}_\xi. \tag{3}$$

The Hamiltonian of Eq. (1) commutes with the angular momentum operators, while the first part of the Hamiltonian with the quadrupole operators gives

$$[\hat{H}_1, \hat{Q}^{(2)}_{SU(3),\nu}] = c\eta[\hat{n}_d, \hat{Q}^{(2)}_{SU(3),\nu}] = c\eta(d^\dagger s - s^\dagger\tilde{d})^{(2)}_\nu. \tag{4}$$

Using

$$\hat{Q}^{(2)}_{\chi,\xi} = \hat{Q}^{(2)}_{SU(3),\xi} + \left(\chi + \frac{\sqrt{7}}{2}\right)(d^\dagger\tilde{d})^{(2)}_\xi, \tag{5}$$

in the second term of the Hamiltonian one gets the intermediate result

$$[\hat{Q}^{(2)}_{\chi}\cdot\hat{Q}^{(2)}_{\chi},\hat{Q}^{(2)}_{SU(3),\nu}] = \sum_{\xi}(-1)^{\xi}\left\{[\hat{Q}^{(2)}_{SU(3),\xi},\hat{Q}^{(2)}_{SU(3),\nu}]\hat{Q}^{(2)}_{\chi,-\xi}+\hat{Q}^{(2)}_{\chi,\xi}[\hat{Q}^{(2)}_{SU(3),-\xi},\hat{Q}^{(2)}_{SU(3),\nu}]\right.$$

$$\left.+\left(\chi+\frac{\sqrt{7}}{2}\right)\left\{[(d^{\dagger}\tilde{d})^{(2)}_{\xi},\hat{Q}^{(2)}_{SU(3),\nu}]\hat{Q}^{(2)}_{\chi,-\xi}+\hat{Q}^{(2)}_{\chi,\xi}[(d^{\dagger}\tilde{d})^{(2)}_{-\xi},\hat{Q}^{(2)}_{SU(3),\nu}]\right\}\right\}. \tag{6}$$

In the large N–limit, Eq. (6) can be simplified. In this limit, the ground state band becomes energetically isolated from all other excitations, so SU(3) effectively reduces to a simple rigid rotator. This situation is known as the contraction of SU(3) to R^5[SO(3)] [8, 9]. At the contraction limit, the $Q^{(2)}_{SU(3)}$ operators can be replaced by mutually commuting quantities and terms containing $(d^{\dagger}\tilde{d})^{(k)}$ can be ignored. As seen from Eq. (5), $\hat{Q}^{(2)}_{\chi}$ can be replaced by $\hat{Q}^{(2)}_{SU(3)}$ and in its turn, $\hat{Q}^{(2)}_{SU(3)}$ can be replaced by the intrinsic quadrupole moment q_0, which is $N\sqrt{2}$ in the present case. So, in the large N–limit, the commutator of the second part of the Hamiltonian becomes

$$[\hat{H}_2,\hat{Q}^{(2)}_{SU(3),\nu}] = c(\eta-1)2\sqrt{2}\left(\chi+\frac{\sqrt{7}}{2}\right)(d^{\dagger}s-s^{\dagger}\tilde{d})^{(2)}_{\nu}. \tag{7}$$

In order to get a vanishing commutator, the coefficients of $(d^{\dagger}s-s^{\dagger}\tilde{d})^{(2)}_{\nu}$ in Eqs. (4) and (7) should cancel, leading to the condition

$$\chi(\eta) = \frac{1}{2\sqrt{2}}\frac{\eta}{(1-\eta)}-\frac{\sqrt{7}}{2}. \tag{8}$$

The Alhassid–Whelan arc of regularity [2, 3, 4] is given by the expression [12]

$$\chi(\eta) = \frac{\sqrt{7}-1}{2}\eta-\frac{\sqrt{7}}{2}. \tag{9}$$

Equations (8) and (9) give very similar predictions for values of η between 0 and 0.6, i.e. from the SU(3) vertex until quite close to the critical line, as seen in Fig. 2.

3.2. Matrix Elements

We can also consider the matrix elements of $[\hat{H}_1,\hat{Q}^{(2)}_{SU(3),\nu}]$ and $[\hat{H}_2,\hat{Q}^{(2)}_{SU(3),\nu}]$ within the ground state band, in order to justify the replacement of the quadrupole operator $\hat{Q}^{(2)}_{SU(3)}$ by the intrinsic quadrupole moment in the previous subsection.

Using the standard formalism for treating matrix elements, one finds for the first and second part of the Hamiltonian,

$$\langle[N],(2N,0),\tilde{\chi}=0,L||[\hat{H}_1,\hat{Q}^{(2)}_{SU(3)}]||[N],(2N,0),\tilde{\chi}=0,L\rangle$$
$$= -\frac{2}{3\sqrt{7}}c\eta NR_1\sqrt{2L+1}, \tag{10}$$

$$\langle[N],(2N,0),\tilde{\chi}=0,L||[\hat{H}_2,\hat{Q}^{(2)}_{SU(3)}]||[N],(2N,0),\tilde{\chi}=0,L\rangle$$
$$= \frac{1}{7\sqrt{2}}c(1-\eta)\left(\chi+\frac{\sqrt{7}}{2}\right)q_0R_2\sqrt{2L+1}, \tag{11}$$

where $\tilde{\chi}$ is the Vergados quantum number [13] and R_1, R_2 are complicated expressions [11] of the number of bosons N and the angular momentum L. In the large N limit and for L not too

Journal of Physics: Conference Series **366** (2012) 012025 doi:10.1088/1742-6596/366/1/012025

small, we obtain $R_1 = 1$ and $R_2 = 16/3$. Replacing the intrinsic quadrupole moment q_0 by its value $(N\sqrt{2})$, we see that the two matrix elements vanish when

$$\chi(\eta) = \frac{\sqrt{7}}{8}\frac{\eta}{(1-\eta)} - \frac{\sqrt{7}}{2}. \tag{12}$$

Expressions (12) and (8) give very similar results as seen in Fig. 2.

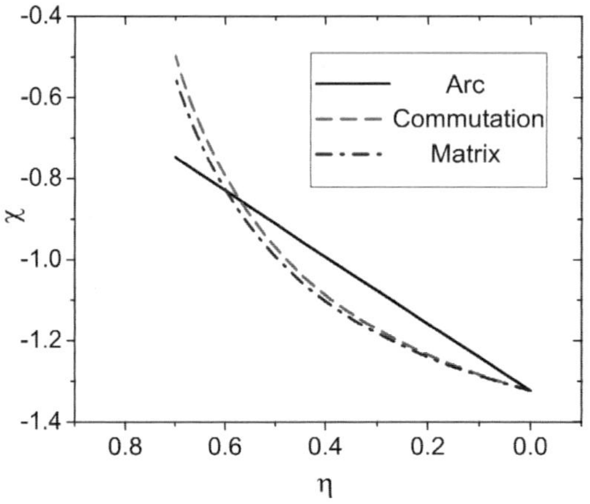

Figure 2. Location of the arc of regularity, as described by the original Eq. (9), and as predicted by the findings of the present work, Eq. (8) and Eq. (12). The η axis has been reversed, in order to correspond directly to Fig. 1.

4. Conclusion

The present work is the first analytical evidence for the existence of an approximate symmetry inside the symmetry triangle of the IBM. The extracted line of SU(3) symmetry follows closely the Alhassid–Whelan arc of regularity, providing thus an explanation for its existence.

References

[1] Iachello F and Arima A 1987 *The Interacting Boson Model* (Cambridge: University Press)
[2] Alhassid Y and Whelan N 1991 *Phys. Rev. Lett.* **67** 816
[3] Whelan N and Alhassid Y 1993 *Nucl. Phys.* A **556** 42
[4] Whelan N 1993 *Ph. D. thesis* (Yale University)
[5] Casten R F 1990 *Nuclear Structure from a Simple Perspective* (Oxford: Oxford University Press)
[6] Bonatsos D, McCutchan E A and Casten R F 2010 *Phys. Rev. Lett.* **104** 022502
[7] İnönü E and Wigner E P 1953 *Proc. Natl. Acad. Sci.* (N.Y.) **39** 510
[8] Le Blanc R, Carvalho J and Rowe D J 1984 *Phys. Lett.* B **140** 155
[9] Rowe D J 1996 *Prog. Part. Nucl. Phys.* **37** 265
[10] Ui H 1970 *Prog. Theor. Phys.* **44** 153
[11] Bonatsos D, Karampagia S and Casten R F 2011 *Phys. Rev. C* **53** 054313
[12] Cejnar P and Jolie J 1998 *Phys. Rev. E* **58** 387
[13] Vergados J D 1968 *Nucl. Phys.* A **111** 681

Journal of Physics: Conference Series **366** (2012) 012026 doi:10.1088/1742-6596/366/1/012026

Nuclear γ-ray line emission induced by energetic ions in solar flares and by galactic cosmic rays

J. Kiener, V. Tatischeff and H. Benhabiles-Mezhoud
CSNSM Orsay, CNRS/IN2P3 and Université Paris-Sud, 91405 Orsay Campus, France

N. de Séréville
IPN Orsay, CNRS/IN2P3 and Université Paris-Sud, 91400 Orsay, France

A. Belhout
USTHB, Faculté de Physique, BP 32, El-Alia, 16111 Bab Ezzouar, Algiers, Algeria

E-mail: Jurgen.Kiener@csnsm.in2p3.fr

Abstract. The γ-ray spectra of the strongest solar flares often show a broad and complex structure in the 0.1-10 MeV region sitting on a bremsstrahlung continuum. This structure is composed of several outstanding narrow lines and of thousands of unresolved narrow and broad lines forming a quasi-continuum. The major part of this emission is due to prompt deexcitation lines following nuclear interactions of accelerated light and heavy ions with the atomic nuclei composing the solar atmosphere. A similar emission is expected from interactions of galactic cosmic rays with the interstellar gas and dust. Experimental nuclear reaction studies coupled with extensive calculations have been done in the last one and a half decade at Orsay for the modelisation of this γ-ray emission. After a description of the nuclear reaction studies the analysis of one solar flare spectrum and predictions for the emission from the inner Galaxy will be presented.

1. Introduction

Nuclear γ-ray line emission from astrophysical sources is a potentially very rich source of information about energetic particle populations and the interacting medium. The different line intensities depend of course on the abundances of the ambient nuclei, but also on the composition and spectra of the energetic particles. Furthermore, most lines from astrophysical sources are Doppler broadened because the ambient media like the solar atmosphere and the interstellar gas have very low densities such that the recoiling nuclei emit almost always in flight. The line shapes may then add another piece of information on the energetic particle properties.

It appears that only lines from interactions of protons and α-particles with the major isotopes of 6 out of the 7 most abundant heavier elements ^{12}C, ^{16}O, ^{20}Ne, ^{24}Mg, ^{28}Si and ^{56}Fe and from α+^{4}He ($\alpha\alpha$) interactions have been unambiguously identified up to now from strong solar γ-ray flares. A first comprehensive study including a compilation and evaluation of cross sections as well as line shape calculations for these reactions has been started in the pioneering work of Ramaty et al. [21]. This work has been continued including many new cross section data measured mainly at the tandem accelerator facilities of Washington [3, 4, 24] and Orsay.

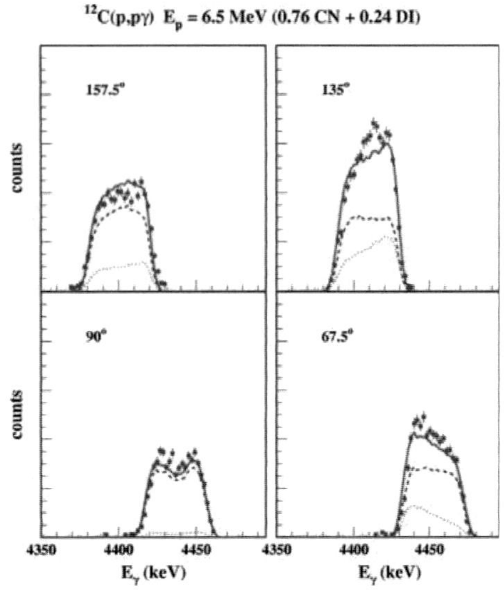

Figure 1. Measured 4.438-MeV line shapes at different angles w.r.t. the beam direction for the inelastic scattering reaction $^{12}C(p,p\gamma)^{12}C$ at $E_p = 6.5$ MeV (symbols). The solid (red) lines show the results of the line-shape calculations, with the direct reaction and CN components indicated by the dotted (green) and dashed (blue) lines, respectively.

We present an overview of experimental nuclear reaction studies coupled with extensive calculations which have been done in Orsay for the modelisation of γ-ray emission from astrophysical sources. A brief description of the solar flare interaction model and the analysis of one solar flare spectrum will be presented in the following. A similar emission is expected from interactions of low-energy cosmic rays with the interstellar gas and dust. We present estimates of the nuclear γ-ray line emission induced by these particles in the inner Galaxy and discuss its detectability with future space-borne γ-ray telescopes.

2. Nuclear reaction studies

The strongest nuclear deexcitation lines in solar flares are, with the exception of two lines from $\alpha\alpha$ reactions, the lines resulting from the deexcitation of the first few excited states in the above cited nuclei. For the latter γ-ray production cross sections in proton and α-particle induced reactions have been measured in three experimental campaigns at the tandem Van-de-Graaff accelerator of the IPN Orsay [18, 1, 2]. In a fourth experiment we studied ^{3}He-induced reactions with ^{16}O and ^{24}Mg [5, 6]. Typically 4-5 large-volume Ge detectors with BGO shielding have been employed for the γ-ray detection covering a large angular range in order to accurately determine the γ-ray angular distributions.

Cross section excitation functions have been measured typically in the energy range 5-25 MeV for protons and 5-40 MeV for α-particles. Small steps of $\sim$0.2-0.5 MeV were usually employed at projectile energies below $\sim$15 MeV where often prominent compound-nucleus (CN) resonances show up in the cross sections. The most important reactions for these lines are inelastic scattering reactions, however, for certain lines fusion-evaporation and spallation reactions of a very abundant isotope may significantly contribute.

About 100 different cross section excitation functions have been measured at Orsay during these campaigns, providing an important experimental database for energetic-particle induced γ-ray emission. Many are included in the latest compilation containing 181 excitation functions for γ-ray line production in p, α and ^{3}He-induced reactions [19]. Extrapolations to higher energies and cross sections for other lines are mainly based on calculations with the nuclear reaction code TALYS [13], which generally shows good agreement with experimental γ-ray data.

The shapes of the narrow line components (from energetic light-particle induced reactions) are determined by the fast particle ratios (essentially the α-to-proton ratio α/p) and their energy spectra, and the particle angular distributions, which may be far from isotropic in solar flares. Contrary to cross section data, very few line shape data were published before our work (see e.g. [21]). Therefore we insisted at Orsay on good statistics for the strongest lines and completed the measurements by precise determinations of the beam-induced γ-ray background. Based on these data, complete parameterizations for the calculations of the 4.438-MeV [10] and later the 6.129-MeV line shapes have been obtained [11].

Since then, a new method has been developed, relying on optical-model and resonance calculations specifically to improve the line-shape descriptions in the region dominated by CN resonances [12]. An example is shown in Fig. 1 for the 4.438-MeV line. There, the inelastic scattering is calculated in the coupled-channels (CC) approach. The closest CN state at E_p = 6.5 MeV is the 7.900 MeV, $3/2^+$ states of ^{13}N. The best fit for the CN resonance was found for 60% l=0 and 40% l=2 proton emission to the 4.439-MeV 2^+ state of ^{12}C, representing 76% of the cross section. CN and direct components are added incoherently. Besides the choice of the optical potential for the CC calculation, the only free parameters are the direct-to-CN resonance and the decay angular momentum ratios. It will be applied to proton and α-particle reactions producing the 4.438-MeV and 6.129-MeV lines for which we have extensive sets.

3. Solar flare γ-ray emission

Particle acceleration in impulsive-type solar flares is believed to occur mainly in the low-density solar corona. For the strongest flares, electrons are accelerated to energies exceeding 20 MeV and nuclei can be accelerated to energies above several GeV per nucleon [20]. Part of the energetic particles are injected into magnetic loops where they are trapped and eventually interact with the denser parts of the solar atmosphere, the chromosphere and photosphere. Electromagnetic radiation from the radio to the γ-ray band as well as secondary particle radiation is produced in these interactions. The γ-ray emission below several hundred keV is mostly due to electron bremsstrahlung where the differential photon flux usually follows a power law in energy. At higher photon energies one often observes an additional component superposed on the power-law continuum which is mainly from nuclear γ-ray line emission and sometimes becomes the dominant process in the several MeV range. Beneath the nuclear deexcitation lines, strong narrow lines at 511 keV and 2.223 MeV from positron annihilation and neutron capture on ^{1}H, respectively, are often present. At still higher energies, π^0-decay emission is sometimes observed.

One of the strongest solar flares ever observed occured on Oct. 28, 2003, shortly after 11:00. Its γ-ray emission, lasting about 15 min., has been detected by the Ge spectrometer SPI onboard the Integral spacecraft [7]. Relatively good statistics for the γ-ray spectrum in the 0.6-10 MeV range has been obtained despite a spacecraft orientation which required the radiation to penetrate the satellite bus and the 5cm-thick BGO shield before reaching the Ge detector stack. The observed spectrum (Fig. 2) is dominated by nuclear line emission in the range 1-8 MeV. It shows a very prominent 2.223-MeV line and clearly the 4.438-MeV and 6.129-MeV lines. Assuming power-law spectra in energy, the 2.223/(4.438+6.129) ratio allowed determination of the spectral index s $\sim$ -3.5. The α/p ratio was determined by line-shape analysis (see [11]).

The best overall reproduction of the observed spectrum could be obtained with an accelerated heavy-ion content close to the average composition of impulsive-type flares measured in interplanetary space [22], i.e. strongly enhanced in heavier elements Mg-Ni. In this case, accelerated heavy ions are responsible for more than 80% of the emission, which, because of the large line broadening forms an important continuum in the observed energy range. Compositions with lower enhancements like e.g. the coronal composition are unable to reproduce simultaneously the narrow line and continuum intensities.

In the line shape analysis the 4.438-MeV and 6.129-MeV lines were simultaneously fitted by scanning the parameter space α/p ratio - particle angular distribution around the flare axis. The latter is supposed to be perpendicular to the solar surface at the flare location. The best

Journal of Physics: Conference Series **366** (2012) 012026 doi:10.1088/1742-6596/366/1/012026

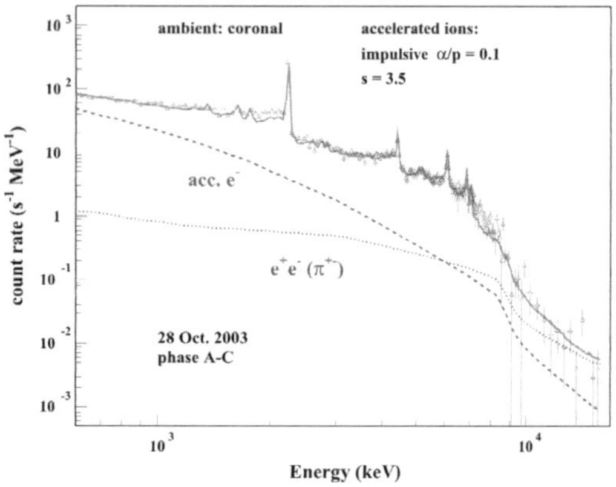

Figure 2. Spectrum of the Oct. 28, 2003 flare as observed by SPI onboard Integral (symbols). The nuclear γ-ray line emission is calculated with particle energy spectra I(E) $\propto$ $E^{-3.5}$ and impulsive-flare-type composition (solid line). Bremsstrahlung emission from accelerated electrons and high-energy leptons from pion decays is shown by the dashed (red) and dotted (magenta) lines, respectively.

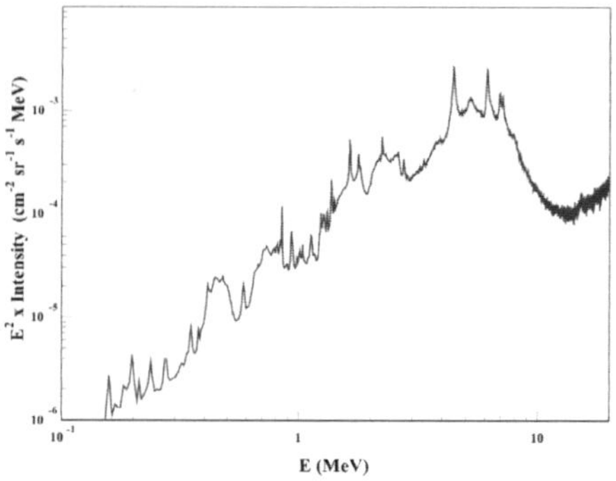

Figure 3. Calculated nuclear γ-ray line emission of the inner Galaxy (300°<l<60°, |b|<10°) induced by CR nuclei including a LECR component.

fit could be obtained with a relatively narrow downward-directed distribution of the interacting particles and $\alpha/p = 0.12$. These results could be obtained with spectra resulting from very unfavorable observing conditions. Better observations can be expected in the future providing probably a detailed picture of the energetic particle properties and eventually their evolution during the flare.

4. Cosmic-ray induced γ-ray emission

The cosmic-ray (CR) spectrum is relatively well established above 1 GeV and up to at least 10^{19} eV. Interstellar particles with energies below 1 GeV per nucleon, however, are strongly deflected by magnetic fields in the heliosphere such that practically nothing is known about them. There is only indirect evidence that our Galaxy contains an important component of low-energy cosmic rays (LECRs). The most compelling is probably the observation of the molecule H_3^+ in diffuse interstellar clouds which requires an H_2 ionization rate $\zeta_2 = 4\times10^{-16}$ s^{-1} [8]. This ionization must be due to CRs because UV photons cannot penetrate these clouds. Standard CRs produce an ionization rate which is ten times smaller than that.

Supposing that this ionization is due to nuclei, one can construct a CR spectrum that satisfies the observed ζ_2 by adding a component below E/A = 1 GeV. Sources of these LECRs could be acceleration in weak shocks at astrospheres as proposed by [23]. We used their model for source spectra and composition and propagated the ions with a leaky-box model. The intensity was normalized to produce $\zeta_2 = 3.6\times10^{-16}$ s^{-1} and then added to the standard CRs. Fermi-LAT recently observed the diffuse γ-ray emission above 100 MeV from the inner Galaxy [25]. From their preliminary data they extracted the component due to π^0-decay. Calculating the total γ-ray emission from our CR spectra (using the model of [9] for the π^0-decay emission) and adjusting the flux above 100 MeV to the observation provides a robust normalization of the nuclear γ-ray line emission. A γ-ray spectrum for LECRs accelerated at astrospheres with an exponential cutoff of 750 MeV [23] is shown in Fig. 3.

Integrated fluxes of the inner Galaxy for the 4.438-MeV and 6.129-MeV lines are within the reach of γ-ray satellites projects equipped with state-of-the-art detector systems currently used in nuclear and particle physics. Still more favorable is the large bump at $E_\gamma \sim$ 3-10 MeV with an integrated flux of the order of 10^{-4} cm^{-2} s^{-1}. This lies comfortably above the survey broad-band 3-σ sensibility limits of the projects presented to the latest (2010) ESA Cosmic Vision call for a medium-size mission (e.g. $\sim 10^{-5}$ cm^{-2} s^{-1} after 5 years for CAPSiTT).

5. Conclusions

We presented nuclear reaction studies done in the last 15 years contributing to the database for energetic-particle induced nuclear γ-ray line emission. Two examples for application of these data in astrophysical sites have been discussed. The analysis of the Oct. 28, 2003 solar flare illustrated the rich potential of nuclear γ-ray line emission reflecting the energetic particle properties. It has been shown that such an emission could eventually be observed in the near future from the inner Galaxy which would give important clues to low-energy cosmic rays.

6. Acknowledgements

We like to thank the staff at the Orsay tandem accelerator for their efficient help in setting up the experiments and excellent preparation of the ion beams.

References

[1] Belhout A et al. 2007, *Phys. Rev.* **C 76**, 034607
[2] Benhabiles-Mezhoud A et al. 2011, *Phys. Rev.* **C 83**, 024603
[3] Dyer P et al. 1981, *Phys. Rev.* **C 23**, 1865
[4] Dyer P et al. 1985 , *Phys. Rev.* **C 32**, 1873
[5] Tatischeff V et al. 2003, *Phys. Rev.* **C 68**, 025804
[6] Fitoussi C et al. 2008, *Phys. Rev.* **C 78**, 044613
[7] Gros M et al. 2004, *Proceedings of the 5th INTEGRAL Workshop, the INTEGRAL Universe*, 16-20 Feb. 2004, München, Germany, ed. V. Schönfelder, G. Lichti & C. Winkler, ESA SP-552, 669
[8] Indriolo N, Fields B D and McCall B J 2009, *Astrophys. J.*, **694**, 257
[9] Kamae T et al. 2006, *Astrophys. J.* **647**, 692
[10] Kiener J et al. 2001, *Phys. Rev.* **C 64**, 025803
[11] Kiener J et al. 2006, *Astron. & Astrophys.* **445**, 725
[12] Kiener J et al. 2008, *Proceedings of the DAE Symposium on Nuclear Physics Vol. 53*, December 22-26, 2008, Roorkee, India, ed. R. K. Choudhury, A; Saxena, B. J. Roy
[13] Koning A J, Hilaire S and Duijvestijn M C 2008, *Proceedings of the Int. Conf. on Nuclear Data for Science and Technology*, April 22-27, 2007, Nice, France, ed. O. Bersillon, F. Gunsing, E. Bauge, R. Jacqmin, and S. Leray, EDP Sciences
[14] Lang F L et al. 1987, *Phys. Rev.* **C 35**, 1214
[15] Lesko K T et al. 1988, *Phys. Rev.* **C 37**, 1808
[16] Lodders K, Palme H and Gail H-P 2009, *4.4 Abundances of the elements in the solar system.* Trümper J E (ed.) SpringerMaterials - The Landolt-Börnstein Database
[17] Murphy, R.J., Hua, X.-M., Kozlovsky, B., & Ramaty, R. 1990, *Astrphys. J.*, **351**, 299
[18] Kiener J et al. 1998, *Phys. Rev.* **C 58**, 2174
[19] Murphy R J et al. 2009, *Astrophys. J. Suppl. Ser.* **183**, 2009
[20] Ramaty R and Simnett G M 1991, *The Sun in Time*, pages 232-259, The University of Arizona Press
[21] Ramaty R, Kozlovsky B and Lingenfelter R.E. 1997, *Astrophys. J. Suppl. Ser.* **40**, 487
[22] Reames D V 1999, *Space Science Rev.* **90**, 413
[23] Scherer K et al. 2008, *Adv. Space Res.* **41**, 1171
[24] Seamster A G et al. 1984, , *Phys. Rev.* **C 29**, 394
[25] Strong A W et al. 2011, *ICATPP Conference on Cosmic Rays for Particle and Astroparticle Physics*, Villa Olmo (Como, Italy), 7-8 October, 2010, to be published by World Scientific (Singapore)

Journal of Physics: Conference Series **366** (2012) 012027 doi:10.1088/1742-6596/366/1/012027

Fast-timing measurements in ^{95,96}Mo

S Kisyov[1], S Lalkovski[1], N Mărginean[2], D Bucurescu[2], L Atanasova[3], D Balabanski[3], Gh Căta-Danil[2], I Căta-Danil[2], D Deleanu[2], P Detistov[3], D Filipescu[2], D Ghiţă[2], T Glodariu[2], R Mărginean[2], C Mihai[2], A Negret[2], S Pascu[2], T Sava[2], L Stroe[2], G Suliman[2], N V Zamfir[2] and M Zhekova[1]

[1]Faculty of Physics, University of Sofia "St. Kliment Ohridski", Sofia, Bulgaria
[2]National Institute for Physics and Nuclear Engineering "Horia Hulubei", Magurele, Romania
[3]Institute for Nuclear Research and Nuclear Energy, Bulgarian Academy of Science, Sofia, Bulgaria

E-mail: stanimir.kisyov@phys.uni-sofia.bg

Abstract. Half-lives of the $19/2^+_1$ and $21/2^+_1$ states in ^{95}Mo and of the 8^+_1 and 10^+_1 states in ^{96}Mo were measured. Matrix elements for yrast transitions in ^{95}Mo and ^{96}Mo are discussed.

1. Introduction

There are twenty six molybdenum isotopes on the Segré chart, studied by means of γ ray spectroscopy [1]. They span a wide region, where a variety of shapes are observed, from the spherical ^{90}Mo to one of the most deformed nuclei in the $A \approx 110$ mass region. Of particular interest are the ^{95,96}Mo nuclei, which are placed in a region where an offset of the quadrupole deformation takes place [2]. Also, the two nuclei are placed in the third oscillator shell in protons and at the beginning of the fourth oscillator shell in neutrons, where high-j and low-j single-particle orbits with $\Delta j = 3$ enhance the effects of octupole collectivity [3]. Being on the edge between the single-particle and collective modes, the two nuclei represent an excellent laboratory, where different approaches can be tested. In this respect, of particular interest are the nuclear lifetimes of excited states which are directly related to transition matrix elements and hence are sensitive to the underlying structure.

2. Experimental Set Up and Data Analysis

Fast-timing measurements were performed in ^{95,96}Mo. The nuclei were produced in fusion-evaporation reactions, performed at the NIPNE tandem accelerator. A beam of ^{18}O was accelerated up to 62 MeV and focused on ^{82}Se target with a thickness of 5 mg/cm^2 on 2 mg/cm^2 thick Au backing. The cross section for the ^{82}Se(^{18}O, 5nγ)^{95}Mo channel was estimated at 400 mb, while that for the ^{82}Se(^{18}O, 4nγ)^{96}Mo channel at about 100 mb. The beam intensity was of the order of 20 pnA.

The gamma ray detector system consisted of eight $LaBr_3$:Ce detectors and eight HPGe detectors, working in coincidence [4, 5]. The $LaBr_3$:Ce detectors issue positive dynode and negative anode signals. The dynode signals were used for energy measurements, while the anode signals were used for timing. The time signals were shaped by Constant Fraction Discriminators

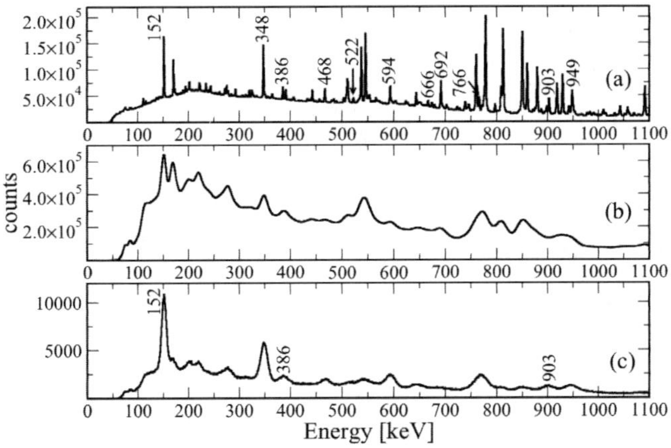

Figure 1. ^{95}Mo energy spectra. HPGe total projection (a), $LaBr_3$:Ce total projection (b), $LaBr_3$:Ce energy spectrum gated on 692-keV transition with HPGe detectors (c).

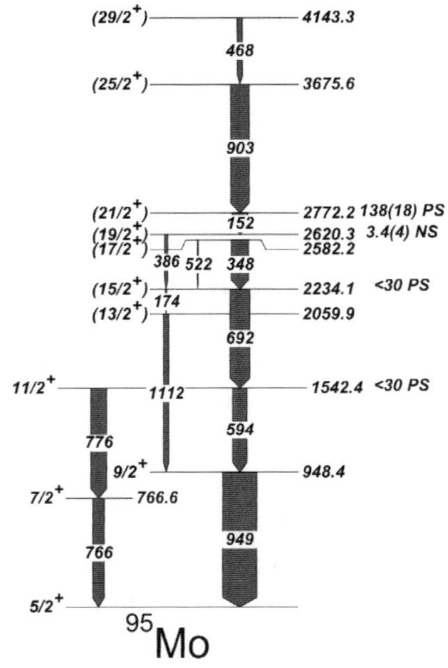

Figure 2. Partial level scheme of ^{95}Mo.

(CFDs) and sent to Time-to-Amplitude-Converters (TACs), operating in a common STOP mode. The energy and time signals were digitized and the data were stored in event files, collected for approximately two hours each. The system was triggered by coincidences between two $LaBr_3$:Ce and one HPGe detector. The data were analyzed with GASPware [6] and RadWare [7] packages.

Three dimensional matrices were sorted with $LaBr_3$:Ce energy on two of the axes and a relative time, defined as $T_{1,2} = t \pm (t_1 - t_2)$, on the third axis. Here, t_1 is the moment of interaction of the preceding in the time γ ray and t_2 is the moment at which the delayed γ ray was detected, while t is an arbitrary offset. If the two γ rays feed and de-excite a particular nuclear level, and $E_{\gamma 1}$ denotes the energy of the feeding transition, while $E_{\gamma 2}$ is the energy of the delayed transition, then for each event two matrix elements $(E_{\gamma 1}, E_{\gamma 2}, T_1)$ and $(E_{\gamma 2}, E_{\gamma 1}, T_2)$ were incremented. Then, from the time projected $(E_{\gamma 1}, E_{\gamma 2})$ and $(E_{\gamma 2}, E_{\gamma 1})$ two-dimensional gates, two time spectra, symmetric with respect to the arbitrary offset t, were obtained. If the lifetime of the level of interest is of the order of the detectors time resolution, the centroids $C_D =< t >= \int tD(t)dt / \int D(t)dt$ of the time distributions will be displaced by 2τ one from the other, where τ is the lifetime of the level. This procedure is known as the centroid shift method [8,9]. It should be noted that the two centroids will overlap in the cases where $E_{\gamma 1}$ and $E_{\gamma 2}$ feed and de-excite nuclear level with a lifetime shorter than the binning of the converter, which was of 6 ps/channel. For lifetimes longer than the detector time resolution, the slope method was used to determine level lifetimes. To reduce the background in the $LaBr_3$:Ce spectra, the (E_γ, E_γ, T) matrices were incremented after an (E_γ, T) gate was implied on any of the HPGe detectors.

Sample energy spectra, obtained with HPGe and $LaBr_3$:Ce detectors, are presented in figure 1. Figure 1(a) shows a total energy projection for all HPGe detectors. Peaks corresponding to transitions in ^{95}Mo are denoted with the transition energies. Figure 1(b) presents the total energy projection obtained with all $LaBr_3$:Ce detectors. A $LaBr_3$:Ce spectrum, gated on the 692-keV transition from ^{95}Mo in the HPGe detectors, is shown in figure 1(c).

Table 1. Energy levels and transitions in ^{95}Mo.

$J^{\pi\dagger}$	E^a_{level}[keV]	E^b_γ [keV]	I_γ	$\lambda M^\dagger$	$T_{1/2}$
$5/2^+$	0.0				stable
$7/2^+$	766.6(7)	766.3	0.80(16)	M1+E2	4.4(7) ps†
$9/2^+$	948.4(7)	948.7	2.4(5)	E2(+M3)	2.58(11) ps†
$11/2^+$	1542.2(10)	593.6	1	M1+E2	≤30 ps
$11/2^+$	1542.2(10)	775.7	1.09(22)	E2	≤30 ps
$(13/2^+)$	2060.2(17)	1112.0		(E2)	
$(15/2^+)$	2234.2(15)	173.8	0.086(10)	(M1+E2)	≤30 ps
$(15/2^+)$	2234.2(15)	692.0	1.39(25)	E2	≤30 ps
$(17/2^+)$	2582.4(16)	348.1	1.23(11)	M1(+E2)	d
$(17/2^+)$	2582.4(16)	522.1	0.062(12)	[E2]^c	d
$(19/2^+)$	2620.5(16)	38.1†	0.056	[M1+E2]^c	3.4(4) ns
$(19/2^+)$	2620.5(16)	386.4	0.23(2)	E2	3.4(4) ns
$(21/2^+)$	2772.4(17)	152.3	1.1(5)	M1	138(18) ps
$(25/2^+)$	3675.8(20)	903.4	1.3(3)	E2	d
$(29/2^+)$	4143.5(23)	467.7	0.37(5)	E2	d

†from NNDC [10], unless otherwise noted; ^{a}from a least-squares fit to E_γ; ^{b}uncertainty of 1.0 keV; ^{c}from the J^π difference; ^{d}no enough statistics in the present experiment.

The partial level scheme of ^{95}Mo, presented in figure 2, is based on the coincidence measurements performed in the present study and is consistent with the level scheme presented in Ref. [11]. Nuclear level spins and parities J^π, level energies E_{level}, γ-ray energies E_γ, intensities I_γ and multipolarities λM along with the half-lives $T_{1/2}$ of the respective levels in ^{95}Mo are listed in Table 1. Gamma-ray intensities for the strongest yrast transitions were deduced from a spectrum gated on the 949-keV transition and normalized to the 594-keV transition. In the cases where a particular level decays by more than one transition, spectra gated on the feeding transition were used to normalize the intensity of the weaker de-exciting transition to the intensity of the strongest one. Energy and intensity balance was performed.

The lowest lying levels in ^{95}Mo (figure 2) have half-lives shorter than the electronics time binning of 6 ps/channel. The centroids of the symmetric time distributions constructed for the $9/2^+$ state (figure 3a) overlap within the time binning, which is consistent with the half-life of 2.58(11) ps quoted in NNDC [10].

Figure 3(b) shows the time distributions sorted for the $19/2^+$ state in ^{95}Mo. The time spectrum is obtained by gating on the 152-keV and 386-keV transitions with the $LaBr_3$:Ce detectors. In addition, to reduce the background, gates on the 949-keV, 594-keV, 692-keV or 766-keV transitions were applied with the HPGe detectors. A half-life of 3.4(4) ns was obtained from the slope of the time distribution. According to NNDC, the $19/2^+$ state decays via two branches with energies of 38 keV and 386 keV. The 38-keV transition is highly converted, and its energy is outside of the detectors range of sensitivity. Hence, the transition was not observed experimentally in the present experiment. Therefore, the intensity of the 38-keV transition was estimated from the intensity balance performed for the 2582-keV level and after evaluation of the levels side feeding. Using the branching ratios obtained from the present data, $B(E2) = 0.12(9)$ W.u. was obtained.

Figure 3(c) shows symmetric time spectra for the $21/2^+$ state in ^{95}Mo. The spectra are gated on the 903-keV and 152-keV transitions in the $LaBr_3$:Ce detectors. To reduce the background, the same HPGe gates as for the $19/2^+$ state were imposed. A half-life of 138(18) ps was obtained

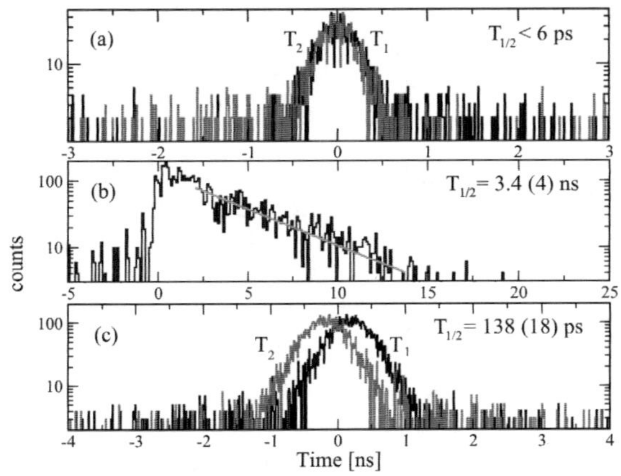

Figure 3. Time spectra for the $9/2^+$ state (a), for the $19/2^+$ state (b), and for the $21/2^+$ state (c) in ^{95}Mo.

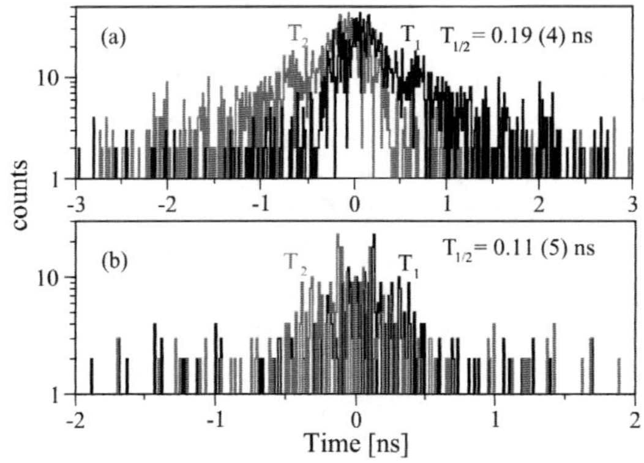

Figure 4. Time spectra for the 8_1^+ state (a), and for the 10_1^+ state in ^{96}Mo.

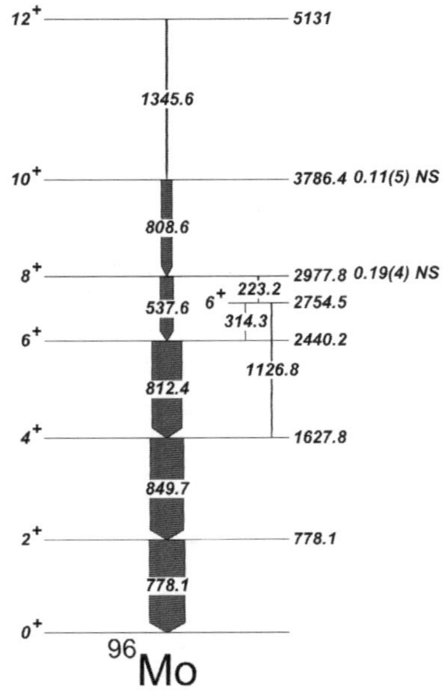

Figure 5. Partial level scheme of ^{96}Mo [12].

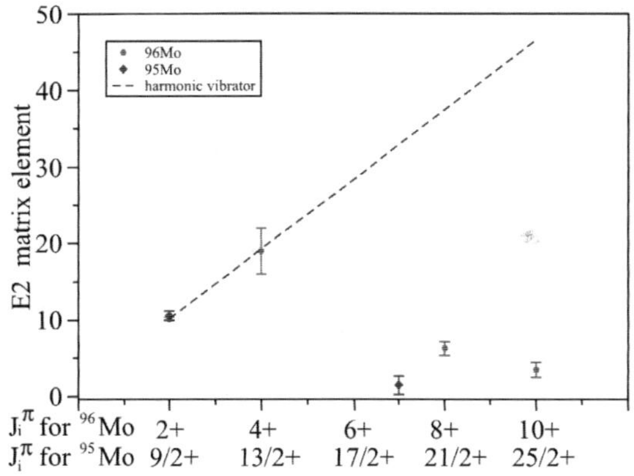

Figure 6. Matrix elements calculated for the $E2$ transitions de-exciting the first five yrast states in ^{95}Mo and ^{96}Mo and compared to the matrix elements, calculated for the harmonic vibrator and normalized with respect to the first phonon in ^{96}Mo.

from the centroid shift method.

Another reaction channel leads to the even-even ^{96}Mo nucleus. An analysis similar to the one performed for ^{95}Mo was carried out. Two levels with picosecond half-lives were observed. Figure 4(a) shows the time distributions obtained for the 8_1^+ state in ^{96}Mo. The time spectrum is gated on the 809-keV and 538-keV transitions with $LaBr_3$:Ce detectors and on the 812-keV transition with HPGe detectors (figure 5). A half-life of 0.19(4) ns was obtained with the centroid shift method. Hence, the $B(E2; 8_1^+ \rightarrow 6_1^+)$ is 2.3(5) W.u.

Time spectra for the 10^+ state in ^{96}Mo are presented in figure 4(b). The same gate as in the case of the 8_1^+ state was set on the HPGe detectors. Gates on the 1346-keV and 809-keV transitions imposed on $LaBr_3$:Ce detectors were used to increment the symmetric time spectra. A half-life of 0.11(5) ns was obtained using the centroid shift method. The transition strength for the 809-keV transition results as $B(E2; 10_1^+ \rightarrow 8_1^+)$ =0.6(3) W.u.

3. Discussion

The ground state of ^{95}Mo is a $J^\pi = 5/2^+$ state, which is consistent with the ordering of the neutron single-particle levels in the $A \approx 100$ mass region [13] where $\nu d_{5/2}$ single-particle orbit appears at the beginning of the fourth oscillator shell. Next to the $\nu d_{5/2}$ single-particle orbit is $\nu g_{7/2}$. Indeed, the first $J^\pi = 7/2^+$ level, placed at 767 keV in ^{95}Mo, decays via a $M1 + E2$ transition to the ground state with $B(E2; 7/2^+ \rightarrow 5/2^+) = 0.96$ W.u [10]. The next yrast $J^\pi = 9/2^+$ level in ^{95}Mo has an energy of 948 keV, which is close to the 2_1^+ level energy of the neighboring even-even nuclei. Also, the $9/2^+$ state decays via a strong $E2$ transition with $B(E2) = 11.3(6)$ W.u [10]. Such a state can be interpreted as a $\nu d_{5/2}$ single-particle state fully aligned with the first 2_1^+ phonon of an even-even core [14]. In this respect, it is interesting to compare the matrix elements obtained for the $E2$ transitions in the even-even ^{96}Mo with the $E2$ matrix elements obtained from the yrast transitions in the even-odd ^{95}Mo nucleus. The matrix elements, calculated from $|< \psi_f||E2||\psi_i > | = \sqrt{(2J_i + 1) \times B(E2; J_i \rightarrow J_f)}$ for the yrast transitions in ^{95,96}Mo, are shown in figure 6. The experimental matrix elements are also compared to the harmonic vibrator matrix elements, which were parametrized from the first excited state in ^{96}Mo. The $< 2^+||E2||4^+ >$ matrix element, calculated for ^{96}Mo [15], remarkably coincides with the harmonic vibrator $< 2^+||E2||4^+ >$ matrix element. Also, the $< 5/2^+||E2||9/2^+ >$ matrix element, calculated from the ^{95}Mo data, is consistent with its interpretation as a fully aligned particle-core coupled state. Further, with the increase of the angular momentum the experimental matrix elements deviate from the harmonic vibrator description. In fact, the experimental matrix elements are even smaller than the $< 0_1^+||E2||2_1^+ >$ matrix element. In this spin range the odd- and even-mass molybdenum matrix elements are again in the same range. Considering the above and the high-J values, these excited states more probably involve considerable contribution from high-j single-particle excitations.

4. Acknowledgments

This work is supported by the Bulgarian Science Fund - contracts DMU02/1, DRNF02/5, DID-02/16, Romanian UEFISCDI - contract IDEI 115/2011 and by a Bulgarian-Romanian partnership contract, numbers DNTS-02/21 and 460/PNII Module III.

References

[1] NNDC data base (www.nndc.bnl.gov)
[2] Casten RF, 2000, *Nuclear Structure from a Simple Perspective* Oxford University Press
[3] Lalkovski S *et al.* 2007 *Phys. Rev.* C**75** 014314
[4] Mărginean N *et al.* 2010 *Eur. Phys. J.* A **46** 329336
[5] Kisyov S, *et al.* 2011 *Phys. Rev.* C **84** 014324
[6] Mărginean N 2009 Private Communication
[7] RadWare package (http://radware.phy.ornl.gov/)
[8] Andrejtscheff W *et al.* 1982 *NIM* **204** 123-128
[9] Régis J-M *et al.* 2010 *NIM* A **622** 83-92
[10] Basu S K *et al.* 2010 *Nuclear Data Sheets* **111** 2555
[11] Zang Y H *et al.* 2009 *Phys. Rev.* C **79** 044316
[12] Chatterjee J M *et al.* 2000 *Nucl. Phys.* A **678** 367-381
[13] Heyde KLG 1994 *The Nuclear Shell Model*, Springer-Verlag
[14] de-Shalit A *Phys. Rev.* **122** 1530
[15] Abriola D *et al.* 2008 *Nuclear Data Sheets* **109** 2501

g-factor measurements of short lived states in the mass region 50<A<110.

G J Kumbartzki

Department of Physics and Astronomy, Rutgers University, New Brunswick, NJ, USA

E-mail: kum@physics.rutgers.edu

Abstract. Magnetic moments provide important nuclear structure information. This contribution will present the experimental tools and challenges of measuring magnetic properties of excited nuclear states, especially of the short lived picosecond states. Our understanding of the transient field and its applicability will be discussed using examples and results of recent measurements on medium heavy nuclei (A∼100).

Magnetic moments provide sensitive information on the individual contributions of protons and neutrons to the wave functions of nuclei and their excited states. This fact is a direct consequence of the large magnitude and opposite signs of the free proton and neutron g factors. A good illustration can be found in the g factors of the 2_1^+ states in the isotopic chains of Sr and Zr around the closed neutron shell at N = 50. Large positive g factors for N = 50 are followed by negative g factors in ^{92}Zr and ^{94}Zr.

Magnetic moments are usually measured by observing the spin precession of the nuclei in a known magnetic field. The spin of the nuclear state will precess with the Larmor frequency

$$\omega_L = \frac{d\theta}{dt} = g\frac{\mu_N}{\hbar}B \tag{1}$$

Depending on the lifetime of a state, the required field strength for a measurable$\Delta\ \theta$ may be in the 10^3 Tesla range. Such fields cannot be provided in the laboratory, but they do exist as hyperfine fields in atoms. For instance, the spin of a single electron in the $1s$ orbit gives rise to a magnetic field at the site of the nucleus of

$$B_{1s} = 16.7Z^3[\text{Tesla}] \tag{2}$$

Fully occupied shells do not contribute. Such fields are indeed observed in highly ionized atoms under conditions which occur when atoms move through a solid medium. Outer electrons are stripped off, electrons are rapidly lost and recaptured, and deep inelastic scattering creates inner shell vacancies. If such ions traverse a polarized ferromagnetic medium, which acts as a reservoir of electrons with a preferred spin direction, this polarization is transferred into the moving ions, giving rise to a net field direction of the hyperfine fields. In this picture the field strength is a dynamic changing sum of the products of hyperfine fields, the charge state distributions of ions moving in a ferromagnetic medium, and a polarization factor, which may

Journal of Physics: Conference Series 366 (2012) 012028 doi:10.1088/1742-6596/366/1/012028

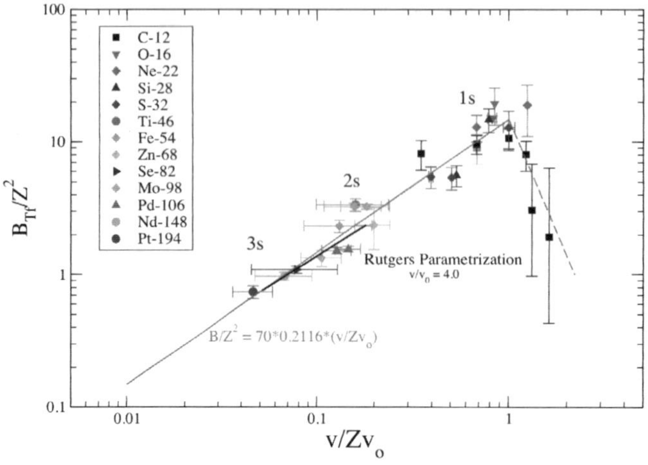

Figure 1. Compilation of transient fields, B_{TF}, measured for a set of calibration points. The velocity of the ions is given in terms of v/Zv_0, a measure of the electronic occupation of the ion in a given charge state. Three different regions can be assigned corresponding to the hyperfine field being dominated by $1s$ ($v/Zv_0 \approx 1$), $2s$ and $3s$ ($v/Zv_0 \leq 0.2$) electrons.

depend on the probe ions, their velocities and the ferromagnetic host. This description is of little practical use, since none of the details are known. The fields were therefore calibrated with a set of known g factors. It was found that the fields vary smoothly with the charge, Z, and velocity, v, of the ions. A compilation of fields derived from transient field measurements using gadolinium for states with independently measured g factors is presented in figure 1. This presentation implies a linear dependence on Z. The v dependence is only related to charge state distributions.

Figure 2. Setup at WNSL Yale Tandem.

Journal of Physics: Conference Series **366** (2012) 012028 doi:10.1088/1742-6596/366/1/012028

Above the 1s-orbital velocity of Zv_0 (Bohr velocity $v_0 = c/137$), as more ions are bare, the fields decrease and vanish. The polarization transfer is the least understood part of this approach.

In practice, transient field measurements rely on parametrizations for absolute g factor values. The Rutgers parametrization [1] is applicable in the 2s and 3s regions and for ion velocities between 2 and 8v_0. The 1s region was parametrized for high velocity application of the transient field by Stuchbery [2].

The standard experimental setup is an angular correlation table (figure 2) with four γ detectors close to the vacuum chamber. The vacuum chamber contains the usual triple-layer target mounted in a magnet. The target should be coolable to liquid nitrogen temperature or lower. The chamber also contains particle detectors for particle-γ coincidence measurements.

The probe nuclei are excited and spin aligned by projectile Coulomb excitation in inverse kinematics. The target is typically a combination of carbon, where the reaction takes place, gadolinium, the ferromagnetic medium, and copper, where the probe ions come to rest and decay in an environment free of internal fields. The knock-on carbon nuclei have sufficient energy to penetrate both the target and a thin-foil beam stop to reach a particle detector. At energies near the Coulomb barrier the projectiles tend to pick up an α particle from the carbon. This α-transfer reaction makes it possible to measure g factors of some unstable nuclei in inverse kinematic conditions.

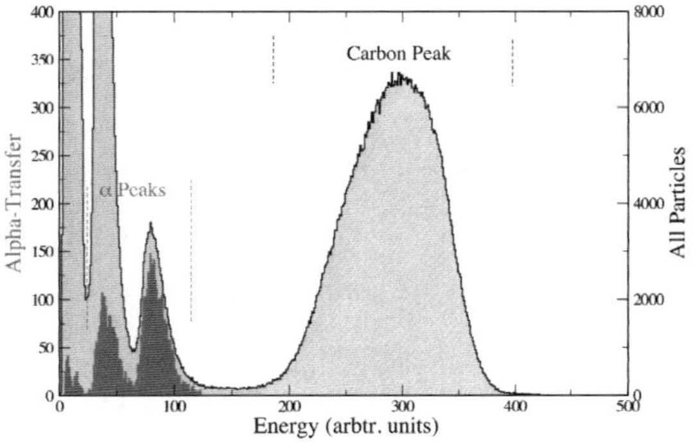

Figure 3. Particle spectrum in a 100μm thick silicon detector.

Figure 3 shows a particle spectrum obtained with a 345 MeV ^{96}Ru beam. The energy loss in the carbon target layer is ~30 MeV. The Coulomb barrier is at 333 MeV. At higher energies light particles (not stopped in the detector) from other reactions dominate while the Coulomb excitation is suppressed.

Well-separated γ spectra are generated by gating on specific particle groups. Individual spectra are shown in figure 4 for ^{96}Ru and ^{100}Pd.

Requiring a particle-γ coincidence selects an ensemble of aligned nuclei which decay with a characteristic angular distribution. The spin rotation in the transient magnetic field causes a rotation of the angular correlation pattern. This rotation is measured as a rate change in the stationary γ-detectors. The precession effect $\epsilon = (\rho - 1)/(\rho + 1)$ is calculated from quadruple ratios involving four detectors, with the same quadrant angle θ_γ:

$$\rho = \sqrt{\rho_{1,4}/\rho_{2,3}}\ , \text{where} \quad \rho_{i,j} = \sqrt{(N_i^\uparrow N_j^\downarrow)/(N_i^\downarrow N_j^\uparrow)}\ . \tag{3}$$

N↑ and N↓ are the integrated γ-peak rates for two magnetic field directions. The polarizing field

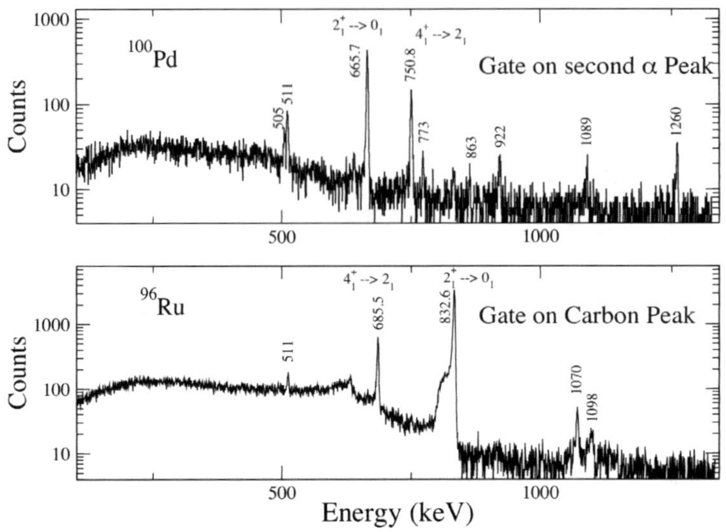

Figure 4. Gated γ spectra.

at the target is reversed every few minutes. The logarithmic slope, $S(\theta_\gamma) = \frac{1}{W(\theta_\gamma)} \cdot \frac{dW(\theta_\gamma)}{d\theta}$, of the angular correlation is needed to transform the rate change at the detector positions into the rotation angle$\Delta\ \theta = \epsilon / S(\theta_\gamma)$. The particle-$\gamma$ angular correlation, using the usual notation,

$$W(\theta) = 1 + A_2^{exp} Q_2 P_2(cos\theta) + A_4^{exp} Q_4 P_4(cos\theta) \tag{4}$$

can be determined from anisotropy ratios obtained from the same precession data in the individual clover segments, making use of the approximation

$$A_k^{exp} = A_k^{th}(1 - \eta k(k+1)) \quad \text{for} \quad \eta \ll 1. \tag{5}$$

When measuring g factors across an isotope chain, the$\Delta\ \theta$ values themselves represent the relative g factors. They can be measured with high statistical accuracy. For absolute g factors the actual transient field strength has to be known. This field strength is taken from a parametrization (calibration). In the medium-heavy nuclei region the Rutgers parametrization

$$B_{TF} = 96.7 M Z^{1.1} (v/v_0)^{0.45} \tag{6}$$

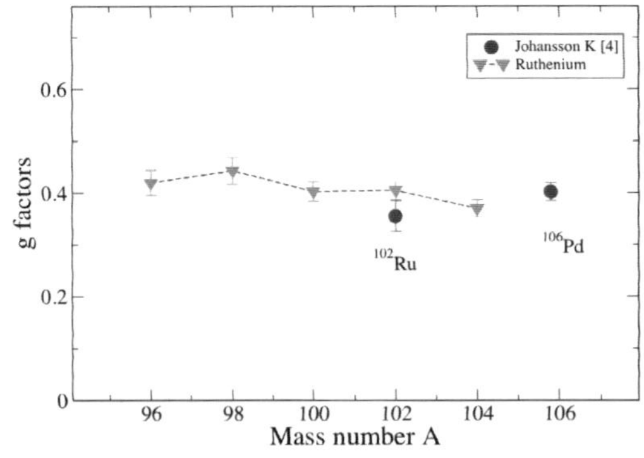

Figure 5. Relative $g(2_1^+)$ of the stable ruthenium isotopes calibrated against ^{106}Pd.

Table 1. Results from an α-transfer reaction measured simultaneously with the Coulomb excitation of the projectiles [5].

Nucleus	E_x(keV)	I^π	Transition	τ(ps)	$\|S(67°\|$	g
^{100}Pd	665.5	2_1^+	$2_1^+ \to 0_1^+$	9.0(4)	0.324(54)	+0.30(14)
	1415.9	4_1^+	$4_1^+ \to 2_1^+$	3.6(3)	0.550(91)	+0.45(14)
^{96}Ru	832.6	2_1^+	$2_1^+ \to 0_1^+$	4.24(9)	1.85(5)	+0.46(3)

is used. Here, M is the magnetization of the ferromagnetic layer. It is measured offline as a function of the temperature and care has to be taken that it does not vary in beam. At beam energies of 300 MeV and higher and at beam currents of ~1pnA, the local beam spot temperature can reduce the magnetization and therefore the effective field.

As example, results for the 2_1^+ states in the stable ruthenium isotopes, measured with one target and calibrated against the $g(2_1^+, ^{106}\text{Pd})$, are shown in figure 5 [3]. The results for ^{100}Pd simultaneously measured with ^{96}Ru are listed in table 1.

Conclusion

- The transient field together with Coulomb excitation in inverse kinematics is a powerful tool to measure magnetic moments of the very short lived excited states in nuclei. The technique is applicable to rare and radioactive nuclei if they are provided as accelerated beams.
- The α-transfer reaction allows measurements of magnetic moments using the transient field in inverse kinematic conditions on certain unstable nuclei, for which otherwise a radioactive beam is needed.
- It has proven difficult to measure higher excited states because of low Coulomb excitation cross-sections.
- At beam energies above the Coulomb barrier, as new reaction channels open, the Coulomb excitation becomes minute and the detectors are swamped with undesirable particles and γ rays.
- A better knowledge of the transient field itself is desirable.

Acknowledgments
Special thanks to my collaborators from Rutgers University, Benczer-Koller N, Torres D A, Gürdal G and Sharon Y Y. The experimental work was done at the Yale Tandem Accelerator; the help of the Yale students, postdocs, faculty and staff is greatly appreciated. The work was supported in part by the U.S. National Science Foundation and the U.S. Department of Energy under Grant No. DE-FG02-91ER-40609.

References

[1] Shu N K B, Melnik B, Brennan J M, Semmler W and Benczer-Koller N 1980 *Phys. Rev.* C **21** 1828
[2] Stuchbery A E 2004 *Phys. Rev.* C **69** 044302
[3] Taylor M J, Gürdal G, Kumbartzki G, Benczer-Koller N, Stuchbery A E, Sharon Y Y *et al.* 2011 *Phys. Rev.* C **83** 044315
[4] Johansson K, Karlsson E, Norlin L-O, Windahl R A and Ahmed M R 1972 *Nucl. Phys.* A 188 600
[5] Torres D A, Kumbartzki G J, Sharon Y Y, Zamick L, Manning B, Benczer-Koller N *et al.* 2011 *Phys. Rev.* C **84** 044327

Journal of Physics: Conference Series **366** (2012) 012029 doi:10.1088/1742-6596/366/1/012029

Single-particle isomeric states in ^{121}Pd and ^{117}Ru

S Lalkovski[1,2,†], A M Bruce[2], A M Denis Bacelar[2], M Górska[3], S Pietri[3,4], Zs Podolyák[4], P Bednarczyk[3], L Caceres[3], E Casarejos[5], I J Cullen[4], P Doornenbal[3,6], G F Farrelly[4], A B Garnsworthy[4], H Geissel[3], W Gelletly[4], J Gerl[3], J Grębosz[3,7], C Hinke[8], G Ilie[9,10], G Jaworski[11,12], S Kisyov[1], I Kojouharov[3], N Kurz[3], S Myalski[7], M Palacz[12], W Prokopowicz[3], P H Regan[4], H Schaffner[3], S Steer[4], S Tashenov[3], P M Walker[4,13], H J Wollersheim[3] and M Zhekova[1]

[1]Faculty of Physics, University of Sofia "St. Kliment Ohridski", Sofia 1164, Bulgaria
[2]School of Computing, Engineering and Mathematics, University of Brighton, Brighton BN2 4JG, UK
[3]Gesellschaft für Schwerionenforschung mbH, Planckstr 1, D-64291 Darmstadt, Germany
[4]Department of Physics, University of Surrey, Guildford GU2 7XH, UK
[5]Facultad de Física, Universidad de Santiago de Compostela, Santiago de Compostela 15782, Spain
[6]Institut für Kernphysik, Universität zu Köln, Zülpicher Straße 77, D-50937 Köln, Germany
[7]Niewodniczański Institute of Nuclear Physics, Polish Academy of Science, ul. Radzikowskiego 152, Krakow 31-342, Poland
[8]Physik-Department E12, Technische Universität München, D-85748 Garching, Germany
[9]Wright Nuclear Structure Laboratory, Yale University, New Haven, Connecticut 06520, USA
[10]National Institute for Physics and Nuclear Engineering, P.O. Box MG-6, Bucharest, Romania
[11]Heavy Ion Laboratory, Warsaw University, ul. Pasteura 5A, 02-093 Warszawa, Poland
[12]Faculty of Physics, Warsaw University of Technology, Koszykowa 75, 00-662 Warszawa, Poland
[13]CERN, CH1211 Geneva 23, Switzerland

E-mail: †stl@phys.uni-sofia.bg

Abstract. Neutron-rich nuclei were populated in a relativistic fission of ^{238}U. Gamma-rays with energies of 135 keV and 184 keV were associated with two isomeric states in ^{121}Pd and ^{117}Ru. Half-lives of 0.63(5) μs and 2.0(3) μs were deduced and the isomeric states were interpreted in terms of prolate deformed single-particle states.

1. Introduction

Nuclear shell structure is one of the milestones in Nuclear Physics. It is related to enhanced robustness of nuclei with respect to excitations, when a particular number of nucleons is present [1]. A key element of the nuclear shell model is the spin-orbit force, which lifts the j-degeneracy and pushes down the orbits with higher-j towards orbits of opposite parity. One of the major successes of the model is the ability to reproduce the nuclear magic numbers and to explain the islands of isomerism emerging on the Segré chart when approaching the magic medium-mass and heavy nuclei.

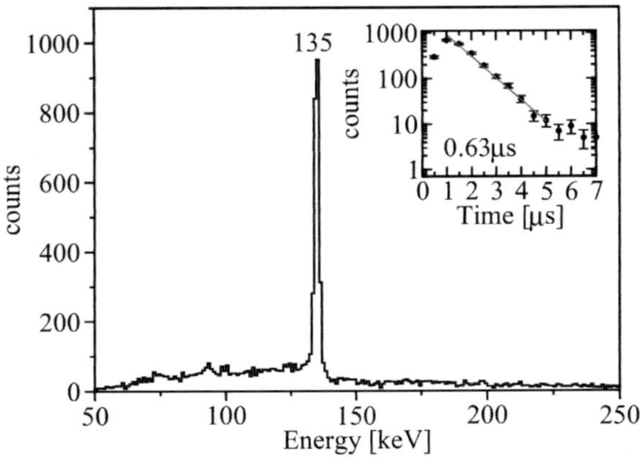

Figure 1. Isomeric transition in ^{121}Pd and time spectrum (inset) gated on it.

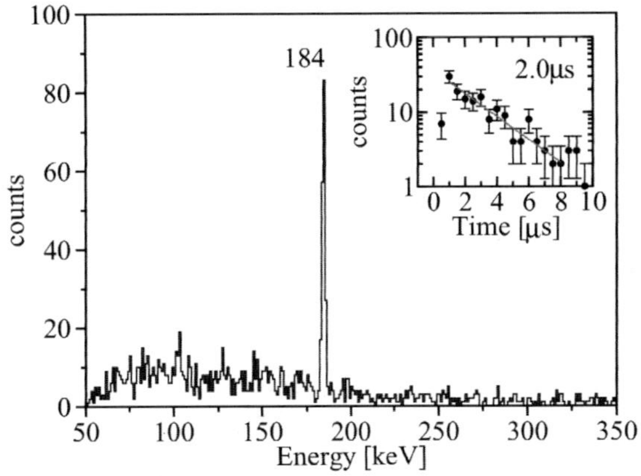

Figure 2. Isomeric transition in ^{117}Ru and time spectrum (inset) gated on it.

In addition to the spin-orbit force, it is the nuclear quadrupole deformation, which splits the high-j states into a number of orbits with a different third projection of the total angular momentum [2], causing new sub-shell closures to emerge [3] and high-K isomeric states to appear in the deformed regions on the Segré chart [4].

The present article addresses the structure of $^{121}_{46}$Pd$_{75}$ and $^{117}_{44}$Ru$_{73}$, which are the most neutron-rich nuclei in the palladium and ruthenium isotopic chains studied by means of γ-ray spectroscopy. They are placed in a region of the Segré chart where variety of shapes are expected, involving large degree of triaxiality and γ-softness [5, 6]. Also, because of the vicinity of the $N = 82$ magic number, a deformed-to-spherical shape transition [5, 6] is expected to take place. Due to the nature of the shell structure, isomeric states were observed in the two nuclei, which can be related to specific single-particle configurations, and hence can shed light on the nuclear shapes in the mass region.

2. Experimental set up and data analysis

Neutron-rich palladium and ruthenium isotopes were produced in a relativistic fission reaction. A ^{238}U beam was accelerated up to 750 MeV/n by the GSI SIS accelerator and impinged on 1g/cm^2 ^{9}Be target. The fission products were separated by the GSI Fragment Separator (FRS) and implanted in a passive stopper. Isomeric delayed transitions were detected by the RISING detector array [7], comprising 105 HPGe detectors. The data were processed by XIA DGF modules providing signals with an energy resolution of 3 keV at 1.3 MeV and a time resolution of 25 ns. The data acquisition was triggered by a particle detector, placed at the final focal plane of the FRS. The data were stored in event-by-event mode. Experimental details were previously published elsewhere [8].

Fig. 1 presents the γ-ray energy spectrum in singles, obtained in coincidence with the ^{121}Pd nuclei. A γ transition with an energy of 135 keV is observed. The inset of the figure presents the time distribution, gated on the 135-keV γ-ray. The half-life, obtained with a single exponential curve, is 0.63(5) μs.

Fig. 2 shows the energy spectrum, gated on ^{117}Ru ions, where a single transition with an energy of 184 keV was observed. The inset of the figure presents the time distribution of the 184-keV transition and a half-life of 2.0(3) μs was obtained from the slope of the curve. The two γ rays in ^{121}Pd and ^{117}Ru were previously observed [9], but no half-life information was published.

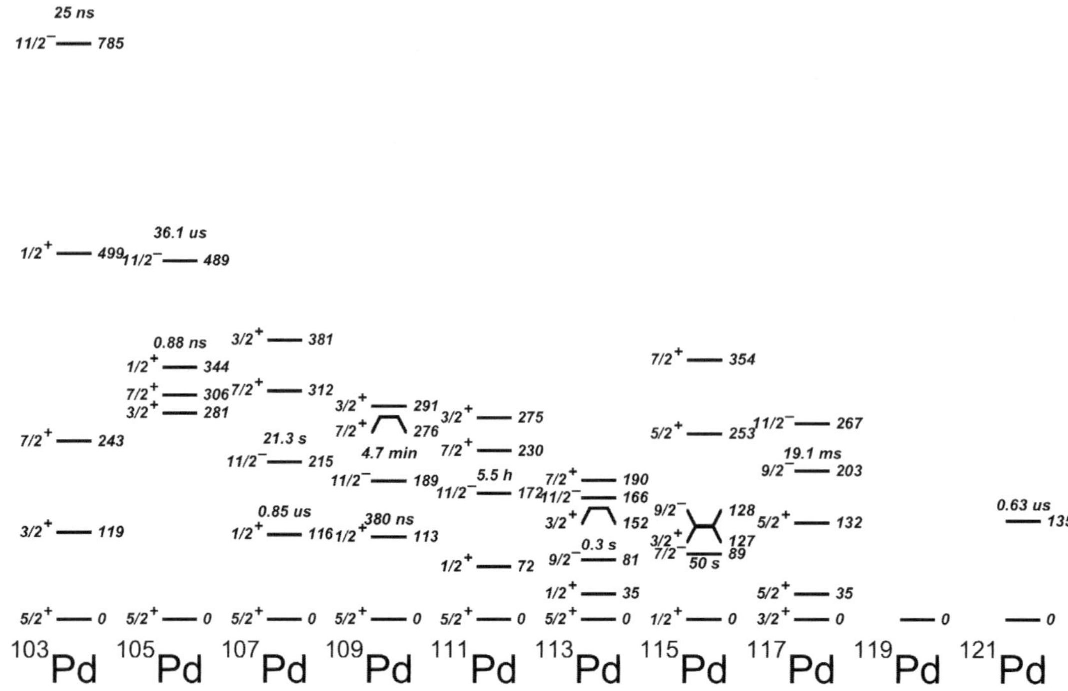

Figure 3. Systematics of the low-lying excited states in the neutron-rich palladium isotopes: $^{103-113}$Pd [10], ^{115}Pd [11], ^{117}Pd [12] and ^{121}Pd (present work).

3. Discussion

Fig. 3 presents the systematics of the low-lying excited states in the neutron-rich odd-A palladium isotopes. The ground state in ^{103}Pd to ^{113}Pd is a $J^{\pi}(g.s.) = 5/2^+$ state [10], while the J^{π} value for the ground state in ^{115}Pd and ^{117}Pd are subject of discussion in the literature [10, 11, 12]. Here, we adopt the J^{π} assignments for ^{115}Pd and ^{117}Pd as given in [11] and [12], respectively. The extrapolation of the systematics towards the extremely neutron-rich odd-A palladium nuclei suggests that $J^{\pi} = 1/2^+$ to $7/2^+$ and $J^{\pi} = 7/2^-$ to $11/2^-$ states will be present close to the ^{121}Pd ground state, but not all candidates give rise to a sub-microsecond isomer. The low-lying isomeric states, observed in the odd-A palladium isotopes are related to the decay of the negative-parity states as well as to the decay of the $J^{\pi} = 1/2^+$ excited state via stretched $E2$ low-energy transitions. Hindrance factors (HF), defined as $F_W = T_{1/2,\gamma}/T_{1/2,W.e.}$ where $T_{1/2,\gamma}$ is the partial half-life and $T_{1/2,W.e.}$ is the single-particle estimate, were calculated for the isomeric $E1$, $E2$, $E3$ and $M2$ transitions and plotted on figure 4 as a function of the neutron number.

The neutron-rich odd-A palladium ($Z = 46$) isotopes show structure close to that of the odd-A ruthenium ($Z = 44$) nuclei (figure 5). The HF, calculated for the low-lying isomeric transitions in the neutron-rich odd-A ruthenium nuclei are also plotted on fig. 4. To increase the statistics on the systematics in figure 4, HF were calculated for a number of low-lying isomeric transitions in the odd-A Cd nuclei [10, 13]. Inside the encircled regions on figure 4 HF for the 184-keV and 135-keV isomeric transitions in ^{117}Ru and ^{121}Pd are presented.

On figure 4, the $E1$ HF are in the $10^4 < F_W < 10^8$ range. If the 135-keV and the 184-keV isomeric transitions, observed in ^{121}Pd and ^{117}Ru nuclei, are of $E1$ nature the respective HF will follow the systematics presented in figure 4. The $E2$ HF shows a steady behaviour with the neutron number having $F_w \approx 1$. The ^{121}Pd and ^{117}Ru data points would follow the systematics

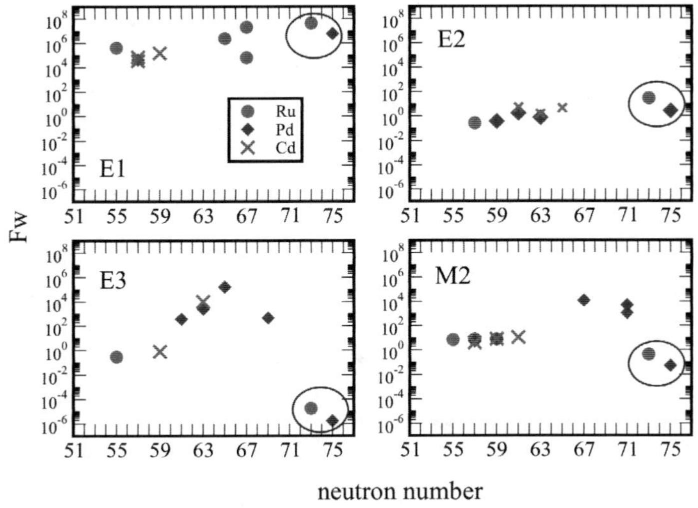

Figure 4. Hindrance-factor systematics for the low-lying isomeric states in the neutron-rich odd-A ruthenium, palladium and cadmium nuclei. The encircled regions present the hindrance factors for the 135-keV and 184-keV transitions studied here.

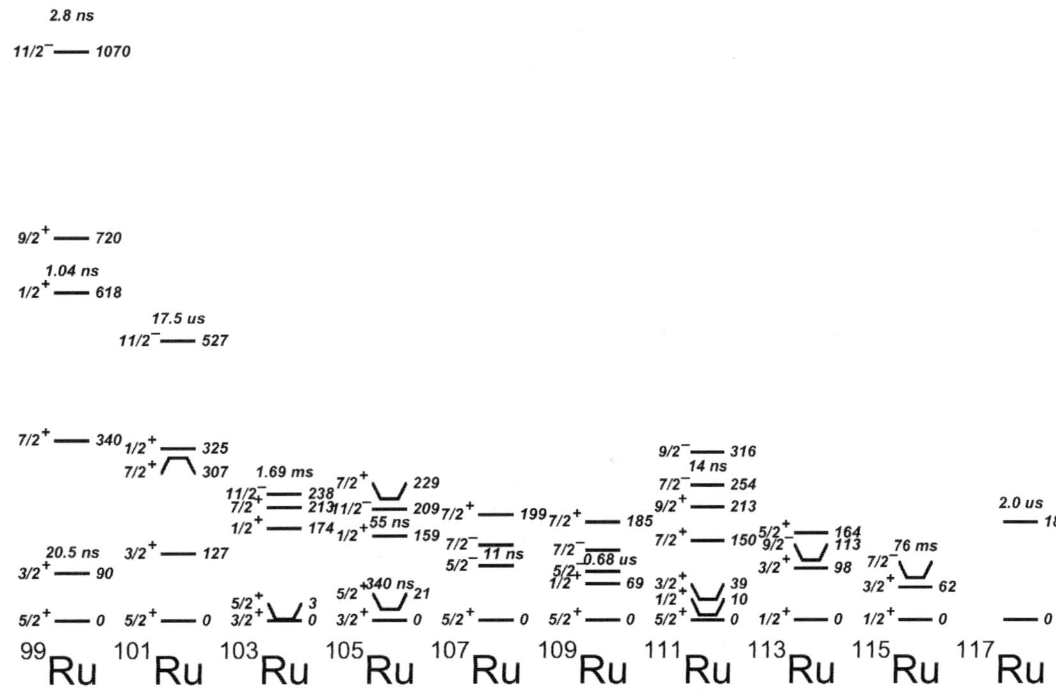

Figure 5. Systematics of the low-lying excited states in the neutron-rich ruthenium isotopes. Data from [10] for $^{99-115}$Ru and from the present work for ^{117}Ru.

in figure 4 if the two transitions are of $E2$ nature. In contrast to the $E2$ HF, the $E3$ F_W data points show an abrupt change varying six orders of magnitude. If the 135-keV and 184-keV transitions are of $E3$ nature, then they will be six orders of magnitude more enhanced than the other $E3$ transitions. Hence, the $E3$ nature for the two isomeric transitions in ^{117}Ru and

^{121}Pd can be ruled out. The $M2$ transitions in the region have $F_w \geq 1$. If 135-keV and 184-keV transitions are of $M2$ nature, than they would be enhanced and would not follow the systematics of the $M2$ transitions. Therefore, $M2$ nature is unlikely.

The single-particle levels, calculated with a Woods-Saxon potential for the nuclei in the $A \approx 110$ mass region region [3], show that the ^{121}Pd Fermi surface involves $7/2^+[404]$, $1/2^+[400]$, $3/2^+[402]$, and $9/2^-[514]$ Nilsson orbits, where $\beta_2 = 0.19$ was calculated from the energy of the first excited state in ^{120}Pd. Given the HF systematics (fig. 4), $9/2^- \rightarrow 7/2^+$ $E1$ and $3/2^+ \rightarrow 7/2^+$ $E2$ transitions would cause isomeric states in the sub-microsecond time range. However, $J^\pi = 9/2^-$ assignment for the isomeric state can be ruled out because of the direct β^- feeding observed to the $9/2^+$ state in ^{121}Ag [14] suggesting the $J^\pi = 9/2^-$ isomeric state in ^{121}Pd has a half-life longer than the half-life of the observed in the present study nanosecond isomer. Therefore, we tentatively interpret the 135-keV transition as a prolate $3/2^+[402] \rightarrow 7/2^+[404]$ $E2$ transition.

The 73th neutron of ^{117}Ru, which has β_2 =0.25 estimated from the first excited state in ^{116}Ru, is placed on $1/2^+[400]$ Nilsson orbit [3]. At this deformation, the $1/2^-[541]$ down-sloping orbit is also present close to the ground state, therefore we tentatively interpret the isomeric transition as a $1/2^- \rightarrow 1/2^+$ $E1$ transition. It should be noted, however, that at this deformation the $3/2^+[402]$, $7/2^+[404]$ and $9/2^-[514]$ Nilsson orbits are also present close to Fermi level, which can give rise to a situation similar to the ^{121}Pd scenario. The analysis of ^{121}Pd and ^{117}Ru is further complicated by the fact that the two nuclei are in a region where triaxiality, shape transitions and shape co-existence [6] are expected to occur, which may result in unexpected deviations from the hindrance factors systematics presented here. Therefore, in order to draw firm conclusions on the structure of these neutron-rich nuclei more experimental information and deeper theoretical analysis are needed.

4. Conclusions

Neutron-rich nuclei were populated in a relativistic fission of ^{238}U. Gamma rays with energies of 135 keV and 184 keV were associated with isomeric states in ^{121}Pd and ^{117}Ru, respectively. Half-lives of $0.63(5)\mu$s and $2.0(3)\mu$s were deduced and the isomeric states are interpreted in terms of prolate deformed single-particle states.

Acknowledgments

This work is partially supported by the UK STFC, AWE plc., Royal Society, and by the Bulgarian National Science Fund under contract No: DMU02/1. S.L. acknowledges constructive discussions with Dr. F.G. Kondev and Prof. F.R. Xu.

References

[1] Goeppert-Mayer M, 1948 *Phys. Rev.* **74** 235, 1949 *Phys. Rev.* **78** 16
[2] Nilsson S V, 1955 *Dan.Mat.Fys.Medd.* 29
[3] Xu F R, Walker P M, Wyss R, 2002 *Phys. Rev.* C **65** 021303(R)
[4] Walker P and Dracoulis G, 1999 *Nature* **399** 35
[5] Skalski J, Mizutori S and Nazarewicz W, 1997 *Nucl. Phys.* A**617** 282
[6] Nomura K, Shimizu N and Otsuka T, 2010 *Phys. Rev.* C**81** 044307
[7] Pietri S *et al.*, 2007 *Nucl. Instr. Meth.* B**261** 1079
[8] Bruce A M *et al.*, 2010 *Phys. Rev.* C**82** 044312
[9] Tomlin B E, Mantica P F, Walters W B, 2007 *Eur. Phys. J* ST **150** 183
[10] www.nndc.bnl.gov/ensdf
[11] Kurpeta J *et al.*, 2010 *Phys. Rev.* C**82** 027306
[12] Urban W *et al.*, 2004 *Eur. Phys. J.* A**22** 157
[13] Kisyov S *et al.*, 2011 *Phys. Rev.* C**84** 014324
[14] Stefanescu I *et al.*, 2009 *Eur. Phys. J.* A**42** 407

Complete γ-spectroscopy of neutron-rich nuclei around ^{48}Ca by Heavy-Ion Transfer reactions

S. Leoni

University of Milano, Department of Physics, Milano, Italy
INFN, Sez. Milano, Milano, Italy

silvia.leoni@mi.infn.it

Abstract. The gamma decay of neutron-rich nuclei around ^{48}Ca was measured at Legnaro National Laboratory with the PRISMA-CLARA setup, using the heavy-ion transfer reaction ^{48}Ca on ^{64}Ni at ~6MeV/A. The analysis shows the feasibility to perform complete in-beam γ-spectroscopy with heavy-ion transfer reactions (in terms of angular distributions, polarizations of the γ rays and lifetimes of the nuclear states), providing a method that can be further exploited in the future with heavy targets and radioactive beams. In the one neutron transfer channels, ^{49}Ca and ^{47}Ca, states arising by coupling a single particle to the 3$^-$ phonon of ^{48}Ca are observed, showing the robustness of nuclear collectivity in rather light systems.

1. Introduction

One of the most interesting issues in nuclear physics is the study of nuclei far from stability, which represent a test bench for nuclear models and a challenge for the experimental investigation. In recent years, it has been shown that deep inelastic collisions between heavy ions are a valuable tool to populate moderately neutron rich nuclei in different mass regions, allowing to perform spectroscopic studies of exotic systems [1,2].

In this contribution we focus on neutron-rich systems around ^{48}Ca, in particular on ^{49}Ca and ^{47}Ca, for which no firm spectroscopic information are available, being hard to reach by standard fusion-evaporation reactions. These nuclei are only one neutron away from the ^{48}Ca doubly magic core, therefore they are good candidates to search for states arising from coupling a single particle to a phonon, such as the 3$^-$ state in ^{48}Ca. The understanding of particle–phonon and phonon–phonon couplings is indeed a very important issue, because this phenomenon at the basis of fermionic many-body interacting systems, both in solid state and nuclear physics. In particular, the coupling of a single particle to the vibrational motion is the key process at the origin of anharmonicities of the vibrational spectra and a building block of phonon–phonon interactions. In nuclear physics, particle-phonon coupled states are responsible for the damping of collective excitations (as giant resonances), effective masses and for the quenching of spectroscopic factors [3,4]. Experimentally, several indications have been found of discrete states of particle–phonon nature, mostly in medium-heavy nuclei [3], but it is still an open question whether such states can be considered a general nuclear property, down to the region of medium-light systems with reduced collectivity. In this work we give evidence for particle–phonon coupled states in ^{49}Ca and ^{47}Ca, which were populated by the heavy-ion transfer reaction ^{48}Ca+^{64}Ni, at ~6 MeV/nucleon [5,6]. The experiment was performed at Legnaro National Laboratory of INFN (Padova) using the large acceptance magnetic spectrometer PRISMA [7,8] combined with the high efficiency Ge array CLARA [9]. The work demonstrates the feasibility of performing complete γ-spectroscopy with heavy-ion binary reactions at low energy, providing a method which can be further exploited with heavy targets and radioactive beams.

2. The experiment

The ^{48}Ca beam, impinging on a 0.98 mg/cm^2 thick ^{64}Ni target, was provided by the Tandem-Alpi accelerators at 282 MeV, with an average current of 1 pnA. The reaction products were measured by the magnetic spectrometer PRISMA [7,8] and the coincident γ-rays by the CLARA HPGe-array [9].

PRISMA is among the largest acceptance (≈ 80 msr) magnetic spectrometers currently operating. It is based on one quadrupole and one dipole magnets and a system of entrance and focal plane detectors which allow mass and charge identification of the detected ions after a trajectory reconstruction, event by event. In the current experiment PRISMA was placed at the grazing angle for this reaction, i.e. 20°, with an angular acceptance of ± 6°. The Ge array CLARA consisted of 23 composite Ge detectors of EUROBALL CLOVER type, each equipped with Anti-Compton shield and characterized by a total absolute efficiency of ~3% at 1.3 MeV. The CLOVER detectors were arranged in a hemisphere opposite to PRISMA with most of the Ge crystals placed in 3 rings at average azimuthal angles θ_{CLA} = 100°, 130° and 150° with respect to the entrance direction of the spectrometer. In this way reaction products detected in the spectrometer focal plane, in coincidence with the γ-rays, have a forward trajectory with respect to the array.

The analysis of the γ rays measured by the CLARA array was performed in coincidence with some of the most intense reaction products. In particular, we have studied the angular distributions of the most intense transitions taking advantage of the ring geometry of the Ge detectors. For several known E2 transitions in ^{46,50}Ca and ^{51}Sc nuclei, the intensities measured in the three rings at 100°, 130° and 150° have been normalized to the one at 100°, and fitted by the angular distribution function $W(\theta) = 1 + a_2P_2(\cos\theta)$ (with P_2 the Legendre polynomial and a_2 the fitting parameter). It is found that the ratio $\alpha_2 = a_2/a_{2max}$ is approximately 70%, where a_2 the measured coefficient and a_{2max} the maximum value allowed for the corresponding γ decay with fully aligned spin. As a consequence, linear polarization measurements can be also performed, as it has been demonstrated making use of the most efficient CLOVER detectors at 100° [6] . The analysis shows the possibility to firmly establish spin and parity of the excited states by performing γ-spectroscopy studies with low energy heavy-ion transfer reactions [6,10]. In the following Sections we apply these techniques, together with lifetime measurements, to the study of ^{49}Ca and ^{47}Ca nuclei.

3. γ–Spectroscopy of ^{49}Ca and ^{47}Ca: Particle-vibration coupling in medium-light nuclei

Nuclei only one nucleon away from doubly magic systems are ideal candidates to search for states arising by coupling a particle/hole to a phonon. In the case of the doubly magic nucleus ^{48}Ca, the 3^- state at 4.507 MeV exhibits a sizable collectivity, with a measured B(E3) of 6.8±1 Wu [11]. On the contrary, the 2^+ state at 3.832 MeV is less collective, with a B(E2) value of ~2 Wu [12]. The rather strong collective character of the 3^- suggests the existence of multiplets of positive parity states arising by coupling the 3^- phonon with the unpaired $p_{3/2}$ neutron of ^{49}Ca and the $f^{-1}_{7/2}$ neutron hole of ^{47}Ca. Experimentally, candidates for this type of states have been found both in ^{49}Ca and ^{47}Ca by low resolution particle spectroscopy studies: their common features are largely reduced spectroscopic factors and "anomalous" angular distributions of the detected particles [13,14]. For these states, no firm spin and parity assignments have been obtained, so far, and no lifetime analysis has been performed in order to establish their collectivity.

Figure 1 shows in the bottom panel the γ-spectrum measured in coincidence with ^{49}Ca ions detected in PRISMA. All strong transitions correspond to the decays from states identified as single particle excitations, besides the 660- and 3357-keV lines which depopulate the levels at 4017 keV and 3357 keV, with the tentative spin assignment $9/2^+$ and $7/2^-$, respectively [2]. In particular, the 4017 keV state is expected to originate from the $p_{3/2}$ neutron coupled to the 3^- phonon of ^{48}Ca. The present experiment has proven these spin assignments by measuring the angular distribution and polarization of both transitions, resulting in a_2 coefficients and asymmetry values consistent with electric stretched dipole for the 660-keV line and electric stretched quadrupole for the 3357-keV γ ray (cf. Fig. 1). In addition, to firmly establish the nature of the states, a lifetime analysis has been performed by applying a differential plunger technique to the heavy-ion transfer reaction ^{48}Ca+^{208}Pb, measured with PRISMA-CLARA [15]. As shown in Fig. 1 b), the lifetime of the 4017 keV level (depopulated by the 660-keV γ-ray) is τ = 8.5±2.0 ps. Taking into account the branching from the 4017 keV state (of the order of

9% for the direct decay to the ground state) one obtains the reduced transition probability B(E3) = 7.9 ± 2.0 Wu, which agrees with the measured strength of the 3^- phonon of ^{48}Ca.

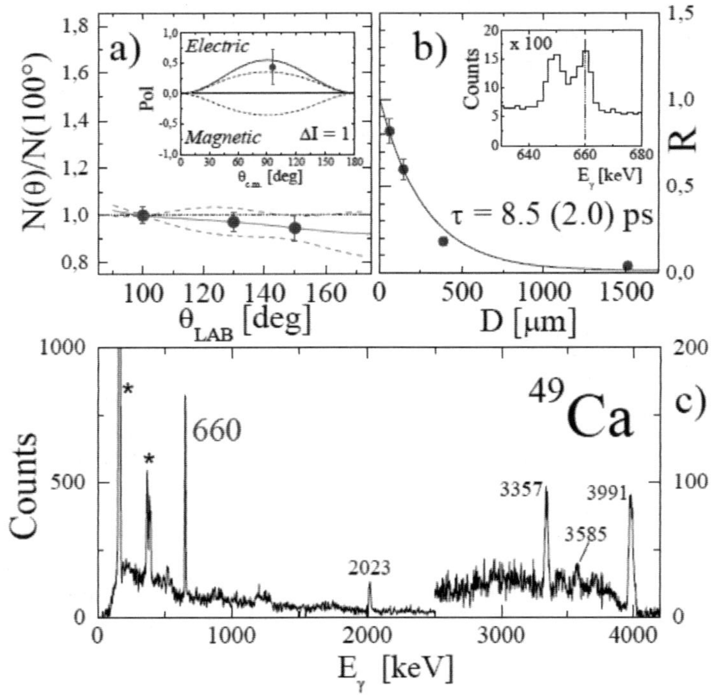

Figure 1. Bottom panel c): γ-spectrum of ^{49}Ca, with contaminant lines from ^{63}Ni marked by stars. Panels a): Angular distribution of the 660-keV transition. The lines represent the fit by the function W(θ), with parameter a_2 = 0.072±0.05 (consistent with a stretched dipole with ~4% admixture of quadrupole decay). The inset of panel a) shows the polarization of the 660-keV γ ray, in comparison with prediction for a pure electric or magnetic dipole. Panel b) shows, in the inset, the Doppler corrected γ-ray spectrum around 660 keV, as measured in the differential plunger experiment [15]. From the relative intensity of the high- and low-energy peaks, corresponding to the decay after and before the degrader, as a function of the target-to-degrader distances, the lifetime of the state is obtained [6].

A similar analysis has been performed for the ^{47}Ca reaction product. Figure 2 shows in the bottom panel the γ-spectrum, with labels indicating transitions de-exciting states of simple particle-hole nature [2]. In addition, a very weak transition at 437 keV is also observed, which has been established to depopulate a state at 3999 keV. Such state has been proposed to be the $11/2^+$ or $13/2^+$ member of the multiplet arising by coupling the 3^- phonon of ^{48}Ca with the $f^{-1}_{7/2}$ neutron hole of ^{47}Ca. Due to the limited statistics collected for this transition no angular distribution and polarization measurements could be performed to firmly determine the spin and parity of the 3999 keV level, however an attempt was made to measure its lifetime using the differential plunger technique applied to the case of ^{49}Ca (cfr. panels a) and b) of Fig. 2). It is found that the lifetime is τ = 64±16 ps. Therefore, by taking into account the decay branch of ~58% to the ground state, the reduced transition probability becomes equal to B(E3)=7.4±1.9 Wu, again very similar to the 3^- phonon strength of ^{48}Ca.

Theoretically, a weak coupling model [3] can be applied to predict the excitation energy spectrum arising from coupling a particle/hole to a phonon. This type of calculation has been performed for both ^{47}Ca and ^{49}Ca. As shown in Fig. 3, the model predicts, at the lowest perturbative order, multiplets of states with energies shifted with respect to the 3^- phonon. In particular, in the case of ^{49}Ca the levels are found to be quite closely spaced and moved down in energy, with very good agreement with the experimental energy for the $9/2^+$ state. On the contrary, in the case of ^{47}Ca, a larger energy spread is found within the multiplet, with the $13/2^+$ state being the only one lifted up. This strongly favors the spin assignment of $11/2^+$ to the experimentally observed 3999-keV level. At the same perturbative order, the calculations predict that the reduced transition probability B(EL) to the ground state should be associated to the same multipole of the vibrational state and should have the same value of the core nucleus (B(E3)~7 Wu in this case). Experimentally, the B(E3) values of the ground state decay from the 3999-keV level in ^{47}Ca and from the 4017-keV level in ^{49}Ca are the same (within the error) of the 3^- phonon of ^{48}Ca. This gives strong evidence that these state have a particle-phonon coupled nature. Indication for similar type of states have been recently found also in ^{41,43}Ar nuclei [16].

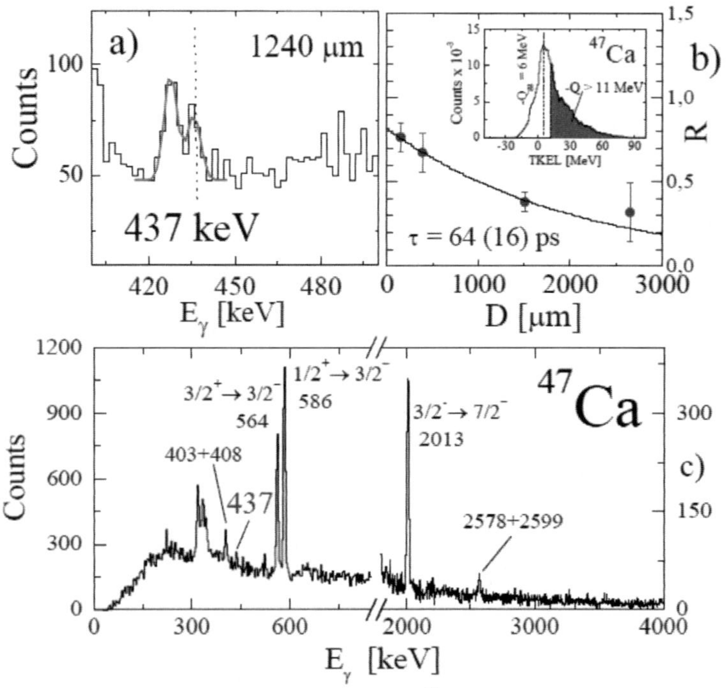

Figure 2. Bottom panel c): γ-spectrum of ^{47}Ca. Panels a) and b): Lifetime analysis of the 3999-keV level, based on the relative intensity of the high- and low-energy component of the 437-keV γ–ray de-exciting the state (panel a)), as a function of the target-to-degrader distances (panel b)). To improve the selectivity to γ decays from states around 4 MeV, only events with large Total Kinetic Energy Loss were considered, as shown in the inset of panel b) [10].

4. Conclusion

Based on the study of the binary reaction ^{48}Ca + ^{64}Ni at ~6 MeV/A, we have demonstrated the possibility of performing complete–γ spectroscopy studies (i.e. angular distribution, polarization and lifetime analysis) with low energy heavy-ion transfer reactions. This opens up the possibility to investigate in great details the properties of moderately neutron rich systems moving away from stability and it provides a method that can be further exploited in the future with heavy targets and radioactive beams. In the specific cases of ^{49}Ca and ^{47}Ca, evidence is given for particle-phonon coupled states based on the 3^- excitation of ^{48}Ca. These are among the few fully established examples of particle-phonon coupled states in rather light nuclear systems. They are, in general, of great importance for the understanding of the origin of the anharmonicity of vibrational spectra, for damping phenomena and for the quenching of spectroscopic factors.

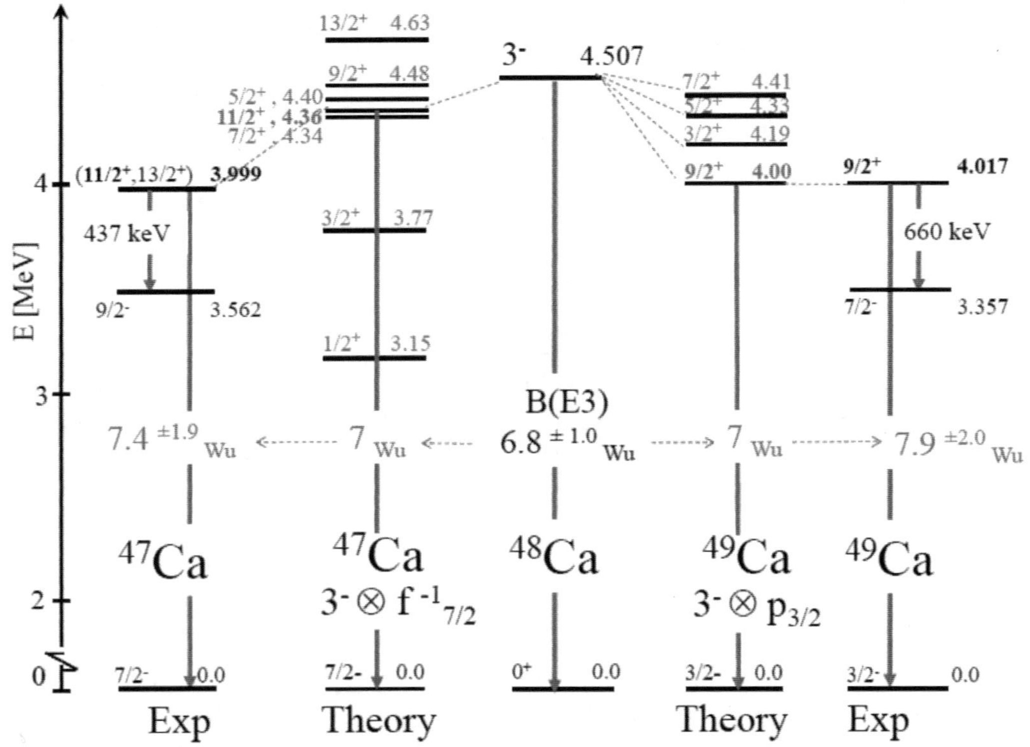

Figure 3. Partial level scheme of ^{49}Ca (right) and ^{47}Ca (left) showing the positive parity states arising by coupling one neutron ($p_{3/2}$) or one neutron-hole ($f^{-1}_{7/2}$) to the collective 3^- phonon of ^{48}Ca (center) . Experimental data refer to this work, theoretical prediction are obtained by the weak coupling model of Ref. [3] using the SkX Skyrme interaction.

References

[1] L. Corradi, G. Pollarolo and S. Szilner, J. Phys. **G36** (2009)113101.
[2] R. Broda, J. Phys. **G32** (2006)R151.
[3] A. Bohr, B.R. Mottelson, Nuclear Structure, vols. I and II, W.A. Benjamin, 1975.
[4] P.F. Bortignon, A. Bracco, R.A. Broglia, Giant Resonances: Nuclear Structure at Finite Temperature, Harwood Academic Publishers, New York, 1998.
[5] D. Montanari et al., Phys. Rev. **C84** (2011)054613.
[6] D. Montanari et al., Phys. Lett. **B697** (2011)288.
[7] A.M. Stefanini et al., Nucl. Phys. **A701**(2002)217c.
[8] D. Montanari et al., Eur. Phys. J. **A47** (2011)4.
[9] A. Gadea et al., Eur. Phys. J. **A20** (2004)193.
[10] D. Montanari et al., to be published.
[11] R.A. Eisenstein, et al., Phys. Rev. **188** (1969)1815.
[12] T. Hartmann, et al., Phys. Rev. **C65** (2002)034301.
[13] Y. Uozumi, et al., Nucl. Phys. **A576** (1994)123.
[14] S. Fortier et al., Nuc. Phys. **A311** (1978)324.
[15] J.J. Valiente-Dobón et al., Phys. Rev. Lett. **102** (2009)242502.
[16] S. Szilner et al., Phys. Rev. **C84** (2011)014325.

Journal of Physics: Conference Series **366** (2012) 012031 doi:10.1088/1742-6596/366/1/012031

Multiphonon Approaches to Complex Spectroscopy

Lo Iudice N, Bianco D, Knapp F† Andreozzi F, Porrino A, P. Vesely

Dipartimento di Scienze Fisiche, Universitá di Napoli Federico II and INFN, Sezione di Napoli, Napoli, Italy

E-mail: loiudice@na.infn.it

Abstract. An equation of motion method for solving the nuclear eigenvalue problem in a basis of microscopic multiphonon states has been upgraded so as to generate states solely composed of quasiparticle Tamm-Dancoff phonons. The method is applied to the neutron rich ^{20}O. A space covering up to three-phonon states is adopted to compute the electric dipole response.

1. Introduction

The most common extension adopted to describe the anharmonic features of collective modes and giant resonances is known as second random-phase approximation (SRPA). It consists in coupling the RPA modes to two-particle two-hole or four quasiparticles configurations [1]. Recent implementations were based on density functional non relativistic [2] and relativistic [3] theories.

Another popular generalization is the quasiparticle phonon model (QPM) [4]. This approach, being based on a Hamiltonian of separable form, is able to extend further the RPA approach by including also a fraction of three-phonon states.

SRPA [3, 5] and QPM [6, 7] were applied to neutron rich nuclei to study the properties of the so called pygmy resonance induced by the oscillation of the valence neutron skin against the core. Such a resonance, whose first evidence was provided about a decade ago [8], is now under intense experimental investigation. A list of references can be found in [9].

We proposed, recently, an equation of motion phonon method (EMPM) [10, 11] for solving the eigenvalue problem in a basis of multiphonon states generated iteratively starting from Tamm-Dancoff phonons. These multiphonon states, however, had a hybrid structure. They were composed of particle-hole (ph) operators coupled to $(n-1)$-phonon states.

This inconsistency is removed in the new formulation, where all quantities are expressed solely in terms of phonons. Moreover, these phonons are constructed in the quasiparticle (qp) scheme so as to allow the study of open shell nuclei.

† Present address: Faculty of Mathematics and Physics Charles University, Prague, Czech Republic

Simple and physically transparent equations generate iteratively a multiphonon basis which can be used to solve the eigenvalue problem for a Hamiltonian of general form. No approximations, apart from the one inherent the *qp* formalism, are involved.

The method was already applied to the neutron rich ^{20}O, chosen as test-ground, to compute the $E1$ response in a space including up to three phonons [12]. A Nilsson+BCS basis in a restricted space was adopted. Here, we use a HF+ BCS basis in the same configuration space.

2. Brief outline of the method

We first solve the *qp* TDA eigenvalue equation

$$\sum_{rs,qt} A_{rsqt} c_{qt}^{\lambda} = \Big\{ \Big[(\epsilon_r + \epsilon_s) - E_\lambda \Big] \delta_{rq} \delta_{st} + \mathcal{V}_{rs,qt} \Big\} c_{qt}^{\lambda} = 0, \tag{1}$$

where ϵ_r are quasiparticle energies and $\mathcal{V}_{rs,qt}$ are the matrix elements of the two-body potential between two-quasiparticle states.

The eigenstates $\mid \lambda\rangle$ of energy E_λ define the phonon operators

$$O_\lambda^\dagger = \sum_{r<s} c_{rs}^{\lambda} \alpha_r^\dagger \alpha_s^\dagger, \tag{2}$$

where $\alpha_r^\dagger$ (α_r) creates (annihilates) a quasiparticle with respect to the BCS vacuum $\mid 0\rangle$.

The main objective of the method is to generate a set of multiphonon states $\mid n; \alpha\rangle$ of the form

$$|n; \beta >= \sum_{\lambda\alpha} C_{\lambda\alpha}^{(\beta)} \, O_\lambda^\dagger \; |n-1; \, \alpha > . \tag{3}$$

The procedure developed to generate these states starts with introducing the equations of motion

$$< n, \beta | [H, O_\lambda^\dagger] | n-1, \alpha >= \left(E_\beta^{(n)} - E_\alpha^{(n-1)} \right) < n, \beta | O_\lambda^\dagger | n-1, \alpha > . \tag{4}$$

We have to expand the commutator on the left-hand side and use closure relations, just as in Refs. [10, 11]. After making additional manipulations, we obtain

$$\sum_{\lambda'\alpha'} \mathcal{A}_{\lambda\alpha,\lambda'\alpha'} X_{\lambda'\alpha'}^{\beta} = E_\beta X_{\lambda\alpha}^{\beta} \, , \tag{5}$$

where

$$X_{\lambda\alpha}^{(\beta)} =< n, \beta | O_\lambda^\dagger | n-1, \alpha > \tag{6}$$

and

$$\mathcal{A}_{\lambda\alpha,\lambda'\alpha'} = (E_\lambda + E_\alpha) \delta_{\lambda\lambda'} \delta_{\alpha\alpha'} + \mathcal{V}_{\lambda\alpha,\lambda'\alpha'}. \tag{7}$$

As shown by the above formula, the matrix $\mathcal{A}_{\lambda\alpha,\lambda'\alpha'}$ is expressed solely in terms of phonons and is in close correspondence with the *qp* TDA matrix $A_{rs,qt}$ defined by the eigenvalue equation (1). Indeed, $\mathcal{A}_{\lambda\alpha,\lambda'\alpha'}$ is formally deduced from $A_{rs,qt}$ by replacing the quasiparticle energies ϵ_i with the phonon energies E_λ and the quasiparticle two-body potential $\mathcal{V}_{rs,qt}$ with the phonon-phonon potential $\mathcal{V}_{\lambda\alpha,\lambda'\alpha'}$.

Journal of Physics: Conference Series **366** (2012) 012031 doi:10.1088/1742-6596/366/1/012031

Equation (5) is not yet an eigenvalue equation. This is obtained once we express the X amplitudes (6) in terms of the C coefficients by making use of the expansion (3). Once this is done, we obtain from (5) the generalized eigenvalue equation within the n-phonon subspace

$$\sum_{\lambda'\alpha'} (\mathcal{AD})(\lambda\alpha, \lambda'\alpha') C^{\beta}_{\lambda'\alpha'} = E_{\beta} \sum_{\lambda'\alpha'} \mathcal{D}_{\lambda\alpha,\lambda'\alpha'} C^{\beta}_{\lambda'\alpha'} , \qquad (8)$$

where

$$\mathcal{D}_{\lambda\alpha,\lambda'\alpha'} = < n-1, \alpha' | O_{\lambda'} O^{\dagger}_{\lambda} | n-1, \alpha > \qquad (9)$$

is the metric matrix. $\mathcal{A}$ and $\mathcal{D}$ are expressed in terms of quantities defined in the $(n-1)$-phonon subspace. The recursive character of these formulas allows to solve Eqs. (8) iteratively. The redundant states are eliminated by a procedure [10, 11] based on the Cholesky decomposition method. A set of orthonormal multiphonon states $\{|0>, |\lambda>, \ldots |n, \alpha> \ldots\}$ is generated. The Hamiltonian is diagonal in each n-phonon block. The terms coupling different blocks are also expressed solely in terms of phonons and have a simple structure. The Hamiltonian matrix can therefore be brought easily to diagonal form yielding the eigenfunctions

$$| \Psi_{\nu} \rangle = \sum_{n,\alpha} C^{(\nu)}_{\alpha} \mid n; \alpha \rangle . \qquad (10)$$

As shown by the formula, the ground state is treated at the same footing of the other states and is explicitly correlated. Indeed, our approach does not rely on any approximation, except for the one inherent the Bogoliubov qp transformation.

3. Application of the method to ^{20}O

A test calculation of the $E1$ response in ^{20}O was already performed [12] using a Nilsson basis. The lowest three major shells, up to (s, d), were used to generate the constituent qp TDA phonons. All multiphonon states spanning up to the entire three-phonon subspace were taken into account. The space considered is slightly more restricted than the one used in a shell model calculation [13].

Here, we replace the Nilsson with a HF basis. The HF states incorporate high energy configurations even if the calculation is performed in a restricted space. They are used to generate the quasiparticle states in BCS. As in [12], a Brueckner G-matrix derived from the CD-Bonn nucleon-nucleon potential are adopted.

The spurious admixtures induced by the center of mass motion and by the particle number violation, caused by the Bogoliubov transformation, are removed in the preliminary TDA stage. We have just to adopt the Gramm-Schmidt procedure to orthogonalize the $2qp$ states to $\propto \vec{R}_{CM}|0>$ and $\propto \hat{n}|0>$ and use this spurious free basis to generate the TDA phonons. The multiphonon states resulted to be spurious free as well, since their phonon structure remains unchanged all along the procedure.

As shown in Figure 1, the $E1$ reduced strength gets dramatically damped and spread as the two-phonon space is included. The three-phonons induce further quenching and fragmentation.

Journal of Physics: Conference Series **366** (2012) 012031 doi:10.1088/1742-6596/366/1/012031

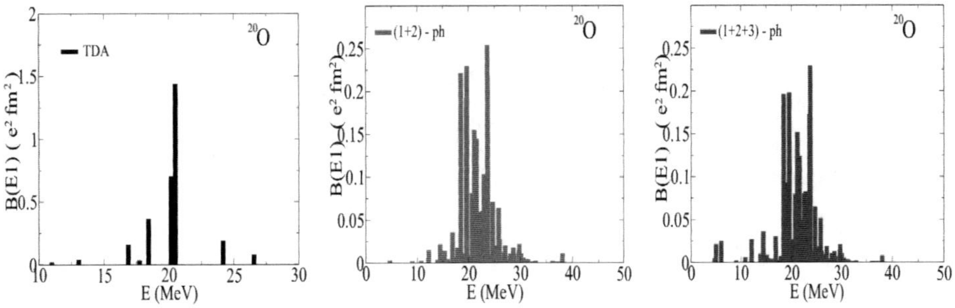

Figure 1. $E1$ reduced transition strengths using a HF basis in different multiphonon spaces

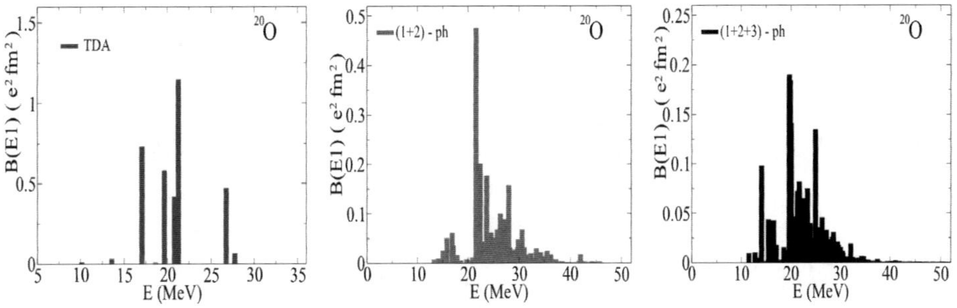

Figure 2. $E1$ reduced transition strengths obtained from using a Nilsson basis

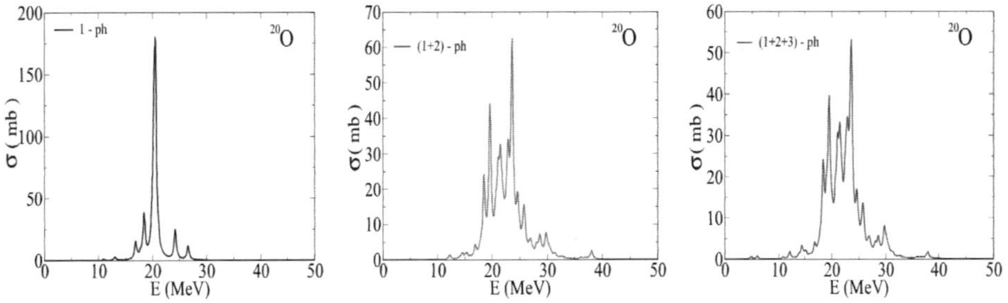

Figure 3. $E1$ cross sections computed in a HF+BCS basis

The phonon coupling induces damping and quenching also in the Nilsson basis (Figure 2). Though HF and Nilsson spectra are seen to differ in several details, in both cases, few low-lying peaks appear below 16 MeV, consistently with early experiments [8], especially when three-phonon states are included.

The effects of the phonon coupling are visible also in the cross sections, shown in Figure 3. Due to the smoothing action of the width ($\Gamma = 1MeV$), however, the cross section emphasizes the damping and spreading induced by the two phonons, while the role of the three phonons is attenuated.

The transition densities shown in Figures 4 disclose the nature of two typical states. The plot in the left panel clearly indicates that the peak in the GDR region describes an out of phase oscillation of protons against neutrons. On the other hand,

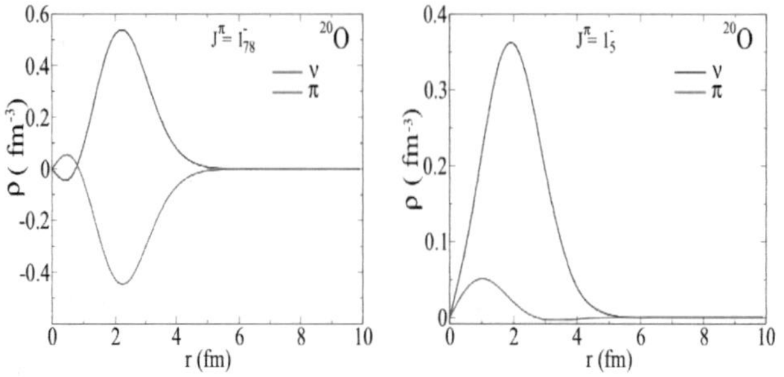

Figure 4. Transition densities for some selected 1^- states in ^{20}O

the transition density pertaining to a low-lying state plotted in the right panel is typical of an oscillation of neutron skin versus the core indicating the occurrence of a pygmy resonance.

No evidence of such a collective mode emerged from the calculation using the Nilsson basis [12]. The low-lying states obtained in this basis were simply non collective. Apparently, the high energy configurations incorporated in the HF basis play an important role in determining the relative weight of proton and neutron components.

The discrepant results obtained in the HF and Nilsson basis strongly suggests that a larger configuration space must be used in generating the constituent TDA phonons.

On the other hand, the number of multiphonon states becomes prohibitively large as the number of majors shells increases. A severe truncation of the multiphonon space is necessary. The phonon formalism is naturally suited to such a task. It allows, in fact, to select the TDA phonons according to their energy and collectivity. A work along these line is being completed.

References

[1] Wambach J 1988 *Rep. Prog. Phys.* **51** 989
[2] Gambacurta D, Grasso M and Catara F 2010 *Phys. Rev.* C **81** 054312
[3] Litvinova E, Ring P and Tselyaev V 2010 *Phys. Rev. Lett.* **105** 022502
[4] Soloviev V G 1992 *Theory of atomic nuclei : Quasiparticles and Phonons* (Bristol: Institute of Physics)
[5] Paar N, Vretenar D, Khan E and Colò G 2007 *Rep. Prog. Phys.* **70** 691 for review and references
[6] Tsoneva N, Lenske H and Ch. Stoyanov 2004 *Phys. Lett.* B **586** 213
[7] Tsoneva N and Lenske H 2008 *Phys. Rev.* C **77** 024321
[8] Leistenschneider A *et al.* 2001 *Phys. Rev. Lett.* **86** 5442
[9] Tamii A *et al.* 2011 *Phys. Rev. Lett.* **107** 062502
[10] Andreozzi F, Knapp F, Lo Iudice N, Porrino A and Kvasil J 2007 *Phys. Rev.* C **75** 044312
[11] Andreozzi F, Knapp F, Lo Iudice N, Porrino A and Kvasil J 2008 *Phys. Rev.* C **78** 054308
[12] Lo Iudice N, Bianco D, Knapp F, Andreozzi F, Porrino A and Vesely P 2011 *Proceedings 14th International Symposium on Capture Gamma-Ray Spectroscopy and Related Topics (CGS14)(Guelph)* (Singapore: World Scientific) to be published
[13] Sagawa H and Suzuki T 1999 *Phys. Rev. C* **59** 3116

Journal of Physics: Conference Series **366** (2012) 012032 doi:10.1088/1742-6596/366/1/012032

Analysis of ^{6}He+^{12}C elastic scattering and breakup reactions using a microscopic optical potential model

V K Lukyanov[1], D N Kadrev[2], E V Zemlyanaya[1], A N Antonov[2], K V Lukyanov[1] and M K Gaidarov[2]

[1] Joint Institute for Nuclear Researches, Dubna, 141980, Russia
[2] Institute for Nuclear Research and Nuclear Energy, Bulgarian Academy of Sciences, Sofia 1784, Bulgaria

E-mail: vlukyanov@jinr.ru

Abstract. Theoretical analysis is made for the ^{6}He+^{12}C elastic scattering data at three different beam energies. The breakup effect of the ^{6}He at higher energies is also studied. Calculations were performed using microscopic optical potentials (OP's) obtained by a double-folding procedure and also those inherent in the high-energy Glauber-Sitenko approximation. The problem of ambiguity of the adjusted depths of these potentials is resolved. The role of breakup processes in formation of the imaginary potential in elastic channel is also discussed.

^{6}He is a typical nucleus having the weak binding energy and extended neutron halo in its periphery. The latter is the reason why in collisions with the proton and nuclear targets the projectile nucleus ^{6}He is breaking up with a comparably large probability that causes the flux loss in the elastic channel. Therefore, the study of elastic scattering of ^{6}He on protons or light targets is a powerful tool to get information on peculiarities of the mechanism of such processes. We perform an analysis of the ^{6}He+^{12}C elastic scattering data at three beam energies $E = 3$ [1], 38.3 [2] and 41.6 MeV/N [3] using the microscopically calculated OP. The main expressions for the real and imaginary parts are

$$U(r) = V^{DF}(r) + iW(r), \qquad V^{DF}(r) = V^D(r) + V^{EX}(r). \tag{1}$$

The real part V^{DF} consists of the direct and exchange double-folding (DF) integrals that include an effective NN potential and density distribution functions of colliding nuclei. The formalism of the DF potentials is described in Refs. [4, 5]. In the considered case V^D and V^{EX} include only the isoscalar part of the NN-potential while its isovector part is omitted because $Z = N$ in the target nucleus ^{12}C. All details of the mathematical treatments and calculations are given in Ref. [6].

Concerning the imaginary part of our OP, we take it in two forms. The first one has the same form as the real part of OP but with different strength. Also we test another shape of the imaginary part that corresponds to the full microscopic OP derived in Refs. [7, 8] within the HEA [9, 10] whose imaginary part is as follows

$$W^H(r) = -\frac{1}{2\pi^2}\frac{E}{k}\bar{\sigma}_N \int_0^\infty j_0(qr)\rho_p(q)\rho_t(q)f_N(q)q^2dq. \tag{2}$$

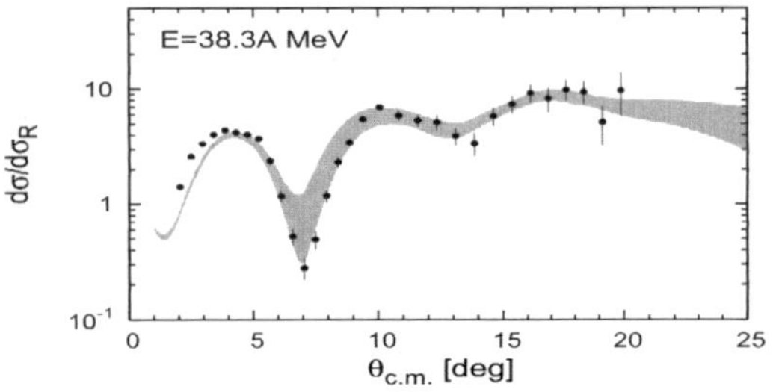

Figure 1. Cross sections of ^{6}He+^{12}C elastic scattering at E = 38.3 MeV/N calculated by fitting the N_R, N_I, N_I^{sf} parameters of the microscopic OP [Eqs. (3)-(4)] (gray area). The experimental data are taken from Ref. [2].

Here $\bar{\sigma}$ is the isotopically averaged total NN cross section, and f_N is the form factor of the NN amplitude. The final microscopic volume OP has the following form:

$$U(r) = N_R V^{DF}(r) + iN_I W(r), \tag{3}$$

where $W(r)$ is taken to be equal either to $V^{DF}(r)$ or to $W^H(r)$. The parameters N_R and N_I entering Eq. (3) renormalize the strength of OP and are fitted by comparison with the experimental cross sections. In the work [11] we attempt to simulate the surface effects caused by the polarization potential by adding to [Eq. (3)] the different surface terms:

$$W^{sf}(r) \to -iN_I^{sf}\frac{dW(r)}{dr}; -iN_I^{sf} r\frac{dW(r)}{dr}; -iN_I^{sf} r^2\frac{dW(r)}{dr}; -iN_I^{sf}\frac{dW(r-\delta)}{dr} \tag{4}$$

where N_I^{sf} is a fitting parameter, and the shift δ is fixed to be $\delta = 1$ fm.

We calculated the ^{6}He+^{12}C elastic scattering differential cross sections using the program DWUCK4 [12] and the microscopically calculated OP [Eq. (3)]. The inclusion of different forms of the surface potential leads to almost similar curves of cross sections displayed in the gray area at Fig. 1. The ambiguity of the values of N's arises. We tried to choose the most physical values of N's by adding the physical criterion that the obtained potentials should obey a determined behavior of the volume integrals dependent on the energy [4]

$$J_V(E) = -\frac{4\pi}{A_p A_t}\int N_R V^{DF}(r) r^2 dr, \qquad J_W(E) = -\frac{4\pi}{A_p A_t}\int N_I W(r) r^2 dr, \tag{5}$$

where the OP's were fitted at different energies. It was shown in Refs. [13, 14, 15]) that the volume integrals J_V decrease with the energy increase at $E < 100$ MeV/N, while J_W increases up to 10-20 MeV/N and then saturates. In Fig. 2 we select those curves whose parameters N's lead to the mentioned behavior of J_V and J_W. In Fig. 3 the deviations of the OP's from their volume parts are presented. In our theoretical scheme they are related to the surface parts of the ImOP in the form of $N_I^{sf} r dW^{H(DF)}/dr$ and $N_I^{sf} r^2 dW^{(H)DF}/dr$. One can suppose that these dynamical polarization potentials arrive due to effects of the breakup of ^{6}He.

We consider the simplest breakup $h\alpha$-model of ^{6}He, where it is suggested consisting of two clusters ^{4}He and h, the correlated pair of two neutrons h=2n [3]. The s-wave function $\varphi_b(\mathbf{s})$ of relative motion of clusters was obtained by fitting to the known binding energy and rms-radius of ^{6}He and defines the density distribution of the latter

$$\varrho_b(s) = |\varphi_b(\mathbf{s})|^2 = (1/4\pi)|\varphi_{l=0}(s)|^2 \tag{6}$$

It then used for calculations of the ground state matrix elements of breakup processes. One can see in Fig.4 that $\varrho_b(s)$ coincides fairly well with $\varrho_L(s)$. Thus, we may apply the 2-cluster $h\alpha$-model for the further calculations.

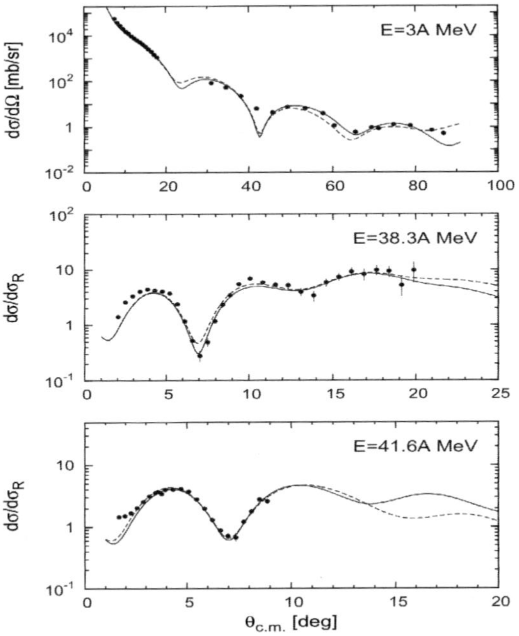

Figure 2. The ^{6}He+^{12}C elastic scattering calculations with the volume (solid for $W = W^H$, dashed for $W = V^{DF}$) and surface part $-iN_I^{sf} r^2 \frac{dW(r)}{dr}$ in OP. For details see Ref. [11].

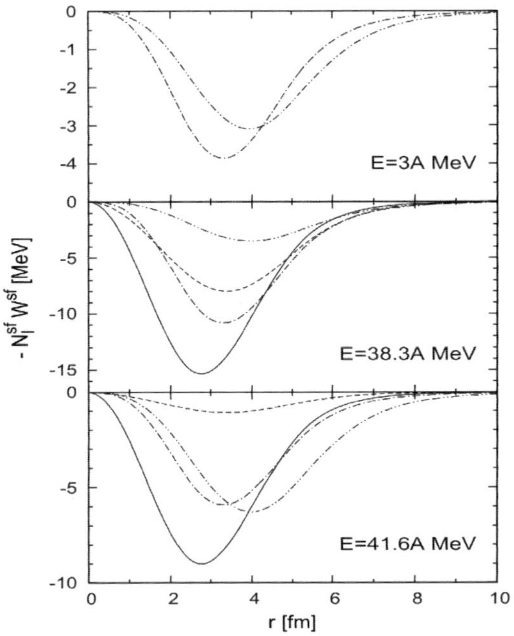

Figure 3. The surface ImOP's used in the calculations of the cross sections. Solid and dot-dashed curves are for $W = W^H$ and others for $W = V^{DF}$, respectively. For details see Ref. [11].

In the framework of the $h\alpha$-model of ^{6}He one can estimate the ^{6}He+^{12}C OP as folding of two OP's of interaction of clusters α and h with ^{12}C:

$$U_{HeC}^{(b)}(r) = V^{DF\,(b)} + iW^{(b)} = \int d^3s\varrho_b(s)\Big\{U_\alpha(\mathbf{r} + (1/3)\mathbf{s}) + U_h(\mathbf{r} - (2/3)\mathbf{s})\Big\} =$$
$$= 2\pi \int_0^\infty \varrho_b(s)s^2ds \int_{-1}^1 dx\Big\{U_\alpha\left(\sqrt{r^2 + (1/9)s^2 + r(2/3)sx}\right) + \tag{7}$$
$$+U_h\left(\sqrt{r^2 + (4/9)s^2 - r(4/3)sx}\right)\Big\}.$$

Here the h-^{12}C potential is taken as the twice neutron-^{12}C OP $U_h = 2U_n$. In turn, potentials U_α and U_n are calculated within the microscopic hybrid model of OP [7]. Doing so, one can see from Fig.5 that angular distributions for different kinds of ImOP in the $h\alpha$-model based on the W^H and V_{DF} forms (solid and dashed curves), as well as in the DF-model (dash-dotted and dotted curves) are closely displayed, and the corresponding total reaction cross sections are almost equal in value. As a result the $h\alpha$-model of ^{6}He seems to be reliable for the further evaluations of total breakup cross sections with a help of the HEA theory. The earlier HEA theory for the breakup processes was developed in Refs.[17],[18]. In recent papers [19],[20] this method was generalized to study breakup reactions of lightest nuclei.

In the Ref.[21] we showed that this method is applicable for the ^{6}He+^{12}C at energies about 40 MeV/N. Thus, we can obtain the total absorbtion cross section for the $h\alpha$-model of ^{6}He withen the HEA method as follows:

$$\sigma_{abs}^{tot} = 2\pi \int_0^\infty b_h db_h \left(1 - |S_h(b_h)|^2\right) (1 - I(b_h)) \tag{8}$$

Journal of Physics: Conference Series **366** (2012) 012032 doi:10.1088/1742-6596/366/1/012032

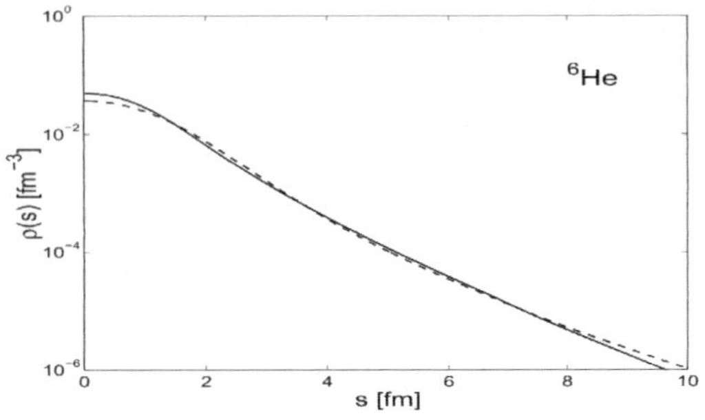

Figure 4. Comparison of the $h\alpha$-model density distribution $\varrho_b(s)$ (solid) with the LSSM density $\varrho_L(s)$ [16] (dashed).

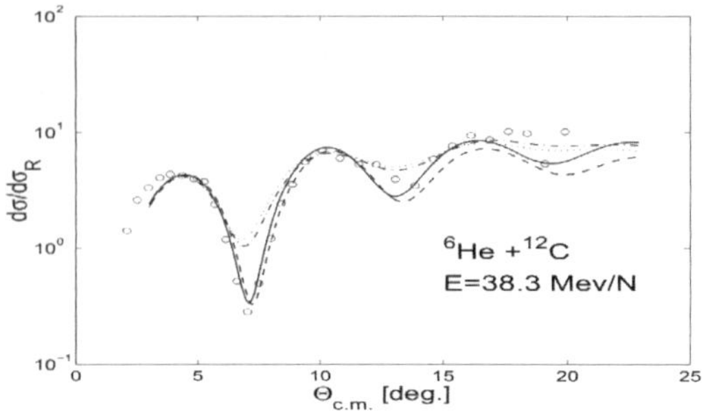

Figure 5. The ^{6}He+^{12}C elastic scattering with ϱ_b (solid, dashed) and $\varrho_L(s)$ densities of ^{6}He (dash-dotted, dotted) [11].

where

$$I(b_h) = \int d^3 s \varrho_b(s) |S_\alpha(b_\alpha)|^2, \quad b_\alpha = \sqrt{s^2 \sin^2\vartheta + b_h^2 - 2sb_h \sin\vartheta\cos\phi}. \tag{9}$$

Here the relation is used of impact parameters $\mathbf{b}_\alpha = \mathbf{b}_h - \mathbf{b}$ with $b = s\sin\vartheta$ being the projection of the $h-\alpha$ vector $\mathbf{s}$ on the plane normal to the 0z-axis along the straight line trajectory of an incident nucleus.

In the case of the stripping reaction with removing h-particle from ^{6}He to the target nucleus, one should use the probability of h to leave the elastic channel $(1 - |S_h(b_h)|^2)$, and for α to continue its elastic scattering with probability $|S_\alpha(b_\alpha)|^2$. Then the probability of the whole process is $|S_\alpha(b_\alpha)|^2 \cdot (1 - |S_h(b_h)|^2)$, and to get the total stripping cross section one must average over $\varrho_b(s)$. In a similar manner the transfer of the α particle can be constructed, and the net contribution of both removal reactions yields the total breakup cross section

$$\sigma_{bu}^{tot} = 2\pi \int_0^\infty b_h db_h \Big\{ |S_h(b_h)|^2 + \Big[1 - 2|S_h(b_h)|^2\Big] \cdot I(b_h) \Big\}. \tag{10}$$

The sum of the absorption (8) and breakup (10) cross sections results in *the total reaction cross section*

$$\sigma_R^{tot} = 2\pi \int_0^\infty b_h db_h \Big(1 - |S_h(b_h)|^2 \cdot I(b_h)\Big). \tag{11}$$

Table 1. The HEA estimations within the $h\alpha$-model of total cross sections of ^{6}He+^{12}C at E=38.3 MeV/N.

potential	σ_{abs}^{tot}, mb	σ_{bu}^{tot}, mb	σ_R^{tot}, mb
ImOP=N$_I$W$^{H(b)}$	392	412	804
ImOP=N$_I$V$^{DF(b)}$	447	383	830

Results and conclusion

The results of the present work can be summarized as follows:

(i) The microscopic OP and cross sections of ^{6}He+^{12}C elastic scattering were calculated at 3 different energies. In contrast to the phenomenological and semi-microscopic models we deal with a fully microscopic approach as a physical ground to account for the single-particle structure of the colliding nuclei.

(ii) At higher energies additional surface terms in OP having a form of a derivative of the imaginary part of the OP became necessary.

(iii) The depths of the real and imaginary parts of the microscopic OP's are considered as fitting parameters. To overcome the ambiguity of these parameters, additional physical constraints were imposed.

(iv) Estimations of the total cross sections were made by means of the preliminary calculated imaginary parts of OP's U_h and U_α for scattering of h- and α-clusters on ^{12}C. The obtained results are shown in Table 1. The breakup cross sections constitute approximately a half of the total reaction cross section. Therefore the significant amount of the imaginary potential of the ^{6}He+^{12}C in elastic scattering channel is formed due to transmittions of the incident flux of ^{6}He into breakup channels.

One can summarize that the elastic scattering data can be explained in a reasonably good agreement by using two folding models. These both models explain fairly well the separation energy and *rms* radius of the nucleus. The $h\alpha$-model can also estimate the other characteristics of the reaction mechanism such as the ^{6}He breakup cross sections. This gives a possibility to understand the significant role of breakup processes in formation of the imaginary part of the ^{6}He+^{12}C elastic scattering OP.

Acknowledgments

The work is partly supported by the Project from the Agreement for co-operation between the INRNE-BAS (Sofia) and JINR (Dubna). Three of the authors (D.N.K., A.N.A. and M.K.G.) are grateful for the support of the Bulgarian Science Fund under Contract No. 02–285 and one of them (D.N.K.) under Contract No. DID–02/16–17.12.2009. The authors E.V.Z. and K.V.L. thank the Russian Foundation for Basic Research (Grant No. 09-01-00770) for the partial support.

References

[1] Milin M, Cherubini S, Davinson T et al 2004 *Nucl. Phys.* **A730**, 285
[2] Lapoux V, Alamanos N, Auger F et al 2002 *Phys. Rev.* C **66**, 034608
[3] Al-Khalili J S, Cortina-Gil M D, Roussel-Chomaz P et al 1996 *Phys. Lett.* B **378**, 45
[4] Satchler G R and Love W G 1979 *Phys. Rep.* **55**, 183
[5] Khoa D T and Satchler G R 2000 *Nucl. Phys.* A **668**, 3
[6] Lukyanov K V 2007 *Comm. JINR* P11-2007-38, Dubna
[7] Lukyanov K V, Zemlyanaya E V, and Lukyanov V K 2004 *JINR Preprint* P4-2004-115, Dubna; 2006 *Phys. At. Nucl.* **69**, 240
[8] Shukla P 2003 *Phys. Rev.* C **67**, 054607
[9] Glauber R J 1959 *Lectures in Theoretical Physics* (New York, Interscience), p.315
[10] Sitenko A G 1959 *Ukr. Fiz. J.* **4**, 152
[11] Lukyanov V K, Kadrev D N, Zemlyanaya E V et al 2010 *Phys. Rev.* C **82**, 024604
[12] Kunz P D and Rost E 1993 *Computational Nuclear Physics*, edited by K. Langanke *et al.* (Springer-Verlag, New York), Vol.2, p.88
[13] Romanovsky E A 1998 *et al., Bull. Rus. Acad. Sci.,* Physics, **62**, 150
[14] Mahaux C and Ngo H 1982 *Nucl. Phys.* **A378**, 205
[15] Brandan M E and Satchler G R 1997 *Phys. Rep.* **285**, 143
[16] Karataglidis S, Dortmans P J, Amos K and Bennhold C 2000 *Phys. Rev.* C **61**, 024319
[17] Glauber R J 1955 *Phys. Rev.* **99**, 1515 (1955); 1955 *ibid.* **100**, 242
[18] Akhiezer A I and Sitenko A G 1955 *Sci. Notes of Kharkov Univ.* **64**, 9 (1955) ; 1957 *Phys. Rev.* **106**, 1236
[19] Hencken K, Bertsch G, Esbensen H 1996 *Phys. Rev.* C **54** 3043
[20] Bertulani C A and Hansen P G 2004 *Phys. Rev.* C **70** 034609
[21] Lukyanov V K, Zemlyanaya E V, Lukyanov K V 2011 *Int. J. Mod. Phys.* E **20** 9

Neutron spectrometry with He-3 proportional counters

M Manolopoulou[1*], M Fragopoulou[1], S Stoulos[1], E Vagena[1], W Westmeier[2] and M. Zamani[1]
[1]School of Physics, Aristotle University of Thessaloniki, Thessaloniki 54124, Hellas
[2]Philipps-Universität, 35032 Marburg, Germany
[*]email: metaxia@auth.gr

Abstract. Helium filled proportional counters are widely used in the field of neutron detection and spectrometry. A review of some parameters and physical effects that influence the He-3 proportional counters response is presented through the experimental study of a commercially available counter. The impact of these effects is taken into account in the description of detector performance when the response is calculated with the Monte Carlo method. A method to determine neutron energy distribution with a bare He-3 counter in an unknown neutron field is presented. The neutron energy range extends up to about 7 MeV. The method does not require a hypothesis about neutron spectrum shape. It is based in the calculated counter response with Geant4, the validity of which is experimentally verified. A detailed description of the steps needed is presented.

1. Introduction

Application of He-3 counters in neutron detection and spectrometry has been long proposed mainly due to high cross section of ^{3}He interaction with neutrons though for spectrometry applications several causes like recombination, space charge and wall effect make response function calculation a difficult task [1-6]. In this work the response of a proportional He-3 counter is studied experimentally and simulated with the Geant4 code [7]. The calculated response of the system is used to determine neutron energy distribution in an unknown neutron field.

2. Experimental technique

Irradiations were performed with nearly mono-energetic neutron beams in the energy range from 230 keV to 10.7 MeV provided by a Tandem, Van de Graff accelerator facility at the Institute of Nuclear Physics, NCSR Demokritos, Athens, Hellas. Neutrons with energies up to 3.3 MeV were obtained via the ^{7}Li(p,n)^{7}Be reaction, from 3.75 to 10.7 MeV were produced via ^{2}H(d,n)^{3}He reaction while higher energies up to 22 MeV via ^{3}H(d,n)^{4}He reaction. Details about the irradiations can be found elsewhere [8]. The description of the counter and the operational characteristics obtained during the above irradiations are presented in Table 1.

Irradiations were also performed in a calibrated neutron field, in the low-scatter measurement hall of the Physikalish-Technische Bundesanstalt (PTB, Braunschweig) irradiation facility, in order to determine the response function of the counter. Neutron beam energies were 0.5, 1.2 and 2.5 MeV while irradiations were carried out additionally with a shadow cone between the target and the counter to determine the influence of the room returned neutrons to the spectra acquired. Irradiation time was scheduled to attain sufficient statistics in the spectra, around 1000 counts in the centroid of the full energy peak. The count rate was kept low, around 10^3 cps to avoid space-charge effect and/or pulse pile-up events in the counter.

Table 1. Description and operational characteristics of the counter.

Pressure [atm]	Gas content	Cathode material / thickness	Anode material / diameter	Effective length [cm]	Effective diameter [cm]	Operating Voltage [V]	Shaping Time [μs]
6	^{3}He 64.7% Kr 33.3% CO_2 2.0%	Stainless Steel 304 / 0.089 cm	Tungsten / 0.025 mm	15	5	1700	8

3. Results and Discussion

3.1 Pulse height distributions

Pulse height distributions obtained during irradiations for 8 neutron beam energies are presented in figure 1. This complex response is made up of (i) The "Full Energy" peak, at the end of each spectrum, formed when total kinetic energy of the ^{3}He(n,p)T reaction products is deposited in the effective gas volume. The reaction is exothermic, with Q=764 keV. The neutron kinetic energy is added to the Q value of the reaction, and the total kinetic energy is distributed among the two reaction products according to their emission angles in the lab frame. For neutrons with energies above around 7 MeV, as it is calculated by the SRIM/TRIM code [11], the proton/triton range becomes larger than the dimensions of the effective volume of the counter. Therefore the full energy peak cannot be formed (e.g. 7.7 MeV in figure 1).

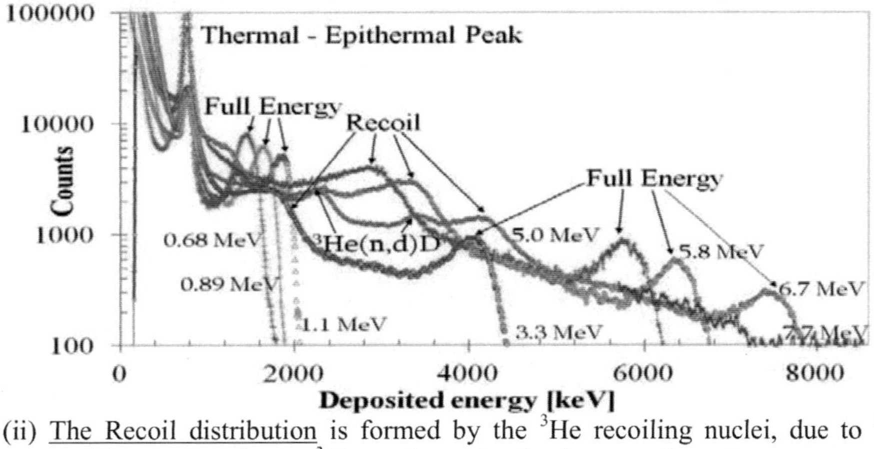

Figure 1. Pulse height distributions obtained with He-3 counter for several neutron beam energies. For the neutron beam energy of 7.7 MeV, only the last part of the spectrum is presented for clarity reasons.

(ii) The Recoil distribution is formed by the ^{3}He recoiling nuclei, due to ^{3}He(n,elastic) reaction. The energy E_R transferred to a ^{3}He nucleus via elastic scattering of a neutron of energy E_n is $E_R \approx 4E_n \cos^2\theta$ where θ is the scattering angle of the recoiling nucleus in the lab coordinate system [1]. As all scattering angles are allowed, in principle, the result in the spectrum is a continuum between a minimum of zero energy and a maximum of whenever the recoiling nucleus is emitted at 0° with respect to the neutron direction. This maximum energy is calculated to be 75% of the neutron energy for the ^{3}He nucleus and 29%, 22% and 5% for the C, O and Kr nuclei, respectively. It should be pointed out that although the cross-section of elastic scattering for fast neutrons is larger than the ^{3}He(n,p)T reaction, the result in the pulse height distribution is not so prominent as compared with the full energy peak, due to the distribution of the recoil events over a wide energy region of the spectrum (figure 1). (iii) Gamma-ray distribution. γ – rays interact mainly with the counter walls and Kr atoms producing electrons. Due to their long range, electrons can deposit a limited amount of energy (usually less than about 200-300 keV) in the counter, producing pulses with small amplitudes and long rise times. These pulses are recorded in the low-energy region of the pulse height distributions, presenting an exponentially decreasing function of which the exponent is a function of the γ −ray field intensity [5,8,9] (iv) Finally, in all spectra the Thermal-Epithermal peak is formed from scattered, low energy, neutrons. It should be noted here that as the counter was covered with 1 mm Cd foil, in the thermal-epithermal peak are registered neutrons with energies above about 1 eV up to several tenths of keV due to system resolution. (iv) Finally, for neutron energies above 4.3 MeV, the $^3He(n,d)D$ reaction contributes to the pulse height distributions, with Q=-3.27 MeV (figure 1).

3.2 Energy calibration and Resolution

The deposited energy of the $^3He(n,p)^3H$ reaction products which is registered in the full-energy peak area of the spectrum exhibits a linear relationship with the pulse height although a nonlinear behaviour is observed for recoil peak events as presented in figure 2(a). Moreover, when the recoil nuclei pulse height is converted to energy via energy calibration of the full energy peak, a significantly lower energy than the one expected from kinematics is calculated, mainly as a result of a recombination effect. The energy calibration equation for recoil events has been reported previously [10]. Resolution of the full energy peak is in the range of 4%-8% (figure 2(b)).

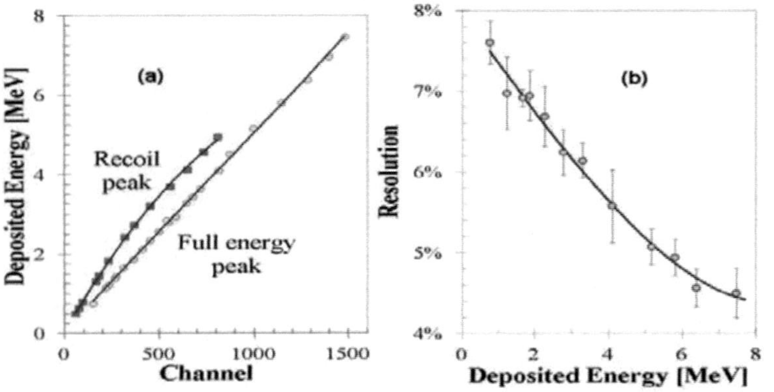

Figure 2. (a) Energy calibration of the system for the full energy and recoil peaks and (b) resolution as a function of neutron energy.

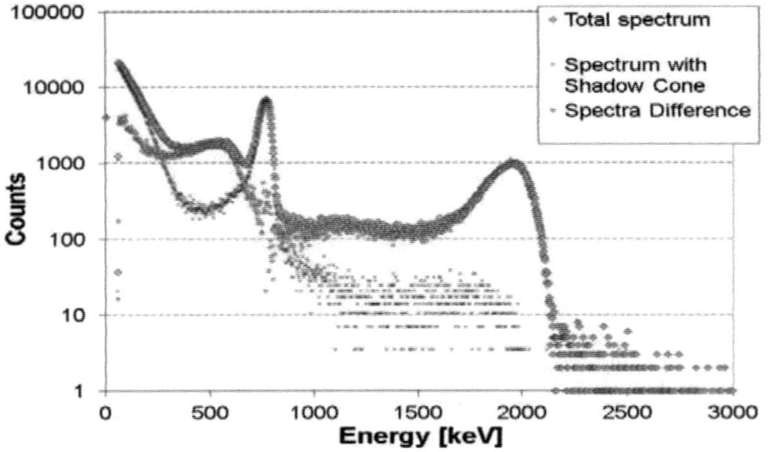

Figure 3. Pulse height distributions obtained during irradiations at PTB facility with 1.2 MeV neutrons. The original spectrum (Total) and the additional spectrum with a shadow cone between the target and the counter mainly due to room-returned neutrons. The difference is the calculated counter response from direct neutrons only.

3.3 Efficiency calibration measurements

The response of the system was measured for neutron energies: 0.5, 1.2 and 2.5 MeV. Irradiations were carried out at the low-scatter measurement hall of the PTB irradiation facility. In figure 3 the pulse height distributions obtained with and without the shadow cone, as well as the difference of the above two, are presented. Neutron fluencies were certified with uncertainties 4% to 8%. The efficiency of the system for the full energy peak is determined $1.64 \cdot 10^{-2}$ ±9.6%, $1.29 \cdot 10^{-2}$ ±5.2% and $7.2 \cdot 10^{-3}$ ±7.2% for neutron energies 0.5, 1.2 and 2.5 MeV respectively, for the irradiations conditions, thus when neutrons are crossing the face of the counter and are travelling parallel to the counter axis.

3.4 Simulation of counter response with GEANT4

Counter response was calculated using the Geant4 simulation kit (v.4.9). An accurate description of the counter regarding its geometrical characteristics, gas mixture and construction materials was provided by the manufacturer (LND Inc.). In order to take into account the displacement of the recoil events, namely the two different energy calibrations according to the origin of the events, during the calculation the particles that are found to deposit energy in the effective volume of the counter are identified and their deposited energy is scored to the corresponding calculated spectrum according to their identity. An example of the different calculated spectra coming from (n,p), (n,elastic) or other reactions with the above procedure is presented in figure 4. The spectrum originating from recoil events is first treated with the correction function due to pulse height reduction (figure 4). Subsequently, each spectrum is folded with a Gaussian distribution according to the experimentally determined resolution function. It should be noted here that the resolution measured at recoil edges is about 3 times larger than that of the full-energy peaks. Finally, the spectra are added together with an exponentially decreasing function to account for gamma ray contribution in the lower part of the pulse height distribution. The parameters of this function are calculated from the counts registered in the first channels of the experimental spectra [8]. Comparison of the experimental spectra with the simulated ones is presented in figure 5.

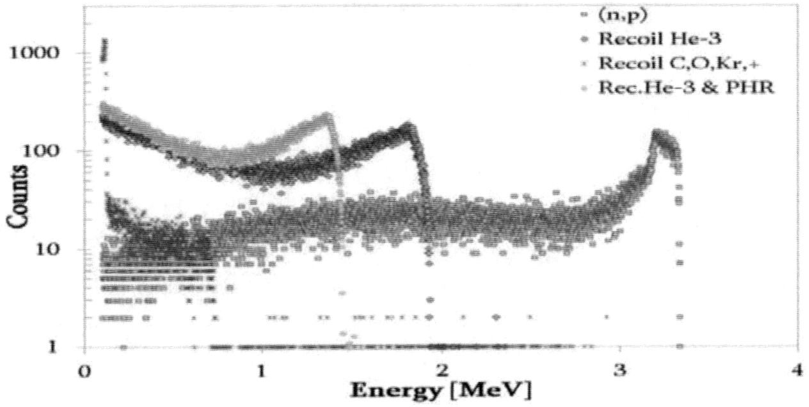

Figure 4. Calculated spectra originating from (n,p), (n,elastic) and other reactions for incoming neutron beam energy of 2.5 MeV. With dots the He-3 recoil spectrum after the application of the Pulse Height Reduction correction function.

3.5 Neutron spectrum unfolding.

The calculated response of the system can be used to unfold a spectrum in an unknown neutron filed. An example is provided below. Details about the neutron source and the irradiation are described in reference [12].Neutron energy range from 0.1 to 7 MeV is divided to several isolethargic bins. Counter response is calculated for each bin. Multiple regression analysis is performed between the experimental pulse height distribution data (independent variables) and the calculated counter response for each bin (dependent variables). Comparison of the experimental pulse height distribution with the simulated one using the determined neutron fluencies is presented in figure 6(a). The determined neutron fluencies are found in acceptable agreement with the expected ones obtained from the simulation with DCM-DEM code, figure 6(b) [13,14].

4. Conclusions

The response of a proportional He-3 counter was studied experimentally and the observed performance was implemented to the simulated response of the counter with Geant4. A satisfactory agreement between the calculated spectra and the experimental pulse height distributions obtained in calibrated neutron fields is observed. The calculated response of the counter was used to unfold spectra from an unknown neutron filed. The calculated neutron fluence distribution is in acceptable agreement with the expected from simulation for that field.

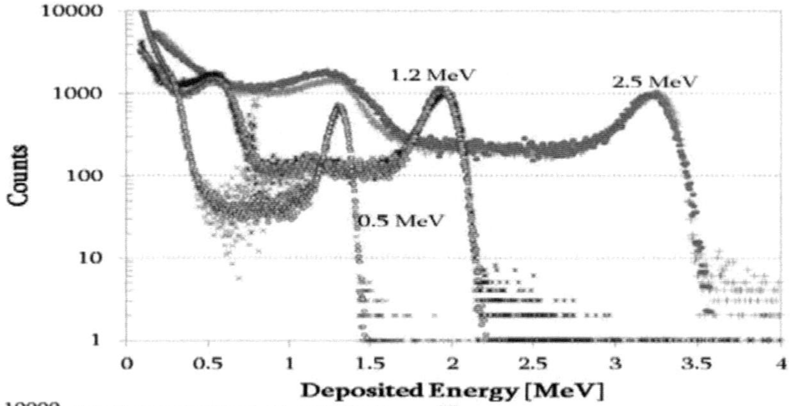

Figure 5. Experimental pulse height distributions (cross) and simulated spectra (dots) with GEANT4 for neutron beam energies 565 keV, 1.2 MeV and 2.5 MeV.

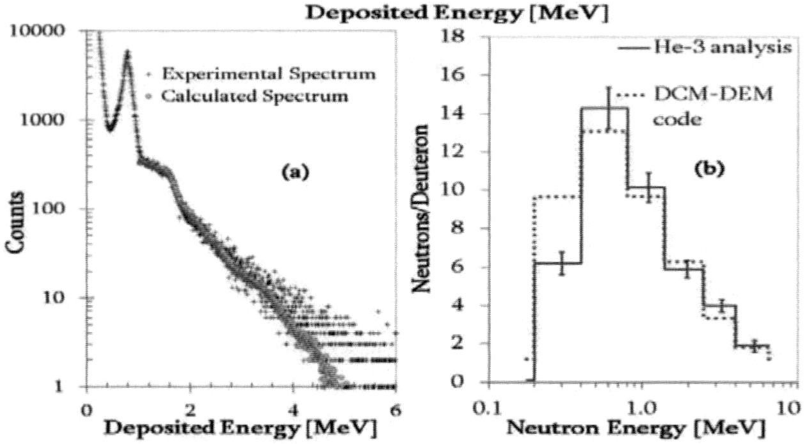

Figure 6. (a) Experimental (cross) and simulated with Geant4 (dots) pulse height distributions. (b) Energy distribution of neutron yield as it is figured out from the analysis of He-3 counter spectra and comparison with the calculated one by DCM-DEM code.

5. References

[1] Knoll G F 1999 *Radiation Detection and Measurement* (New York: John Wiley & Sons)
[2] Nakazawa M, Iguchi T, Takahashi H and Takada E 1998 *Progr. In Nuclear Energy* **32** 25
[3] Iguchi T, Nakayamada N, Takahashi H and Nakazawa M 1994 *Nucl. Instr. Meth.* A **353** 152
[4] Loughlin M J, Adams J M and Sadler G 1990 *Nucl . Instr. Meth.* A **294** 606
[5] Beimer KH, Nyman G and Tengblad O 1986 *Nucl. Instr. Meth.* A **245** 402.
[6] Matysiak W, Chettle D R, Prestwich W V and Byun S H 2011 *Nucl. Instr. Meth.* A **629** 133
[7] Aggostinelli S et al. 2003 *Nucl. Instr. Meth.* A **506** 250.
[8] Manolopoulou M et al. 2006 *Nucl. Instr. Meth.* A **562** 371
[9] Beddingfield D H, Johnson N H and Menlove H O 2000 *Nucl. Inst. Meth.* A **455** 670
[10] Manolopoulou M et al. 2010 *Nucl. Instr. Meth.* A **618** 284
[11] Ziegler J F Biersack JP and Littmark U *SRIM/TRIM The Stopping and Range of Ions in Matter*
[12] Zamani M, Fragopoulou M, Manolopoulou M, Stoulos S, Jokic S, Sosnin A N and Krivopustov MI 2010 *Annals of Nuclear Energy* **37** 241
[13] Barashenkov V S 2000 *Comput. Phys. Commun.* **126** 28
[14] Sosnin A N et al. 2002 *Izvestiya RAS, Phys. Ser.* **66** 1494

Journal of Physics: Conference Series **366** (2012) 012034 doi:10.1088/1742-6596/366/1/012034

Projectile Coulomb Excitation of the nucleus ^{194}Pt

T Möller[1], C Bauer[1], R V F Janssens[2], C J Lister[2], E A McCutchan[2], N Pietralla[1], G Rainovski[1,3], D Seweryniak[2], C Stahl[1] and S Zhu[2]

[1] Institute for Nuclear Physics, Darmstadt Technical University, Germany
[2] Physics Division, Argonne National Laboratory, USA
[3] Faculty of Physics, St. Kliment Ohridski University of Sofia, Bulgaria

Abstract.
Low-lying collective excited states of ^{194}Pt have been studied via the ^{12}C(^{194}Pt,^{194}Pt*) projectile Coulomb excitation reaction at 85% of the Coulomb barrier (850 MeV) using the Gammasphere Ge-detector array at Argonne National Laboratory. Absolute $E2$ transition strengths have been obtained from the Coulex cross sections that were deduced from the relative γ-ray yields. They are discussed with respect to the structure suggested by the O(6) symmetry of the Interacting Boson Model.

In nuclear physics, the three dynamical symmetries [1; 2] of the Interacting Boson Model (IBM), U(5) [3], SU(3) [4], and O(6) [5] provide valuable benchmarks for the description of nuclear quadrupole collectivity at low and medium angular momenta. These three symmetries correspond to analytically solvable cases of the Bohr Hamiltonian [6] – the harmonic vibrator, the quadrupole-deformed axial rotor, and the γ-unstable rotor [7].

The O(6) symmetry of the sd-IBM-1 is based on the chain $U(6) \subset O(6) \subset O(5) \subset O(3)$ of nested sub-algebras with quantum numbers N, σ, τ, and L, respectively [1; 2]. The empirical evidence for the existence of nuclei at the O(6) dynamical limit of the IBM is based on energy level patterns, branching ratios and, more convincingly, on selection rules for $E2$ transitions. Within the Consistent Q-Formalism (CQF) [8], they are such that $E2$ transitions are allowed and collective only between states with $\Delta\sigma = 0$ and $\Delta\tau = \pm 1$ [2]. It is the $\Delta\sigma = 0$ selection rule that is definitive of pure O(6) symmetry while the $\Delta\tau = \pm 1$ selection rule actually stems from the O(5) symmetry and is therefore rather ubiquitous for all nuclei between U(5) and O(6) dynamical symmetries. Observation of O(6) symmetry in nuclei has first been reported in the case of ^{196}Pt [9]. This claim was based on energy level patterns and $E2$ decay branching ratios that closely follow the O(6) selection rules. It was, later on, supported by establishing a lower limit for the lifetime of the 0^+_3 state, the lowest state of the $\sigma = N - 2$ representation [10]; the resulting upper limits for the absolute $B(E2)$ values were small, in agreement with the pure O(6) dynamical symmetry [10]. Another, even more extensive region of O(6)-candidate nuclei was found in the Xe-Ba-Ce region [11] around mass number $A = 130$. It has been shown that the low-spin structures of the nuclei ^{128}Xe [12], ^{126}Xe [13] and ^{124}Xe [14] manifest O(6)-like arrangements of energy levels and $E2$ branching ratios which reflect the selection rules for the $\sigma = N$ states of O(6). Lately, the results of new Coulex experiments with the nuclei ^{124,126,128}Xe [15; 16] showed moderate collective character in their respective $0^+_{\sigma=N-2} \to 2^+_{1,2}$ transitions, which can be regarded as a severe violation of the $\Delta\sigma = 0$ selection rule. Quantitative analyses showed a complete dissolution of the O(6) symmetry in these nuclei, while the O(5) symmetry is preserved

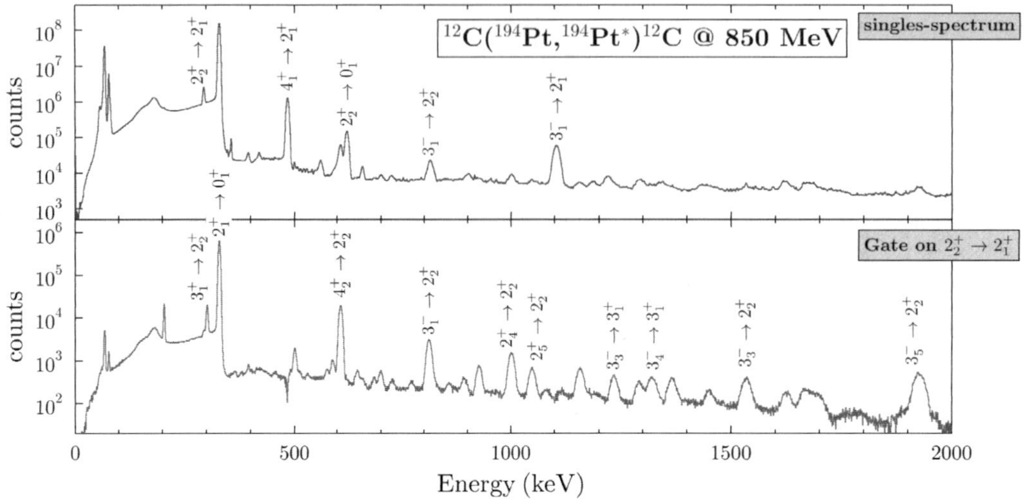

Figure 1. Background subtracted, Doppler-corrected γ-ray spectra of ^{194}Pt after Coulomb excitation on a Carbon target observed with the Gammasphere spectrometer. In the top panel, the sum singles spectrum is shown. In the bottom panel, a spectrum of the γ-rays coincident to the $2_2^+ \to 2_1^+$ transition is shown. The strongest transitions are indicated.

to a large extent. The amount of symmetry breaking has been quantified by a newly developed analysis of wave-funtion amplitudes.

In the light of the recent findings it is intriguing to investigate to what extent the O(6) symmetry is preserved or broken in the Pt isotopes. In this work, we report on an experiment of projectile Coulomb excitation of ^{194,196}Pt nuclei, performed in order to study the preservation of the selection rules characteristic for the O(6) symmetry on the basis of the measurement of absolute transition strengths. In the following we will concentrate on the nucleus ^{194}Pt.

The experiment was carried out at Argonne National Laboratory. A pulsed (12 MHz) ^{194}Pt beam was delivered by the ATLAS accelerator. The 850 MeV beam was incident on a 1 mg/cm^2 ^{12}C target. The deexcitation γ-rays, following the Coulomb excitation of the projectile, were detected with the Gammasphere array [17] which consisted of 100 HPGe detectors arranged in 16 rings. In a first run the beam intensity was ~1 pnA. Gammasphere was used in singles mode resulting in an average counting rate of $1.5 \cdot 10^4$ counts-per-second (cps). A total of $1.6 \cdot 10^9$ events of γ-ray fold 1 or higher was collected in about 28 hours. In a second run, the trigger of the data aquisition was set to a gamma-ray fold of 2 or higher, allowing to increase the beam intensity to ~5 pnA. The average count rate was $\sim 6 \cdot 10^3$ counts-per-second. In this run, γ-ray coincidence data of $2.3 \cdot 10^8$ events was collected in about 12 hours of beam time. The data from the first run was used to create singles spectra which have been Doppler corrected with respect to the velocity of the recoiling projectile ions of $v/c \approx 8.1\%$. The contribution of the room background was eliminated in the offline sort by correlating the γ-rays with the accelerator radio-frequency (rf) signal. The final spectrum, which is a difference between the 'beam-on' (with respect to the rf) spectrum and the 'beam-off' spectrum, scaled to eliminate the 1461 keV room background transition from ^{40}K, is shown in the top panel of Figure 1.

The coincidence data from the second run was used to create a $\gamma\gamma$-coincidence matrix. As an example a spectrum of γ-rays coincident to the transtion $2_2^+ \to 2_1^+$ is shown in the lower panel of Figure 1. A total of 67 transitions could be observed in the coincidence spectra. Due to this large number of ovserved transitions, only the coincidence data allowed the reconstruction of the level scheme from the observed γ-rays. A part of the deduced level scheme is shown in

Journal of Physics: Conference Series **366** (2012) 012034 doi:10.1088/1742-6596/366/1/012034

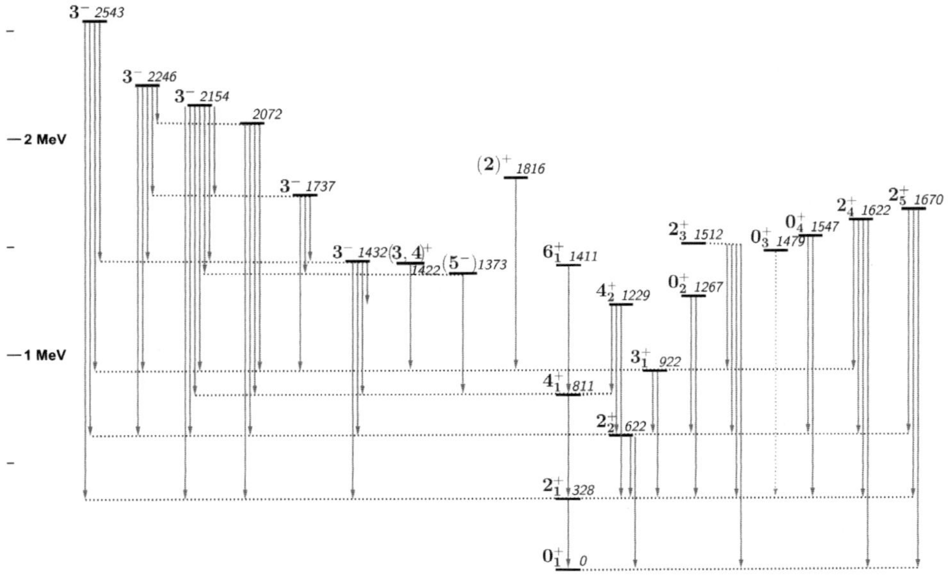

Figure 2. Part of the deduced level scheme of ^{194}Pt together with the placements of some of the γ-ray transitions observed in our experiment.

Figure 2.

All observed γ rays originate from ^{194}Pt nuclei. Many of these γ rays have already been identified in ^{194}Pt [18]. Some transitions, however, have not been reported so far, e.g. the transitions depopulating a state at an excitation energy of 2072 keV and the transitions depopulating the 3^- states at excitation energies of 2145 keV, 2246 keV and 2543 keV. These states have previously been observed in (p,p') reactions only [19; 20].

Efficiency-corrected transition intensities were determined in the singles spectrum where possible. For most of the transitions, the intensity had to be determined from the coincidence data, using transitions visible both in the singles spectrum and in the coincidence data for normalization. The transition intensities were then used to calculate Coulomb-excitation yields, normalized to the yield of the state 2_1^+ at 328 keV. These relative Coulex yields are proportional to the relative Coulex cross sections.

At this stage of our analysis, the relative Coulomb excitation yields of 11 states have been fitted to the Winther-de Boer theory [21] using a multiple CE code [22] and taking into account the energy loss of the beam in the target which can be estimated to about 142 MeV from a Bethe-Bloch approach. The matrix elements involved in the calculations were constrained to the experimental branching ratios and multipole mixing ratios. Adopted quadrupole moments [18] for the states 2_1^+, 2_2^+ and 4_1^+ were included into the calculation. The quadrupole moments of the remaining states have been varied within the values of the rotational limit, contributing on average by $\sim 4\%$ to the uncertainties of the results. An additional degree of freedom in the Coulex calculation is introduced by the choice of the signs of transition matrix elements. In the case of states with more than one possible excitation path, different combinations of the signs for the transition matrix elements involved in the excitation lead to an increase in the uncertainty of additional $10-20\%$ or to ambiguous results beyond their respective uncertainties, like in the case of the state 0_2^+ at 1267 keV. The scale for the calculation is set by the strength of the transition of the state 2_1^+ to the ground state of $B(E2; 2_1^+ \rightarrow 0_1^+) = 49.2(8)$ W.u. [18].

The calculations resulted in 23 absolute $E2$ transition strengths. The strengths of the

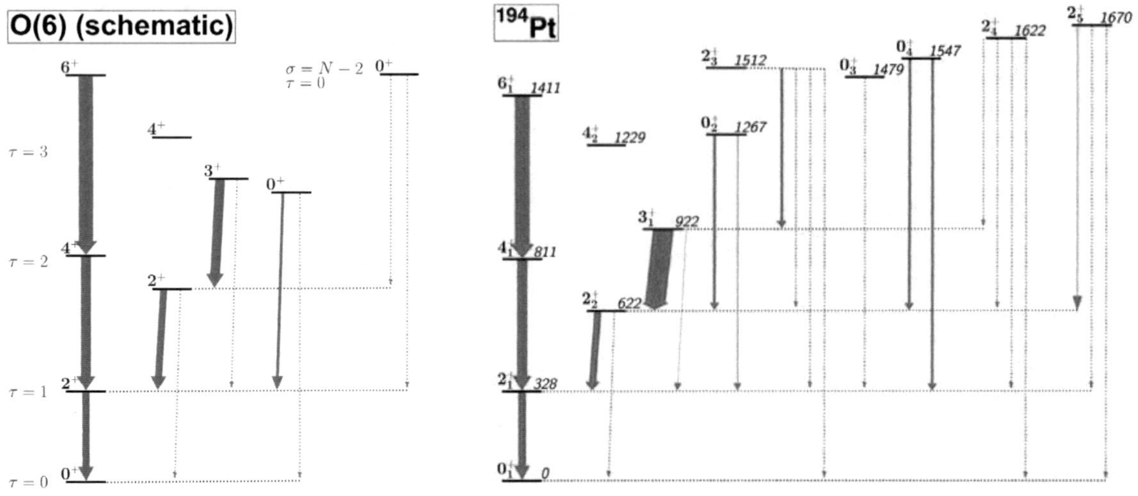

Figure 3. Part of the level scheme, for which transition strengths have been deduced compared to a schematic O(6) level scheme. The widths of the arrows indicating $E2$ decay transitions are proportional to their $B(E2)$ values deduced from our experiment. For $B(E2) < 1\,\mathrm{W.u.}$ we use dashed lines. Decay transitions of the bandhead of the $\sigma = N-2$ group to the $2^+_{1,2}$ states are forbidden in the O(6) limit. In ^{194}Pt, a clear assignment of the $\sigma = N-2, \tau = 0$ state is difficult due to the existence of two close-lying 0^+ states in the energy range in question.

transitions depopulating the state 3^+_1 at 922 keV, and the $2^+_{3,4,5}$-states at 1512 keV, 1622 keV and 1670 keV, respectively, could be deduced for the first time. The resulting strengths of the transitions depopulating the 0^+_2 state at 1267 keV are very sensitive to the choice of the signs of the transition matrix elements involved in the excitation process. Two different possible solutions have been found. At an excitation energy of 1479 keV a 0^+ state is reported, e.g. from beta decay studies [23; 24]. Transitions depopulating this state 0^+_3 could not be observed directly in our data. An estimate of an upper limit for the intensity of the transition to the state 2^+_1 was made from which a small, non-collective upper limit for the value of $B(E2; 0^+_3 \to 2^+_1)$ could be deduced.

In terms of the selection rules of the O(6) symmetry, a clear conclusion cannot be drawn. The $\Delta\tau = \pm 1$ selection rule is satisfied for transitions depopulating the states 2^+_2, 3^+_1, 0^+_2 and 2^+_3 if the τ quantum numbers are assigned to these states according to the typical O(6) scheme [5]. The more difficult question is whether the $\Delta\sigma = 0$ selection rule is preserved. To address this question, one first has to identify states belonging to a representation with different σ quantum number than the yrast states. The most promising candidate to study is the 0^+ bandhead of the $\sigma = N-2$ family (cf [15]), which will only be populated strongly in the case of a broken O(6) symmetry. In the neighbouring nucleus ^{196}Pt, showing a very similar level scheme to ^{194}Pt, this has been identified as the 0^+ state at an excitation energy of 1403 keV. Considering the level scheme of ^{194}Pt in a similar energy range, one finds two close-lying 0^+ states at energies of 1479 keV and 1547 keV [18]. Under the assumption that the state 0^+_3 can be identified as the $\sigma = N-2, \tau = 0$ state, the deduced non-collective strength of $B(E2; 0^+_3 \to 2^+_1)$ would imply a preservation of the $\Delta\sigma = 0$ selection rule. In that case, however, the existence of a close-lying 0^+_4 state with collective $E2$ transitions to the $2^+_{1,2}$ states cannot be accomodated in the O(6) symmetry scheme. If, on the other hand, one assumes the 0^+_4 state to be identified as the state in question, then the collectivity of its decay transitions to the $2^+_{1,2}$ states would imply a severe breaking of O(6) selection rules, similar to the situation observed in the Xe isotopes [15; 16]. In either case, the existence of two 0^+ states lying that close in energy cannot be desribed easily using an IBM-1 model close to the O(6) limit. In fact, their very different transition strengths show, that these two states do not mix and hence have a different character. So the question of the nature of the two $0^+_{3,4}$ states in ^{194}Pt remains to be solved.

In summary, a Coulex experiment on the nucleus ^{194}Pt has been performed. In the analysis new information on transition strengths in this nucleus could be deduced. An interpretation of the results against the background of the O(6) selection rules is difficult due to the unresolved nature of the two close-lying 0^+ states around 1.5 MeV.

Acknowledgements
This work has been funded by the DFG under grant Pi 393/2-2, by the DAAD, and by the Alexander von Humboldt foundation

References

[1] Arima A and Iachello F 1975 *Phys. Rev. Lett.* **35** 1069–1072
[2] Iachello F and Arima A 1987 *The Interacting Boson Model* (Cambridge University Press)
[3] Arima A and Iachello F 1976 *Annals of Physics* **99** 253 – 317
[4] Arima A and Iachello F 1978 *Annals of Physics* **111** 201 – 238
[5] Arima A and Iachello F 1978 *Phys. Rev. Lett.* **40** 385–387
[6] Bohr A 1952 *Mat. Fys. Medd. K. Dan. Vidensk. Selsk.* **26** 14
[7] Wilets L and Jean M 1956 *Phys. Rev.* **102** 788–796
[8] Warner D D and Casten R F 1982 *Phys. Rev. Lett.* **48**(20) 1385–1389
[9] Cizewski J A *et al.* 1978 *Phys. Rev. Lett.* **40**(3) 167–170
[10] Börner H G *et al.* 1990 *Phys. Rev. C* **42**(6) R2271–R2274
[11] Casten R and Brentano P V 1985 *Physics Letters B* **152** 22 – 28 ISSN 0370-2693
[12] Neuneyer U *et al.* 1996 *Nuclear Physics A* **607** 299 – 326 ISSN 0375-9474
[13] Gade A *et al.* 2000 *Nuclear Physics A* **665** 268 – 284 ISSN 0375-9474
[14] Werner V *et al.* 2001 *Nuclear Physics A* **692** 451 – 475
[15] Rainovski G *et al.* 2010 *Physics Letters B* **683** 11 – 16
[16] Coquard L *et al.* 2011 *Phys. Rev. C* **83**(4) 044318
[17] Lee I Y 1990 *Nucl. Phys. A* **520** 641c
[18] Balraj and Singh 2006 *Nuclear Data Sheets* **107** 1531 – 1746
[19] Deason P T *et al.* 1981 *Phys. Rev. C* **23**(4) 1414–1433
[20] Cottle P D, Stuckey K A and Kemper K W 1988 *Phys. Rev. C* **38**(6) 2843–2848
[21] Winther A and de Boer J 1966 *A computer program for multiple coulomb excitation* (Academic Press Inc., New York) p 303
[22] Ower H 1980 *PhD Thesis* Ph.D. thesis Johann Wolfgang zu Goethe Universität zu Frankfurt am Main
[23] Butt D and Lincoln T 1968 *Nuclear Physics A* **121** 673 – 678
[24] Teixeira R R P *et al.* 1997 *Phys. Rev. C* **56**(4) 2320–2323

Viscosity of hot nuclei

Nguyen Dinh Dang
Theoretical Nuclear Physics Laboratory, RIKEN Nishina Center for Accelerator-Based Science, 2-1 Hirosawa, Wako City, 351-0198 Saitama, Japan
and Institute for Nuclear Science and Technique, Hanoi, Vietnam

E-mail: dang@riken.jp

Abstract. The ratio η/s of shear viscosity to entropy density is extracted from the experimental systematics of the giant dipole resonances in copper, tin, and lead regions at finite temperature T. These empirical results are then compared with the predictions by several independent models as well as with almost model-independent estimations. Based on these results, it is concluded that the ratio η/s in medium and heavy nuclei decreases with increasing temperature T to reach $(1.3-4)\times\hbar/(4\pi k_B)$ at $T = 5$ MeV.

1. Introduction

The recent experimental data from the Relativistic Heavy Ion Collider (RHIC) [1] at the Brookhaven National Laboratory and the Large Hadron Collider (LHC) [2] at CERN have revealed that the matter formed in ultrarelativistic heavy-ion collisions is a nearly perfect fluid with extremely low viscosity. This has driven the attention to the calculation of the ratio η/s of shear viscosity η to entropy density s. Using the string theory, Kovtun, Son and Starinets conjectured a universal value $\hbar/(4\pi k_B)$ for this ratio as the lowest bound for all fluids (the KSS bound) [3]. No fluid that violates this lower bound has ever been found experimentally [4]. For finite nuclei, the recent calculations by Auerbach and Shlomo estimated the values of η/s within $(4-19)$ and $(2.5-12.5)$ times the KSS bound for heavy and light nuclei, respectively [5]. These authors used the Fermi liquid drop model (FLDM) [6], applied to the damping of giant collective vibrations. The obtained shear viscosity η increases with temperature T up to almost $T\sim 10$ MeV, then slows down to reach a maximum at $T\sim 13$ MeV. Consequently, within the region $0\leq T\leq 5$ MeV, where giant resonances exist, the damping width predicted by the FLDM increases with T almost quadratically, which is roughly proportional to η [6, 7]. However, the experimental systematics of the width of the giant dipole resonance (GDR) in heavy nuclei formed after heavy-ion fusions and/or inelastic scattering of light projectiles on heavy targets have shown that the GDR width increases with T only within $1\leq T\leq 2.5$ MeV. Below $T\sim 1$ MeV, it remains nearly constant, whereas at $T>3$ - 4 MeV the width seems to saturate [8, 9, 10, 11, 12, 13, 14]. The entropy in Ref. [5] has been calculated by using a linear temperature dependence in the Fermi gas formula $S = 2aT$ with a temperature-independent level density parameter a. This approximation too is rather poor for finite nuclei. Therefore, although by dividing two quantities that increase with T, the obtained result in Ref. [5] for the ratio η/s does decrease qualitatively to a value within 1 order of the KSS bound, a refined quantitative prediction for this ratio still remains a challenge. The present work proposes to extract the shear viscosity η, the entropy density s, and the ratio η/s directly from the most

Published under licence by IOP Publishing Ltd

Journal of Physics: Conference Series **366** (2012) 012035 doi:10.1088/1742-6596/366/1/012035

recent and accurate experimental systematics of the GDR widths in hot nuclei. The extracted empirical values are then confronted with theoretical predictions by four models, which have been developed to describe the temperature dependence of the GDR width, namely, the phonon-damping model (PDM) [15, 16, 17], two thermal-shape fluctuation models (TSFM), as well as the FLDM mentioned above. Finally, the high-T limit of η/s is deduced.

2. Formalism

The shear viscosity η is expressed in terms of the correlation function of the shear stress tensors $T_{xy}(t, \mathbf{x})$ through the Green - Kubo formula [18]. By using the fluctuation-dissipation theorem (FDT), which relates the correlation function to the absorption cross section $\sigma(\omega)$ [19, 20], the shear viscosity $\eta(T)$ can be calculated from the latter as

$$\eta(T) = \lim_{\omega\to 0}\frac{1}{2\omega}\int dtd\mathbf{x}e^{i\omega t}\langle[T_{xy}(t,\mathbf{x}), T_{xy}(0,0)]\rangle = \frac{\sigma(\omega=0,T)}{C}\ , \qquad (1)$$

where C is a normalization constant. The photoabsorption cross section of the GDR is described by the Breit-Wigner distribution as [21] $\sigma_{GDR}(\omega) = \sigma_{int}(GDR)\Gamma/\{2\pi[(\omega - E_{GDR})^2 + (\Gamma/2)^2]\}$, with the integrated cross section $\sigma_{int}(GDR) \simeq (1 + k)S$ ($k \simeq 0.5 - 0.7$), where $S = 60NZ/A$ (MeV mb) is the Thomas-Reiche-Kuhn sum rule, Γ is the GDR full width at half maximum (FWHM), and E_{GDR} is the energy location of the GDR peak (GDR energy). From Eqs. (1) and the expression of $\sigma_{GDR}(\omega)$, one finds the shear viscosity at temperature T as

$$\eta(T) = \frac{\sigma_{int}(GDR)}{\pi C}\frac{\Gamma(T)/2}{E_{GDR}(T)^2 + [\Gamma(T)/2]^2}\ . \qquad (2)$$

which can be applied to any transport process with measurable resonance scattering cross section. By determining the normalization constant C to reproduce the value $\eta(0)$ at $T = 0$, Eq. (2) is rewritten as

$$\eta(T) = \eta(0)\frac{\Gamma(T)}{\Gamma(0)}\frac{E_{GDR}(0)^2 + [\Gamma(0)/2]^2}{E_{GDR}(T)^2 + [\Gamma(T)/2]^2}\ . \qquad (3)$$

The FLDM predicted the value $\eta(0)$ in the range of $(0.5 - 2.5)u$ with $u = 10^{-23}$ Mev s fm^{-3} [5]. Fitting data of giant resonances at $T = 0$ [7] found $\eta(0) = 1.0u$, whereas by fitting nuclear fission [22], one found $0.6u \leq \eta(0) \leq 1.2u$. The present work adopts $\eta(0) = 1.0^{+0.2}_{-0.4}u$ in agreement with these values.

The entropy density is calculated as $s = \rho S/A$ with the nuclear density $\rho = 0.16$ fm^{-3}, and $S = S_F + S_B$, where S_F and S_B are the entropies of the quasiparticle and phonon fields in the PDM Hamiltonian, respectively [See Eq. (1) of Ref. [15]]. The entropy S_α ($\alpha = F, B$) is given in units of Boltzmann constant k_B as

$$S_\alpha = -\sum_j N_j[p_j \ln p_j \pm (1 \mp p_j)\ln(1 \mp p_j)]\ , \qquad (4)$$

where $p_j = n_j$ are the quasiparticle occupation numbers ($\alpha = F$) or phonon occupation numbers $p_j = \nu_j$ ($\alpha = B$), the upper (lower) sign is for quasiparticles (phonons), $N_j = 2j + 1$ and 1 for $\alpha = F$ and B, respectively. For $\alpha = F$, the index j denotes the single-particle energy level, corresponding to the orbital angular momentum j, whereas for $\alpha = B$, it corresponds to that of GDR phonon. In general, n_j has the shape of a Fermi-Dirac distribution, $n_j^{FD} = [\exp(E_j/T) + 1]^{-1}$, smoothed with a Breit-Wigner kernel, whose width is equal to the quasiparticle damping with the quasiparticle energy $E_j = \sqrt{(\epsilon_j - \lambda)^2 + \Delta(T)^2}$ [See Eq. (2) of [17]]. Here ϵ_j, λ, and $\Delta(T)$ are the (neutron or proton) single-particle energy, chemical potential, and pairing gap, respectively. However, because the quasiparticle (single-particle) damping is negligible for heavy nuclei [16], it is neglected in the present calculations of entropy S_F for the sake of simplicity, assuming $n_j = n_j^{FD}$. Regarding the GDR phonon occupation numbers, for $E_{GDR} \gg T$, it is well approximated with the Bose-Einstein distribution as $\nu_{GDR} = [\exp(E_{GDR}/T) - 1]^{-1}$.

For the theoretical description of the GDR width $\Gamma(T)$, the prediction by the PDM is adopted, which employs the single-particle energies obtained within the Woods-Saxon potentials for ^{63}Cu, ^{120}Sn and ^{208}Pb nuclei. In the presence of strong thermal fluctuations, the pairing gap $\Delta(T)$ of a finite nucleus does not collapse at the critical temperature T_c, corresponding to the superfluid-normal phase transition predicted by the BCS theory for infinite systems, but decreases monotonically as T increases [23, 24, 25]. The PDM employs E_j and $\Delta(T)$ by solving the modified BCS equations that include thermal fluctuations of quasiparticle numbers [24]. The GDR width $\Gamma(T)$ is found as the sum of the quantal and thermal widths. The former is caused by coupling of the GDR vibration (phonon) to noncollective particle-hole (ph) configurations with the factors $(1 - n_p - n_h)$, whereas the latter arises due to coupling of the GDR phonon to pp and hh configurations including the factors $(n_s - n_{s'})$ with $(s, s') = (h, h')$ or (p, p'). As the result, the quantal width decreases slightly, whereas the thermal width increases sharply with increasing T and saturates at $T \geq 4$ - 5 MeV in tin and lead isotopes [15]. Thermal pairing gap $\Delta(T)$ causes the GDR width in ^{120}Sn to remain nearly constant or even slightly decrease at $T \leq$ 1 MeV [17] in agreement with the data of Ref. [12]. For ^{63}Cu, the effect of pairing on the GDR width is small, so it is not included in the calculations of GDR width within the PDM, but the finite temperature BCS pairing with blocking by the odd proton is taken into account for the entropy to ensure its vanishing value at low T.

3. Analysis of numerical results

Shown in Fig. 1 are the GDR width, entropy S, shear viscosity η, and ratio η/s as functions of T for copper, tin, and lead regions as predicted by the PDM, the adiabatic thermal shape fluctuation model (AM) [26], the phenomenological thermal shape fluctuation model (pTSFM) [27], and the FLDM, as well as the corresponding results of the empirical extraction from the GDR experimental systematics.

The PDM predictions for the GDR width best fit the experimental systematics for all three nuclei ^{63}Cu, ^{120}Sn, and ^{208}Pb. The AM fails to describe the GDR width at low T for ^{120}Sn because thermal pairing was not included in the AM calculations, while it slightly overestimates the width for ^{208}Pb. (The AM prediction for GDR width in ^{63}Cu is not available.) The predictions by the pTSFM are qualitatively similar to those by the AM, although to achieve this agreement, the pTSFM needs to use $\Gamma(0) = 5$ MeV for ^{63}Cu and 3.8 MeV for ^{120}Sn, i.e., substantially smaller than the experimental values of around 7 and 4.9 MeV, respectively. The widths obtained within the FLDM fit the data fairly well up to $T \simeq 2.5$ MeV. However, they do not saturate at high T but increase sharply with T and break down at $T_c < 4$ MeV.

As for the entropies, the good agreement between the results of microscopic calculations and the empirical extraction indicates that the level-density parameter for ^{63}Cu, within the temperature interval $0.7 < T < 2.5$ MeV, can be considered to be temperature-independent and equal to $a = 63/8.8 \simeq 7.16$ MeV^{-1}, whereas, for ^{120}Sn and ^{208}Pb, the level-density parameter varies significantly with T [10]. The Fermi-gas entropy with a constant level-density parameter a best fits the microscopic and empirical results with $A/a = 8.8$ MeV for ^{63}Cu, and 11 MeV for ^{120}Sn and ^{208}Pb.

The predictions of η by the PDM have the best overall agreement with the empirical results for all three nuclei ^{63}Cu, ^{120}Sn, and ^{208}Pb. The PDM produces an increase in $\eta(T)$ with T up to 33.5 MeV and a saturation in $\eta(T)$ within (2 - 3)u at higher T. The ratio η/s decreases sharply with increasing T up to $T \sim 1.5$ MeV, starting from which the decrease gradually slows down

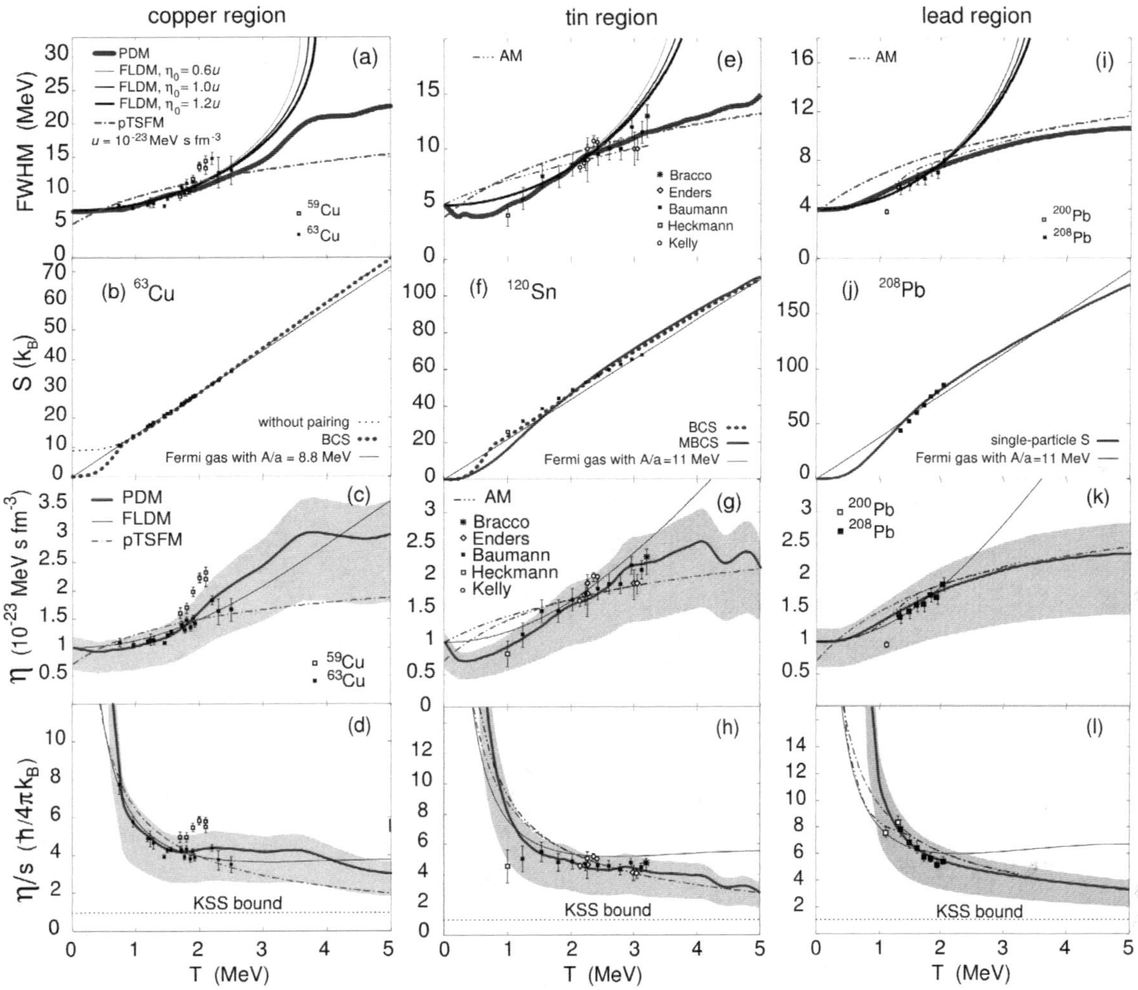

Figure 1. (Color online) Shear viscosity η (a), entropy S (b), and the ratio η/s (c) as functions of T for tin region. The empirical values are extracted from the experimental systematics for for copper (Cu^{59} and Cu^{63} [8]), tin (by Bracco *et al.* [9], Enders *et al.* [10], Baumann *et al.* [11], Heckmann *et al.* [12], and Kelly *et al.* [13]), and lead (Pb^{208} [11] and Pb^{200} [14]) regions.

to reach (2 - 3) KSS units at $T = 5$ MeV. The FLDM has a similar trend as that of the PDM up to $T \sim 2$ - 3 MeV, but at higher T ($T > 3$ MeV for ^{120}Sn or 2 MeV for ^{208}Pb), it produces an increase in both η and η/s with T . At $T = 5$ MeV, the FLDM model predicts the ratio η/s within (3.7 - 6.5) KSS units, which are roughly 1.5 timestwice larger than the PDM predictions. The AM and pTSFM show a similar trend to that predicted by the PDM. However, to obtain such similarity, $\eta(0)$ in the pTSFM calculations has to be reduced to $0.72u$ instead of $1u$. They all overestimate η at $T < 1.5$ MeV.

A model-independent estimation for the high-T limit of η can be obtained under the assumption of GDR width saturation, $\Gamma_{max} \simeq 3\Gamma(0) \simeq 0.9E_{GDR}$ at $T \simeq 5$ - 6 MeV. By using this assumption, one finds $\eta_{max} \simeq 2.551 \times \eta(0)$ from Eq. (3). The high-T limit of the entropy density s is obtained by noticing that $S_F \to 2\Omega\ln 2$ at $T \to \infty$ because $n_j \to 1/2$ where $\Omega = \sum_j (j + 1/2)$ for the spherical single-particle basis or sum of all doubly degenerate levels for the deformed basis. The particle-number conservation requires that $A = \Omega$ since all single-particle occupation

Journal of Physics: Conference Series **366** (2012) 012035 doi:10.1088/1742-6596/366/1/012035

numbers are equal to 1/2. This leads to the high-T limit of entropy density s: $s_{max} = 2\rho \ln 2 \simeq 0.222$ (k_B). Dividing η_{max} by s_{max} yields the high-T limit (or lowest bound) for η/s in finite nuclei

$$\left(\frac{\eta}{s}\right)_{min} \simeq 2.2^{+0.4}_{-0.9} \text{ (KSS units)} , \tag{5}$$

4. Conclusion

In conclusion, by using the Green-Kubo relation and the fluctuation-dissipation theorem, the ratio η/s has been extracted from the experimental systematics of the GDR widths in copper, tin and lead regions at $T \neq 0$, and compared with the theoretical predictions by the PDM, AM, pTSFM, and FLDM. At $T = 5$ MeV, the values of η/s predicted by the PDM reach $3^{+0.63}_{-1.2}$, $2.8^{+0.5}_{-1.1}$, $3.3^{+0.7}_{-1.3}$ KSS units for ^{63}Cu, ^{120}Sn, and ^{208}Pb, respectively. Combining these results with the model-independent estimation for the high-T limit of η/s, which is $2.2^{+0.4}_{-0.9}$ KSS units, one can conclude that the value of η/s for medium and heavy nuclei at $T = 5$ MeV is in between (1.3 - 4.0) KSS units, which is about (3 - 5) times smaller (and of much less uncertainty) that the value between (4 - 19) KSS units predicted by the FLDM for heavy nuclei, where the same lower value $\eta(0) = 0.6u$ was used. This estimation also indicates that nucleons inside a hot nucleus at $T = 5$ MeV has nearly the same ratio η/s as that of QGP, around (2 - 3) KSS units, at $T > 170$ MeV discovered at RHIC and LHC.

The numerical calculations were carried out using the FORTRAN IMSL Library by Visual Numerics on the RIKEN Integrated Cluster of Clusters (RICC) system.

References

[1] K. Adcox *et al.* (PHENIX Collaboration), Nucl. Phys. A **757**, 184 (2005); B.B. Back *et al.*, Ibid. **757**, 28 (2005); J. Arsene *et al.* (BRAHMS Collaboration), Ibid. **757**, 1 (2005); J. Adams *et al.* (STAR Collaboration), Ibid. **757**, 102 (2005).
[2] K. Arnold *et al.* (ALICE Collaboration), Phys. Rev. Lett. **105**, 252302 (2010); G. Aad *et al.* (ATLAS Collaboration), Ibid. **105**, 252303 (2010).
[3] P.K. Kovtun, D.T. Son, and A.O. Starinets, Phys. Rev. Lett. **94**, 111601 (2005).
[4] T. Schäffer and D. Teaney, Rep. Prog. Phys. **72**, 126001 (2009); E. Shuryak, Physics **3**, 105 (2010).
[5] N. Auerbach and S. Shlomo, Phys. Rev. Lett. **103**, 172501 (2009).
[6] V.M. Kolomietz and S. Shlomo, Phys. Rep. **390**, 133 (2004).
[7] N. Auerbach and A. Yeverechyahu, Ann. Phys. (N.Y.) **95**, 35 (1975).
[8] Z.M. Drebi *et al.*, Phys. Rev. C **52**, 578 (1995); M. Kicińska-Habior et al., Phys. Rev. C **36**, 612 (1987); E.F. Garman *et al.*, Phys. Rev. C **28**, 2554 (1983).
[9] A. Bracco *et al.*, Phys. Rev. Lett. **62**, 2080 (1989).
[10] G. Enders *et al.*, Phys. Rev. Lett. **69**, 249 (1992).
[11] T. Baumann *et al.*, Nucl. Phys. A **635**, 428 (1998).
[12] P. Heckmann *et al.*, Phys. Lett. B **555**, 43 (2003).
[13] M.P. Kelly, K.A. Snover, J.P.S. van Schagen, M. Kicińska-Habior, and Z. Trznadel, Phys. Rev. Lett. **82**, 3404 (1999).
[14] D.R. Chakrabarty, M. Thoennessen, N. Alamanos, and P. Paul, and S. Sen, Phys. Rev. Lett. **58**, 1092 (1987).
[15] N.D. Dang and A. Arima, Phys. Rev. Lett **80**, 4145 (1998).
[16] N. Dinh Dang and A. Arima, Nucl. Phys. A, **636** 427, (1998).
[17] N.D. Dang and A. Arima, Phys. Rev. C **68**, 044303 (2003).
[18] R. Kubo, M. Toda, and N. Hashitsume, *Statitiscal Physics II* (Springer, Berlin Heidelberg, 1985).
[19] D.N. Zubarev, Sov. Uspekhi **3**, 320 (1960) [Usp. Fiz. Nauk. **71**, 71 (1960)].
[20] G. Policastro, D.T. Son, and A.O. Starinets, Phys. Rev. Lett. **87**, 081601 (2001).
[21] B.L. Bermann and S.C. Fultz, Rev. Mod. Phys. **47**, 713 (1975).
[22] K.T.R. Davies, A.J. Sierk, and J.R. Nix, Phys. Rev. C **13**, 2385 (1976).
[23] L.G. Moretto, Phys. Lett. B **40**, 1 (1972).
[24] N.D. Dang and A. Arima, Phys. Rev. C **68**, 014318 (2003).
[25] N. Dinh Dang and N. Quang Hung, Phys. Rev. C **77**, 064315 (2008).
[26] W.E. Ormand, P.F. Bortignon, R.A. Broglia, and A. Bracco, Nucl. Phys. A **614**, 217 (1997).
[27] D. Kusnezov, Y. Alhassid, and K.A. Snover, Phys. Rev. Lett. **81**, 542 (1998).

^{182}Pt as a possible candidate for X(5) symmetry

P Petkov[1], K A Gladnsihki[2], A Dewald[3], C Fransen[3], M Hackstein[3], J Jolie[3], Th Pissulla[3], W Rother[3] and K O Zell[3]

[1] Bulgarian Academy of Sciences, Institute for Nuclear Research and Nuclear Energy, 1784 Sofia, Bulgaria

[2] Faculty of Physics, St. Kliment Ohridski University of Sofia, 1164 Sofia, Bulgaria

[3] Institut für Kernphysik, Universität zu Köln, D-50937 Köln, Germany

E-mail: `petkov@inrne.bas.bg`

Abstract.
Recently, a new island of X(5) nuclei has been suggested around A=180 exemplified by some Osmium isotopes. To investigate the limits of its region, a Recoil-distance Doppler shift lifetime measurement has been performed for ^{182}Pt. For the data analysis, the Differential decay curve method has been applied in a newly developed version convenient for low recoil velocities and a non-negligible fraction of nuclei stopped already in the target. The level energies and the newly deduced transition quadrupole moments in the yrast band reveal the persistence of X(5) features in the investigated nucleus, but other spectroscopic data and IBM and GCM calculations indicate shape coexistence and a position of ^{182}Pt close but not at the critical point of the shape-transition.

PACS numbers: 21.10.Tg,21.10.Re,21.60.Ev

As it is well known, the Critical point symmetry (CPS) X(5) [1] describes nuclei at the phase transition between a spherical vibrator and an axially deformed rotor. This model provides a parameter free (up to scaling factors) prediction for the level scheme and B(E2) transition strengths for nuclei at the critical point. Realizations of the X(5) CPS were first found in some isotones with N=90 in the mass region A$\approx$ 150 (^{150}Nd [2], ^{152}Sm [3], ^{154}Gd [4]). Recently, in the region of Os isotopes with A$\approx$ 180, a new island of X(5)-like nuclei has been discovered [5]. It is of interest to establish the limits of this island on the nuclear landscape, as well as to investigate for possible reminiscences of the X(5) structure in neighboring nuclei. In the present work, we concentrate on the nucleus ^{182}Pt whose level scheme is well known from previous investigations (cf. [6, 7] and references therein). Prior to the present study, no lifetimes were known in this nucleus, and thus a precious data set for comparisons with nuclear models, namely electromagnetic transition strengths, was missing. Therefore we decided to perform Recoil distance Doppler-shift (RDDS) measurements in order to determine lifetimes in the yrast band.

The reaction ^{170}Yb(^{16}O,4n) at a beam energy of 87 MeV was used to populate excited states in ^{182}Pt. The beam was provided by the FN Tandem of the Institute für

Kernphysik of the Universität zu Köln. The target consisted of 1.0 mg/cm^2 isotopically enriched ^{170}Yb evaporated onto a 2.1 mg/cm^2 Ta foil serving as a backing and facing the beam. The recoiling nuclei were stopped in a 3.5 mg/cm^2 gold foil after a flight in vacuum with a mean velocity of about 0.63% of the velocity of light, c. The target and stopper foils were mounted in the Cologne plunger apparatus [8]. A setup consisting of five large volume Germanium detectors positioned symmetrically at the backward angle of 143° and a Euroball cluster detector positioned at 0° with respect to the beam axis registered the coincident deexciting γ-rays. The Germanium crystals used were grouped in three rings. Data were taken for 10 target-to-stopper distances x in the range from 3.0 μm to 1500 μm. The data were sorted into 80 8k$\times$8k γ-γ coincidence matrices after corrections for energy shifts and gain matching. The normalization of the data taken at different target-to-stopper distances was performed using coincidence events corresponding to pairs of strong transitions in the yrast band (cf. ref. [8]).

The RDDS method is a well established technique for the determination of picosecond lifetimes of excited nuclear states (for a detailed presentation see e.g. ref. [9] and references therein). For the case of coincidence RDDS measurements where a gate is set on the shifted component of a transition directly feeding the level of interest, a solution of most of the experimental problems was proposed in Ref. [10]. In the present work, because of the relatively small Doppler-shifts of the transitions of interest, it was not possible to use such gating without a significant loss of statistics. Therefore we used the procedure [10] in a variant [11] which is relevant for a case where the gating condition does not influence the timing information for the investigated level and represents a further application of the Differential decay curve method (DDCM) [12, 13] developed in Cologne. Namely, gates were set on the complete line (both shifted and unshifted components included) of the feeding transition and of the transition of interest, respectively. In this way, the problem with the unobserved feeding was solved by using coincidences of a feeding transition with the transition of interest. We summed up the spectra corresponding to gates set in the three independent rings in order to increase statistics. To apply in practice the formalism of Ref. [11], we performed a Monte-Carlo (MC) simulation in three-dimensions of the time evolution of the velocity distribution of the recoils. Further, the "velocity histories" were randomized with respect to the registering detectors. A modified version of the computer code DESASTOP [10, 14] was used for the Monte-Carlo simulation. Details about the code and more specifically, the treatment of the electron and nuclear stoppings can be found in Ref. [15]. In the procedure for the analysis, the background-subtracted line-shapes corresponding to the transition of interest at all distances and the shifted decay curve are fitted simultaneously. For the analysis, the latter is represented by continuously interconnected second-order polynomials over an arbitrarily chosen set of neighboring time-intervals. The fitting problem is linear with respect to the polynomial parameters and the areas of the unshifted peak. To illustrate the application of the procedure, we show in fig. 1 an example of the analysis of the data for the 355 keV transition which depopulates the I^π = 6^+ level of the yrast band. The final result for the lifetime is obtained by averaging

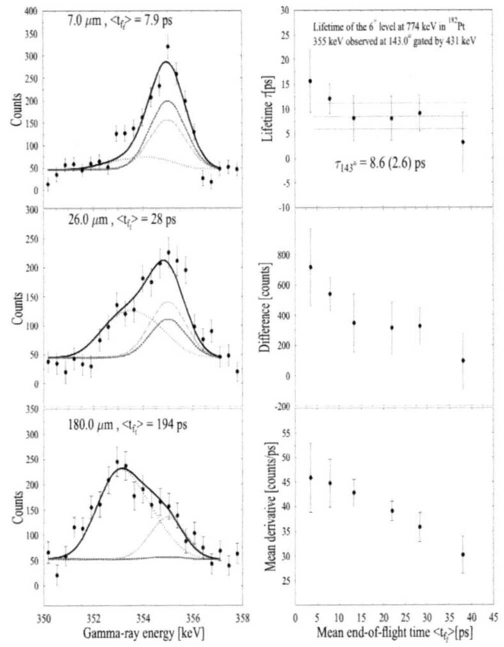

Figure 1. Example of the lifetime analysis of the 355 keV transition. The fits in the left panels illustrate the contributions of the shifted peak (blue dotted line), unshifted peak (green short-dashed line) and decays in the target and DSA-effects (red dot-dashed line). In the top right panel, the τ-curve is displayed. It is a result of the division of the numerator in the middle panel on the r.h.s by the corresponding denominator in the bottom panel on the r.h.s. See also text.

the values derived at the different rings with paying attention to possible systematic errors (see e.g. Ref. [11]). In this way, the lifetimes of five levels in the yrast band till the 10^+ one were derived from the RDDS data giving yield to absolute B(E2) reduced transition probabilities.

The comparison of the X(5) predictions to the data with respect to the level energies and B(E2) branching ratios gives a relatively good agreement with one big exception - the energy of the 0_2^+ state. Experimentally, it lies at about 3.22 times the energy of the 2_1^+ level while according to X(5) it should lie at 5.67 E(2_1^+). Therefore one can conclude that ^{182}Pt is not a perfect candidate for a realization of the X(5) symmetry but possess some of its features, as it will become clear from the discussion below. We performed calculations within two widely exploited models to describe the low-lying states in ^{182}Pt and to make some conclusions about its place in the shape-transitional region. First, we employed the Interacting boson model-1 (IBM-1) [16] in its version called extended consistent Q formalism (ECQF) where the parameters of the Hamiltonian were taken from a systematic study [17] as presented in Ref. [18]. The second model employed in our work is the General collective model (GCM) or Frankfurt model[19, 20]. Both theoretical models describe quite satisfactorily the level scheme and the transitions strengths. The IBM calculation reproduces better the energies in the yrast band and the branching ratios while the GCM gives superior results for the yrast B(E2)'s and the level energies of the side-bands. It is interesting to consider the nuclear shape associated with the fitted parameters of the models, namely with those of the GCM. Since β and γ are dynamic variables in the GCM, the maximum of their probability

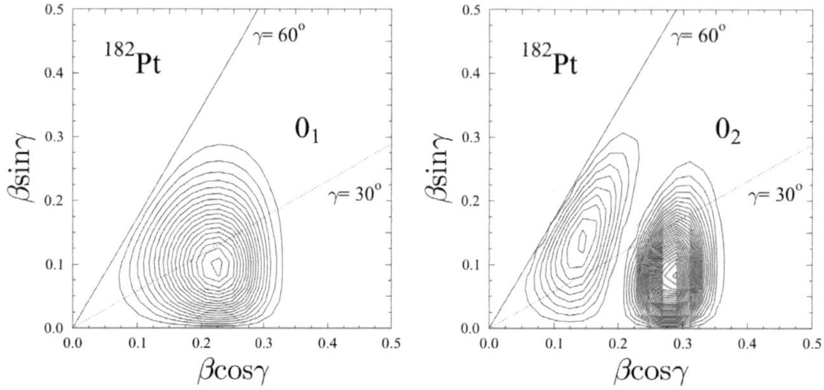

Figure 2. Probability distributions in the β-γ plane of the wave functions of the first two 0^+ states in ^{182}Pt according to the GCM.

distribution (the squared wave function multiplied by the volume element $\beta^4 sin(3\gamma)$) does not necessarily coincide with the potential minimum. Thus, we show in fig. 2 these distributions for the first two 0^+ states. The 0^+_2 state can be interpreted as a β-vibration, its probability distribution has roughly one node in β. On the other hand,

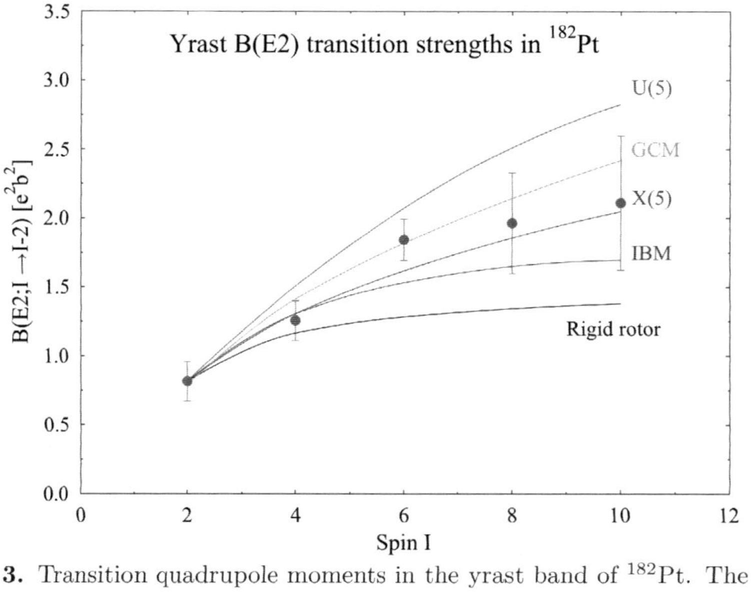

Figure 3. Transition quadrupole moments in the yrast band of ^{182}Pt. The different theoretical models used are indicated. With the exception of the GCM, all models are normalized to the Q_t value for the $2^+_1 \rightarrow 0^+_1$ transition. .

this state is characterized by two composing components: one prolate, more deformed than the 0^+_1 band structure, and one more triaxial to oblate, less deformed structure. Similar observations for a multi-component structure can be made also for the other, higher-lying 0^+ states. In this sense, the GCM predicts a shape coexistence in ^{182}Pt in the framework of a mixing of bands with different quadrupole deformation, confirming earlier experimental findings (e.g. [21]) for the light Pt isotopes. In fig. 3, we present the experimental transition quadrupole moments in the ground-state band compared to model calculations. The best description is provided by the GCM followed by X(5) and IBM. These results for the transition quadrupole moments in the ground-state band may be interpreted as an indication that some X(5) features persist in ^{182}Pt coexisting with excitations corresponding to different nuclear shapes. This is corroborated also by the experimental B(E2) branchings of the transitions from the side bands to the ground-state band. The exact position of ^{182}Pt in the IBM parameter space and its closeness to the shape/phase transitional region may be found in fig.4 of Ref. [17]. In that figure, the Pt isotopes from A=176 to 194 are positioned in the Casten triangle, and ^{182}Pt lies very close, but not in the X(5) transitional region. It deviates somewhat toward the O(6) vertex of the triangle, indicating the importance of the γ-softness effects.

Acknowledgments: K.A.G and P.P. would like to thank the colleagues from the University of Cologne for their kind hospitality. This work was supported by the Deutsche Forschung Gemeinschaft (DFG) within the East-European-Germany collaboration program and contracts No. Jo 391/8-1 and No. Jo 391/11-1. The authors are also indebted for the support by the partnership agreement between the University of Cologne and University of Sofia and by the BgNSF under Contract No. DO 02-219.

References

[1] Iachello F 2001 *Phys. Rev. Lett.* **87** 052502
[2] Krücken R *et al.* 2002 *Phys. Rev. Lett.* **88** 232501
[3] Casten R F and Zamfir N V 2001 *Phys. Rev. Lett.* **87** 052503
[4] Tonev D *et al.* 2002 *Phys. Rev.* C **69** 034334
[5] Dewald A *et al.* 2005 *J.Phys.G* **31** S1427
[6] Popescu D G *et al.* 1997 *Phys. Rev.* C **55** 1175
[7] Singh B and Roediger J C 2010 *Nuclear Data Sheets* **111** 2081
[8] Dewald A *et al.* 1992 *Nucl. Phys.* A **545** 822
[9] Alexander T K and Forster J S 1978 *Adv. Nucl. Phys.* **10** 197
[10] Petkov P *et al.* 1999 *Nucl. Instr. Meth. Phys. Res.* A **431** 208
[11] Petkov P *et al.* 2000 *Nucl. Phys.* A **674** 357
[12] Dewald A, Harissopulos S and von Brentano P 1989 *Z. Phys.* A **334** 163
[13] Böhm G *et al.* 1993 *Nucl. Instr. Meth. Phys. Res.* A **329** 248
[14] Winter G 1983 *Nucl. Instr. Meth.* **214** 537
[15] Petkov P 1998 *et al. Nucl. Phys.* A **640** 293
[16] Iachello F and Arima A 1987 *The Interacting boson model*, (Cambridge University Press, Cambridge, England)
[17] McCutchan E A, Casten R F and Zamfir N V 2005 *Phys. Rev.* C **71** 061301 (R)
[18] Garcia-Ramos J E and Heyde K 2009 *Nucl. Phys.* A **825** 39
[19] G.Gneuss G and Greiner W 1971 *Nucl. Phys.***171** 449
[20] Troltenier D, Maruhn J A and Hess P O 1991, in *Computational Nuclear Physics*, edited by K.Langanke, J.A.Maruhn and S.E.Koonin (Springer, Berlin, 1991) p. 105
[21] G.D.Dracoulis G D *et al.* 1986 *J. Phys. G* **12** L97

Journal of Physics: Conference Series **366** (2012) 012037 doi:10.1088/1742-6596/366/1/012037

Direct Evidence for Proton-Neutron Out-of-Phase Coupling in Mixed-Symmetry States from Scattering Experiments on ^{92}Zr

N Pietralla[1], C Walz[1], V Yu Ponomarev[1], H Fujita[2,3], A Krugmann[1], P von Neumann-Cosel[1], A Scheikh-Obeid[1], J Wambach[1]

[1] Institut für Kernphysik, TU Darmstadt, Schlossgartenstraße 9, D-64289 Darmstadt, Germany

[2] Department of Physics, Osaka University, Toyonaka, Osaka 560-0043, Japan

[3] iThemba LABS, Post Office Box 722, Somerset West 7129, South Africa

E-mail: pietralla@ikp.tu-darmstadt.de

Abstract. The coupling of the giant quadrupole resonance to valence space configurations is shown to dominantly contribute to the formation of low-lying quadrupole collective structures in vibrational nuclei with symmetric and mixed-symmetric character with respect to the proton-neutron degree of freedom. Experimental evidence is obtained from electron- and proton scattering experiments on the nucleus ^{92}Zr that are directly sensitive to the relative phase of valence space amplitudes by quantum interference.

1. Introduction

A general phenomenon of the low-energy structure in heavy atomic nuclei are collective quantum states, in particular of quadrupole nature. The proton-neutron residual interaction in the valence shell is known to represent the dominant origin for the formation and the evolution of nuclear quadrupole collectivity [1]. In order to study the residual proton-neutron interaction in the valence shell of open-shell nuclei it is interesting to determine the properties of those valence shell excitations where valence protons and valence neutrons move out of phase. Such states have been modeled, for example, in terms of mixed-symmetry states in the framework of the interacting boson model [2] with proton-neutron degree of freedom (IBM-2). In the context of nuclear structure physics the Interacting Boson Model (IBM) is an example of an effective field theory (EFT), formulated consistently with the symmetries of the intrinsic shapes of the nucleus. It describes the dynamics of collective low-energy nuclear excitations. Effective field theories for relevant low-energy degrees of freedom of complex systems are a concept of fundamental importance in modern physics with broad applications in many different areas. The underlying principle is a separation of energy (respectively momentum) scales such that the high-energy degrees of freedom are integrated out leading to a low-energy theory consistent with the symmetries and their breaking pattern. The high-energy sector manifests itself in a set of parameters in the low-energy EFT which have to be determined from experiment. Thus each EFT is only capable to describe phenomena at a specific (sufficiently low) energy scale.

In the IBM the relevant low-energy degrees of freedom for the description of quadrupole-collectivity are bosons with intrinsic angular momentum $L = 0$ (s bosons) or $L = 2$ (d bosons).

These lead to an effective Hamiltonian whose parameters are adjusted to data. However, a microscopic and quantitative theory for the derivation of these parameters is still missing. Such an approach could only be hoped for being successful if the dominant origin of the quadrupole collectivity would be included in such a theory. Our recent experiments [3] have provided evidence on how the coupling to cross-shell transitions forming the giant quadrupole resonance (GQR) contributes to the formation of low-energy nuclear collectivity. This has been achieved by an analysis of transition densities in electromagnetic and hadronic scattering reactions that has enabled us for the first time to observe and interpret the interference of the valence-shell components of the lowest-energy one-quadrupole phonon wave functions with those stemming from cross-shell excitations. This lecture aimed at a presentation of our recent material and, hence, this contribution is based on the text published in Ref. [3].

2. The case of ^{92}Zr

To keep the discussion transparent, we studied the formation of quadrupole collectivity in the particularly simple case of a nucleus with a low-energy structure that is dominated by one pair of valence particles each for protons and neutrons. An example is the nucleus ^{92}Zr with 2 neutrons beyond the $N = 50$ shell closure and 2 protons beyond the $Z = 38$ sub-shell closure. The lowest 2-quasiparticle (2qp) states will therefore have $\pi(1g_{9/2})^2$ and $\nu(2d_{5/2})^2$ configurations. Due to the residual proton-neutron interaction two different classes of collective excitations appear at low energy in which the amplitudes of the two most important 2qp configurations are coupled in a symmetric or antisymmetric way, respectively. In ^{92}Zr, these are experimentally identified as the 2_1^+ and 2_2^+ states [4, 5] with some degree of configurational isospin polarization [6].

To shed light on the microscopic origin of the effective coupling strength in the valence shell we consider the quasiparticle-phonon model (QPM) [7]. The QPM starts with interacting particle-hole excitations from which 'phonons' are generated by the Random-Phase Approximation (RPA) approach. These form elementary degrees of freedom from which a more general Hamiltonian can be derived which allows for multi-phonon states. The QPM covers a sufficiently

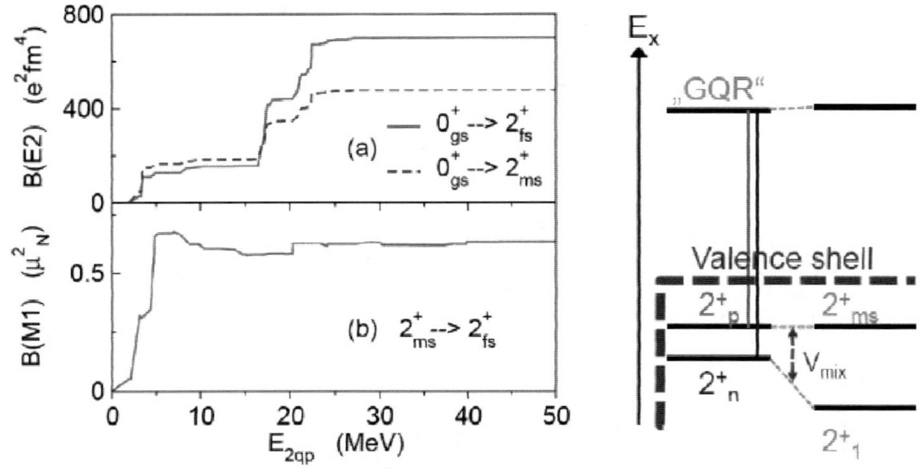

Figure 1. Transition strengths as a function of the maximum 2qp energy included in the QPM calculation for (a) g.s. excitation of the 2_{fs}^+ and 2_{ms}^+ states, and (b) $M1$ transition between both states. The latter are dominated by valence-shell components while the (moderately) collective E2 transitions are dominated by admixtures of cross-shell excitations. For easier interpretation the situation can be simplified to a three-state mixing scenario as sketched on the right.

large single-particle space to satisfy the energy-weighted sum-rules. The QPM approach has proven to account very successfully for the low-energy properties in a large number of vibrational nuclei [8, 9] including ^{92}Zr. In the spirit of an interpretation of the IBM-2 as an EFT, the QPM can be viewed as 'complete theory' including the high-energy degrees of freedom. The QPM wave functions are dominated by the lowest π and ν 2qp components, that show the expected in-phase and out-of-phase behavior for the $2^+_{\rm fs}$ and $2^+_{\rm ms}$ states. The electromagnetic properties and excitation energies are in excellent agreement with the data [9]. The magnetic moments of these states and the strong $M1$ transition between them originate almost entirely from the valence-shell configurations as it is shown in Fig. 1. However, the $B(E2)$ strengths are generated to about 80% from many components beyond the valence shell albeit their total contribution to the wave function norm is small.

The role of the GQR is demonstrated in Fig. 1, which shows the running sum of the $B(M1)$ and $B(E2)$ transition strengths as a function of the maximum 2qp energy included in the calculation. The $M1$ strength saturates at about 6 MeV, i.e., only valence-shell configurations are relevant, while the largest part of the $E2$ strengths exciting the $2^+_{\rm fs}$ and $2^+_{\rm ms}$ phonons is generated by states in the energy region of the GQR around $15 - 20$ MeV. This observation motivates a simple three-state mixing scenario between the proton-valence shell configuration, the neutron-valence shell configuration, and the GQR for a deeper insight in the formation of the one-quadrupole phonon states with symmetric and mixed-symmetry character even on a semi-quantitative level [3]. For the nucleus ^{92}Zr with higher energy for the proton valence-shell component than the neutron valence-shell component at the Z=40 sub-shell closure, this scheme inevitably requires that the neutron valence-shell component flips its phase with respect to the GQR component when going from the proton-neutron symmetric 2^+_1 state to the 2^+_2 state with predominant mixed symmetry. Can we find experimental evidence for this simple scheme, *i.e.*, is it possible to find a signature for the different interference of the dominant valence-shell components with the high-energy mode as predicted in the three-state scheme?

3. Experiment

Apparently, two probes with different sensitivity to protons and neutrons are needed to study this quantum interference experimentally which has not been done so far. Electron scattering at low momentum transfer provides a measure of the charge transition radius. An (e, e') experiment was performed at the high-energy-resolution spectrometer [10] of the Darmstadt superconducting electron linear accelerator (S-DALINAC). An enriched (94.6 %) self-supporting ^{92}Zr target of 9.8 mg/cm^2 areal density was used. Data were taken covering a momentum transfer range

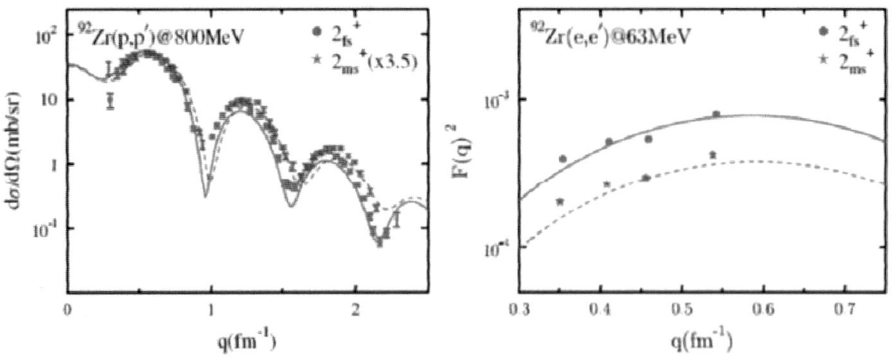

Figure 2. Momentum-transfer dependence of the cross sections of the $2^+_{\rm fs}$ and $2^+_{\rm ms}$ in proton scattering (left) and electron scattering (right). The data are compared to QPM calculations.

between q $\simeq$ 0.3 - 0.6 fm^{-1} indicating no difference between the charge transition radii of the 2^+_{fs} and 2^+_{ms} states within experimental uncertainties (Fig. 2, right). Information about the neutron transition radii can be derived from the proton scattering data of Ref. [11]. At the incident energy of 800 MeV protons interact predominantly via the isoscalar central piece of the effective projectile-nucleus interaction [12]. Clearly, the refraction pattern of the (p,p') cross section for the 2^+_{ms} state are shifted to higher q values as compared to those for the 2^+_{fs} state (Fig. 2, left) corresponding to a smaller transition radius.

4. Analysis

Using the transition densities from Fig. 3, cross sections were calculated in distorted wave Born approximation. The T-matrix parametrization of Franey and Love [12] was used to describe the effective proton-nucleus interaction. The QPM calculation reproduces well both the absolute values of cross sections for both probes and the shift of the refraction pattern to higher q values for the 2^+_{ms} state in the (p,p') reaction as displayed in Fig. 2.

Figure 3 displays the proton and neutron transition densities of the 2^+_{fs} (top) and 2^+_{ms} (bottom) states calculated in the full QPM approach. The full transition densities (solid curves) are decomposed in a collective part stemming from the GQR (dotted curves) and the predominant 2qp $\nu(2d_{5/2})^2$ or $\pi(1g_{9/2})^2$ contributions (dashed curves). The key point is the different

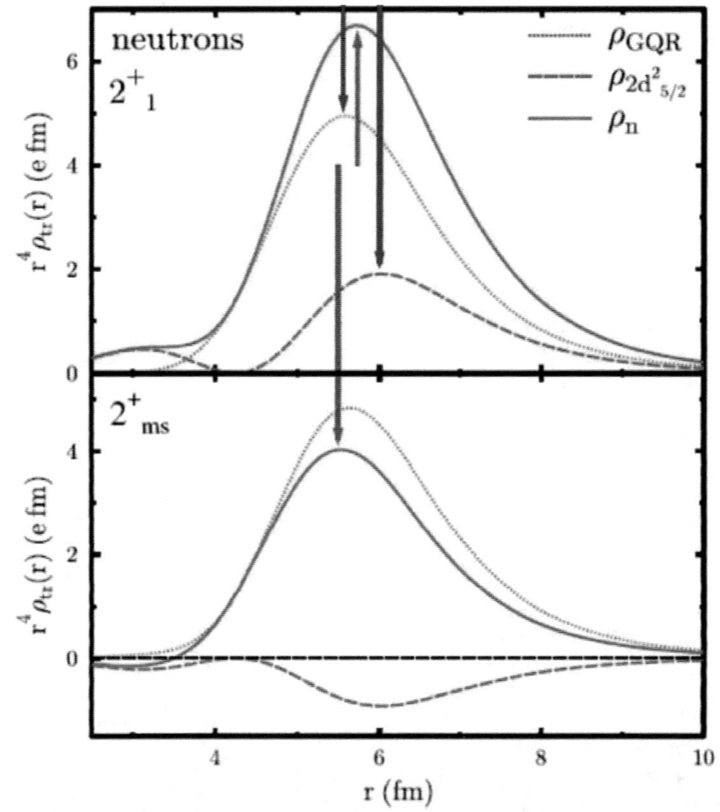

Figure 3. Neutron transition densities of the 2^+_{fs} (top) and 2^+_{ms} (bottom) states of ^{92}Zr from QPM calculations. The full transition densities (solid lines) are decomposed in parts stemming from the GQR (dotted lines) and from the main 2qp configurations (dashed lines). The arrows indicate the maxima of the corresponding full transition densities.

radial behaviour of both parts. An out-of-phase coupling between the neutron valence shell contribution and the contribution from the GQR leads to a destructive quantum interference that predominantly reduces the neutron transition density at large radii (due to the larger radius of the $\nu(2d_{5/2})^2$ orbital) and consequently shifts the maximum of the total neutron transition density to the interior with respect to that one for the 2^+_{fs} state, as indicated by the arrows in Fig. 3. This effect reduces the neutron transition radius of the 2^+_{ms} with respect to the 2^+_{fs} state of ^{92}Zr. In contrast, the proton transition radius remains essentially unchanged since the $\pi(1g^2_{9/2})$ part couples in-phase to the GQR contribution in both states. The combination of both data sets unambiguously demonstrates that the phase of the neutron valence-shell configurations changes its sign between the 2^+_{fs} and the 2^+_{ms} state for the first time [3].

5. Summary

We have presented a simple mechanism to explain the formation of the collective symmetric and mixed-symmetric 2^+ states in vibrational nuclei. The proton-neutron quadrupole interaction responsible for the mixing of the valence space configurations is mediated by their coupling to the GQR, whose contribution dominates the $E2$ strengths of these states. In contrast, magnetic properties can be understood to a large extent within the scope of effective valence-space models. A combination of proton and electron scattering data provides direct evidence for the interference of high-energy configurations from the GQR with leading valence-space configurations. The data are sensitive to the relative signs of the leading valence shell configurations with respect to the GQR-contribution and confirm the symmetric and mixed-symmetric nature of the low-energy 2^+ one-phonon states. In particular, a flip of phase of the neutron valence configuration between the 2^+_{fs} and the 2^+_{ms} state was observed and is understood as a consequence of the lower neutron $2qp$ energy at the $Z = 40$ proton sub-shell closure.

Since typical excitation energies of the low-lying collective states are $1-2$ MeV and that of the GQR $10 - 20$ MeV, a clean separation of scales is given, leading to an interpretation of valence-space approaches like the IBM-2 and shell model as EFTs of microscopic models capable to consistently treat valence-shell excitations and giant resonances. The present findings highlight the peculiar role played by mixed-symmetry states in such efforts and in our understanding of the formation of nuclear collectivity and of collective phenomena in any strongly coupled multi-component quantum system.

Acknowledgments

Discussions with B. Friman, G. Rainovski, R.V. Jolos, T. Otsuka, Ch. Stoyanov, and V. Werner are greatly appreciated. We thank the S-DALINAC crew for preparation of electron beams. This work was supported by the Deutsche Forschungsgemeinschaft under grant No. SFB 634.

[1] Casten R F 1990 *Nuclear Structure from a Simple Perspective* (Oxford: Oxford University Press).
[2] Iachello F and Arima A 1987 *The Interacting Boson Model* (Cambridge: Cambridge University Press).
[3] Walz C, Fujita H, Krugmann A, von Neumann-Cosel P, Pietralla N, Ponomarev V Yu, Scheikh-Obeid A and Wambach J 2011 *Phys. Rev. Lett.* **106** 062501
[4] Werner V *et al.* 2002 *Phys. Lett.* B **550** 140; Fransen C *et al.* 2005 *Phys. Rev.* C **71** 054304
[5] Pietralla N, von Brentano P and Lisetskiy A F 2008 *Prog. Part. Nucl. Phys.* **60** 225
[6] Holt J D *et al.* 2007 *Phys. Rev.* C **76** 034325
[7] Soloviev V 1992 *Theory of Atomic Nuclei: Quasiparticles and Phonons* (Bristol: Institute of Physics Publishing)
[8] Lo Iuduce N and Stoyanov Ch 2000 *Phys. Rev.* C **62** 047302; Lo Iuduce N, Stoyanov Ch and Tarpanov D 2008 *Phys. Rev.* C **77** 044310
[9] Lo Iuduce N and Stoyanov Ch 2006 *Phys. Rev.* C **73** 037305
[10] Lenhardt A W *et al.* 2006 *Nucl. Instrum. Methods Phys. Res.* A **562** 320
[11] Baker F T *et al.* 1983 *Nucl. Phys.* A **393** 283
[12] Franey M A and Love W G 1985 *Phys. Rev.* C **31** 488

Fast-timing measurements performed at ALTO on ^{137,139}Cs

B Roussière[1], M A Cardona[2,3,4], I Deloncle[5], D Hojman[2,4], J Kiener[5], P Petkov[6], D Tonev[6], Ts Venkova[6] and the ALTO collaboration

1. Institut de Physique Nucléaire, IN2P3/CNRS/Université Paris-Sud, F-91406 Orsay, France
2. Departamento de Física, Comisión Nacional de Energía Atómica, 1429 Buenos Aires, Argentina
3. Universidad Nacional de General San Martín, San Martín, Argentina
4. CONICET, 1033 Buenos Aires, Argentina
5. Centre de Spectrométrie Nucléaire et de Spectrométrie de Masse, IN2P3/CNRS/Université Paris-Sud, F-91405 Orsay, France
6. Institute for Nuclear Research and Nuclear Energy, Bulgarian Academy of Sciences, 1784 Sofia, Bulgaria

E-mail: roussier@ipno.in2p3.fr, cardona@tandar.cnea.gov.ar, Isabelle.Deloncle@csnsm.in2p3.fr, hojman@tandar.cnea.gov.ar, Jurgen.Kiener@csnsm.in2p3.fr, petkov@inrne.bas.bg, Dimitar.Tonev@lnl.infn.it, venkova@inrne.bas.bg

Abstract. Fast-timing measurements were performed on ^{137,139}Cs obtained from the β^- decay of ^{137,139}Xe. The neutron-rich Xe nuclei were produced by the ^{238}U photofission induced by the 50 MeV electron beam delivered by the ALTO facility. The half-lives of the first excited state in ^{137}Cs and of six excited states in ^{139}Cs were measured for the first time. These $T_{1/2}$ results give new information on the structure of the $^{137,139}_{55}Cs_{82,84}$ nuclei located near the neutron closed shell N = 82. In particular, it is shown that, in both nuclei, the first excited $5/2^+$ level corresponds mainly to the one-proton state issued from the $\pi 2d_{5/2}$ orbital. From the comparison with other N = 84 isotones, the most probable spin values for the states located at 289.8, 393.5, and 393.8 keV in ^{139}Cs appear to be $5/2^+$, $3/2^+$, and $1/2^+$.

Keywords: Lifetimes, Photofission, Scintillation detectors

1. Introduction

Half-lives of excited nuclear states allow the estimate of transition probabilities. Thus they provide direct insight into the structure of the nucleus and offer one of the most stringent tests of the nuclear models. In the fast-timing technique, the half-life is directly obtained by measuring the time between signals from two fast detectors; the first one, the start, is given by a transition feeding the level and the second one, the stop, by a transition deexciting it. Developed first with BaF_2 crystals [1], this method, well adapted to measure half-lives in the range from 10 ps to some ns, can nowadays take advantage of the $LaBr_3$ scintillators that have timing properties as good as the BaF_2 ones and the best energy resolution among all scintillators. We have developed a fast-timing setup including various scintillators. The aims of the present experiment were to fully characterize our experimental setup by on-line measurements, to demonstrate its ability to measure half-lives in the subnanosecond range and to obtain valuable information on the proton states near the N = 82 closed shell from the transition probabilities deduced from the half-lives.

2. Experimental procedure

The experiment was performed at ALTO, the facility in Orsay dedicated to the production of neutron-rich nuclei from the photofission of ^{238}U induced by the 50 MeV electron beam. The maximum intensity of the incident beam delivered by the linac is 10 μA. But for this study on Cs nuclei, obtained from the β^- decay of the very well produced Xe nuclei, a low intensity ($\sim$ 500 nA) electron beam was used. The 50 MeV electron beam bombarded a thick uranium-carbide target. The fission products were released from the target then ionized with a MK5 ion source. The ions were extracted under 30 kV and mass-separated by the PARRNe magnet. Then the radioactive ions of the selected mass were guided towards the measurement station, where they were collected on a tape. The collected radioactive sources were transported in front of the fast-timing setup composed of three scintillators: one Pilot U plastic for electron detection, one BaF_2 and one $LaBr_3$ crystals for γ-ray detection. All three have 25.4 mm diameters and 3, 30, and 10 mm lengths, respectively and they are coupled to XP2020, XP2020Q, and XP20D0 photomultipliers. A thin electron detector offers a great advantage: high-energy β rays emitted by neutron-rich nuclei with high Q_β values deposit similar energy in the detector, which provides a uniform timing response for the β detector [1]. A HPGe detector has been added to the fast-timing setup; because of its excellent energy resolution, it allows an accurate selection of a decay branch by performing β-$\gamma_{\mathrm{LaBr_3\ or\ BaF_2}}$-$\gamma_{\mathrm{Ge}}$ coincidences.
Fast-timing measurements are high precision experiment, which means that great care has to be taken in the implementation of the experimental setup. The quality of the time spectra obtained can be affected mainly by the accuracy in the position of the radioactive source, the counting rates and the sensitivity of timing electronics and photomultiplier gains to temperature changes. At the collection point, the position of the collected source depends directly on the stability of the radioactive beam. During the experiment, the working conditions of the ion source were very stable and then only small fluctuations in the intensity, size and geometry of the beam could occur. Moreover, the accuracy on the displacements performed by the tape transport system was better than one millimeter. Thus the fluctuations in the source position resulted in increasing only slightly the width of the time spectra. High counting rates can cause dead time problems and pileup effects in the detectors and therefore deformation of the shape of

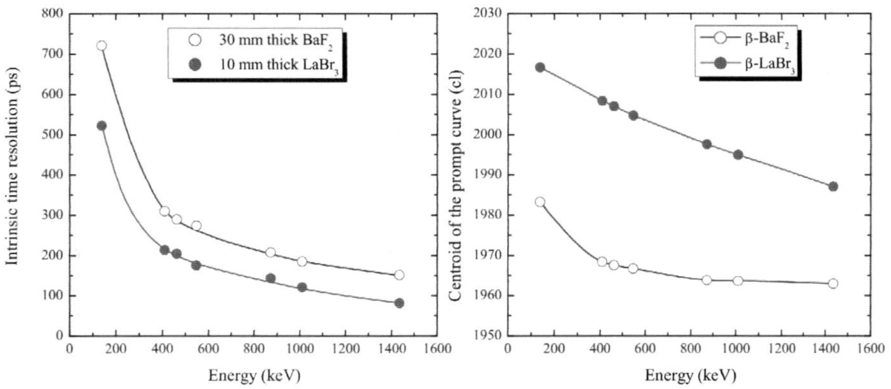

Figure 1. Intrinsic time resolution of the BaF_2 and $LaBr_3$ detectors (left) and centroids of the prompt curves (right) as a function of the γ energy detected in the fast detectors.

the time spectra. To avoid these problems, counting rates below 10000 c/s were used and checked regularly. The temperature variations induce drifts of the centroids of the time spectra that were corrected during the off-line data treatment. More details on the experimental procedure can be found in refs.[2, 3].

3. Experimental results

The measurements performed on the A = 138 mass have been used, since the half-lives in ^{138}Ba are well known [4], to characterize our experimental setup in terms of time resolution, position of the prompt curves and the limits of the half-lives that can be measured by the deconvolution method. Figure 1 shows the intrinsic time resolution obtained for the BaF_2 and $LaBr_3$ as a function of the γ energy detected; the $LaBr_3$ is thinner than the BaF_2 detector then its intrinsic resolution is better. As mentioned above, the response of the thin β detector is independent of energy and its intrinsic time resolution was found to be 244 ps. The position of the centroid of the prompt curve when the time measurement is performed between the β and BaF_2 detectors or beween the β and $LaBr_3$ detectors is also presented in fig. 1. In order to define the limits of the half-lives that can be measured by the deconvolution method, the shape of the prompt curves has been analyzed. Indeed, in the deconvolution method, the fit of the experimental time spectra is done assuming that the prompt curves are described by Gaussian functions. But the prompt curves are not pure Gaussian and exhibit slight asymmetries. In order to quantify these asymmetries, three analyses have performed for each prompt curve: in the first one the fit is done assuming a pure Gaussian function, in the second and the third one the prompt curve are fitted by the function resulting from the convolution of a Gaussian with an exponential function describing an apparent half-life on the right side or on the left side. This apparent half-life should not be mistaken for a true half-life, it is only a parameter used to describe the deviation of the prompt curve from a Gaussian approximation. All the prompt time spectra exhibit

Journal of Physics: Conference Series **366** (2012) 012038 doi:10.1088/1742-6596/366/1/012038

Table 1. Half-lives of the first excited states measured in ^{137,139}Cs.

Nucleus	E_{level} (keV)	$T_{1/2}$	Time spectra	Analysis method
^{137}Cs	455.5	59(4) ps	$T_{\beta\gamma}(\beta\text{-}455_\gamma)$	deconvolution
^{139}Cs	218.6	2.23(5) ns	$T_{\gamma\gamma}(175_\gamma\text{-}218_\gamma)$, $T_{\gamma\gamma}(296_\gamma\text{-}218_\gamma)$	deconvolution
	289.8	96(25) ps	$T_{\beta\gamma}(\beta\text{-}289_\gamma\text{-}225_{Ge})^{a)}$	both
	393.5	24(15) ps	$T_{\beta\gamma}(\beta\text{-}393_\gamma)$	centroid shift
	393.8	1.71(2) ns	$T_{\gamma\gamma}(612_\gamma\text{-}175_\gamma)$	deconvolution
	515.2	6^{+10}_{-6} ps	$T_{\beta\gamma}(\beta\text{-}296_\gamma\text{-}218_{Ge})$	centroid shift
	1006.7	15(10) ps	$T_{\beta\gamma}(\beta\text{-}612_\gamma)$, $T_{\beta\gamma}(\beta\text{-}788_\gamma)$	centroid shift

a) these spectra give access to the sum of the half-lives of the 289.8 and 515.2 keV levels

similar shapes and the values of the apparent half-lives ($\sim$ 20 ps) do not depend on the energy. Applying the criterion proposed in ref. [1] the limit for which the deconvolution method can be used has been fixed to 60 ps; below the centroid shift method has to be used.

The half-lives of the first excited state in ^{137}Cs and of six excited states in ^{139}Cs were measured for the first time in this experiment. The results are presented in table 1. In ^{137}Cs, the ground state and the first excited state located at 455.5 keV are fed respectively by 67 and 31 % of the ^{137}Xe β^- decay. The γ-ray spectrum exhibits almost exclusively the very strong 455.5 keV γ line. Therefore, the half-life of the ^{137}Cs first excited state can be measured by β-γ delayed coincidences only. For ^{139}Cs, it was essential to use the β-γ-γ triple coincidence technique where the timing information is obtained from the β-γ delayed coincidence between fast-timing detectors while the desired decay cascade is selected by a coincident γ-ray in the Ge detector. Two cascades involving γ rays of very close energy, 225-289 keV and 296-218 keV, exist in the level scheme. With scintillators, despite the good energy resolution of $LaBr_3$, the 218-225 keV and 289-296 keV doublets cannot be separated. By setting with the Ge detector a gate on the 225 keV or 218 keV transition, we could select the β-225-289 keV cascade from which we measure the sum of the half-lives of the 515 and 289 keV levels or the β-296-218 keV cascade that leads directly to the half-life of the 515 keV level. The $T_{\beta\gamma}(\beta\text{-}289_\gamma\text{-}225_{Ge})$ time spectrum has been analysed by the deconvolution method and by the centroid shift method: the results obtained, 105(30) ps and 102(15) ps, are quite consistent, which gives confidence in our measurement and analysis procedure. By subtracting the $T_{1/2}$ value of the 515 keV level, we obtained the half-life of the 289 keV level.

4. Discussion

In all N = 82 isotones from Sb to Eu, $7/2^+$ and $5/2^+$ states corresponding to the $\pi 1g_{7/2}$ and $\pi 2d_{5/2}$ configurations are present at very low energy. The transitions linking

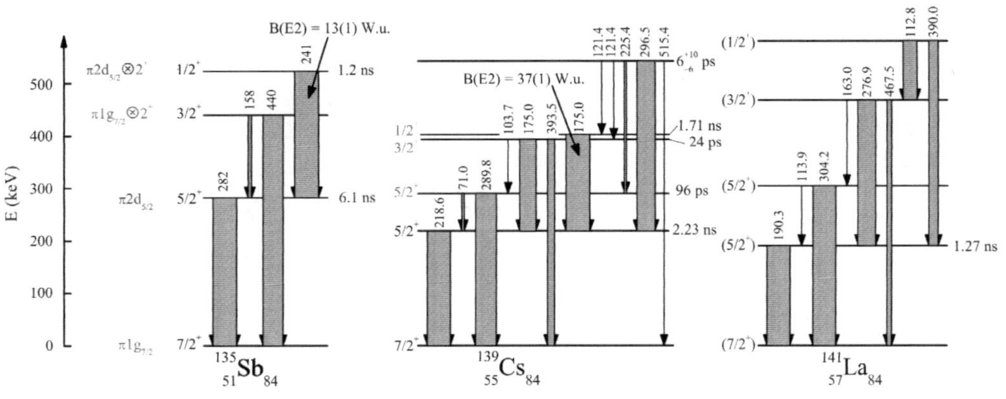

Figure 2. First excited levels in ^{135}Sb [6], ^{139}Cs [7] and ^{141}La [8].

these states is thus a l-forbidden M1 transition. From the half-life measured for the first excited state in ^{137,139}Cs and the transition multipolarities reported in [5], the M1 hindrance factors relative to the Weiskopf estimate for the $\frac{5}{2}^+ \rightarrow \frac{7}{2}^+$ transition have been calculated: the high values found, $F_W(M1) \geq 2440$ and 2880 respectively, show that these proton configurations are quite pure in ^{137,139}Cs.

The other excited states of ^{139}Cs exhibit decay modes very similar to those observed in ^{141}La (see fig. 2). The comparison with ^{135}Sb, another N = 84 isotone, is less easy because, whereas two $5/2^+$ states exist in ^{139}Cs and ^{141}La, there is only one in ^{135}Sb. The B(E2) transition probability for the transition deexciting the $1/2^+$ level in ^{135}Sb is slightly lower than that observed for the 175 keV transition in ^{139}Cs (the E2 multipolarity assumed for the 175 keV transition is discussed in [3]). However the change between these two isotones is quite similar to that observed for the $2^+ \rightarrow 0^+$ in the even-even isotones. All this leads us to propose the spin and parity values $\frac{5}{2}^+$, $\frac{3}{2}^+$, and $\frac{1}{2}^+$ for the levels located at 289.8, 393.5, and 393.8 keV in ^{139}Cs. In order to support these assignments, it would be interesting to measure the half-lives in ^{141}La.

Acknowledgments

This work was supported by the French Ministère des Affaires Etrangères et Européennes via the ECONET program (project 21331RH).

References

[1] Mach H, Gill RL, and Moszyński M, 1989 *Nucl. Instrum. Methods Phys. Res.* A **280** 49.
[2] Deloncle I *et al.*, 2010 *Journal of Physics:* Conference Series **205** 012044.
[3] Roussière B *et al.*, 2011 *Eur. Phys. J.* A **47** 106.
[4] Sonzogni AA, 2003 *Nuclear Data Sheets* **98** 515.
[5] Achterberg E *et al.*, 1972 *Phys. Rev.* C **5** 1759.
[6] Mach H *et al.*, 2007 *Acta Polonica* B **38** 1213.
[7] Burrows TW, 2001 *Nuclear Data Sheets* **92** 623.
[8] Tuli JK, 2001 *Nuclear Data Sheets* **94** 605.

Journal of Physics: Conference Series **366** (2012) 012039 doi:10.1088/1742-6596/366/1/012039

Weak decay rates for waiting-point nuclei involved in the rp-process

P Sarriguren

Instituto de Estructura de la Materia, IEM-CSIC, Serrano 123, E-28006 Madrid, Spain

E-mail: pedro.sarriguren@csic.es

Abstract.
We study weak interaction rates under the stellar density and temperature conditions encountered in the *rp* process. Special attention is paid to neutron-deficient waiting-point nuclei. The nuclear structure part of the problem is described within a deformed Skyrme Hartree-Fock + BCS + QRPA approach, which reproduces not only the beta-decay half-lives but also the available Gamow-Teller strength distributions, measured under terrestrial conditions. The various sensitivities of the decay rates to both density and temperature are discussed. In particular, we study the contributions coming from excited states in the parent nucleus, which are thermally populated, and the competition between beta-decays and continuum electron captures.

1. Introduction

A reliable nuclear input is an essential element in most astrophysical simulations of explosive events, which involve in particular, having a knowledge of the nuclear properties of exotic nuclei. Unfortunately, the experimental information on exotic nuclei is still very poor and thus, the astrophysical models of violent stellar phenomena are mainly based on nuclear model predictions. This is the case of type I X-ray bursts, generated by a thermonuclear runaway in the hydrogen-rich environment of a neutron star that is fed from a red giant binary companion. The explosions typically release $10^{38} - 10^{39}$ erg of energy, producing bursts lasting between seconds and a few minutes. These explosions have recurrence periods ranging from hours to days. They reach peak temperatures (T) and densities (ρ) of about T =1-3 GK and $\rho = 10^6$-10^7 g cm^{-3}, respectively.

This scenario allows the development of the nucleosynthesis rapid proton (rp) capture process [1], which is characterized by proton capture reaction rates that are orders of magnitude faster than any other competing process, in particular β-decay. It produces rapid nucleosynthesis on the proton-rich side of stability toward heavier proton-rich nuclei and explains the energy and luminosity profiles observed in X-ray bursts. The potential impact of X-ray burst nucleosynthesis on the galactic abundances is still an open question. Although ejection of matter is very unlikely, it is not completely discarded. The possible relevance of these events to understand the origin and abundance of some neutron-deficient isotopes is still a matter of debate.

In general, the reaction path follows a series of fast proton-capture reactions until further proton capture is inhibited by photodisintegration. Then, the reaction flow has to wait for a relatively slow β-decay to proceed further and the respective nucleus is called a waiting point (WP). The half-lives of the WP nuclei finally determine the time scale of the nucleosynthesis process and the produced isotopic abundances.

Journal of Physics: Conference Series **366** (2012) 012039 doi:10.1088/1742-6596/366/1/012039

The nuclear structure problem involved in the calculation of these rates must be treated in a reliable way. This means in particular, that the nuclear models should be able to reproduce at least the experimental information available on the decay properties (Gamow-Teller strength distributions and β-decay half-lives) measured under terrestrial conditions. The model used in this work is the quasiparticle random phase approximation (QRPA) based on a deformed Skyrme Hartree-Fock formalism with pairing correlations in BCS approximation. Measured beta-decay half-lives and Gamow-Teller (GT) strength distributions in neutron-deficient medium-mass nuclei [2, 3] are well reproduced by theoretical calculations based on deformed QRPA [4, 5].

Weak decay rates at high ρ and T are different from the decays observed under terrestrial conditions. First, one has to deal with the decay from thermally populated excited states in the parent nucleus. Secondly, because atoms in these scenarios are completely ionized, the existence of an electron plasma opens the possibility for continuum electron captures (cEC). These effects make weak interaction rates in the stellar interior sensitive functions of T and ρ [6, 7], with $T = 1.5$ GK and $\rho = 10^6$ g cm^{-3}, as the most relevant conditions for the rp process [1].

2. Theoretical formalism

The general formalism to calculate weak interaction rates in stellar environments as a function of ρ and T was introduced in Ref. [8]. Improvements on the nuclear structure part were subsequently introduced either from the Shell Model [9] or from QRPA [10].

The decay rate of the parent nucleus is given by [8]

$$\lambda = \sum_i \lambda_i \frac{2J_i+1}{G} e^{-E_i/(kT)} ; \qquad G = \sum_i (2J_i+1) e^{-E_i/(kT)} ; \qquad \lambda_i = \sum_f \frac{\ln 2}{6146} B_{if} \Phi_{if}(\rho, T) , \tag{1}$$

where thermal equilibrium is assumed. In principle, the sum over initial states extends over all populated states in the parent nucleus below the proton separation energy. However, since the range of T for the rp process peaks at $T = 1.5$ GK ($kT \sim 300$ keV), only a few low-lying excited states are expected to contribute significantly in the decay. Specifically, we consider in this work all the (collective) low-lying excited states below 1 MeV.

The rates are decomposed into a nuclear structure part B_{if} and a phase space factor Φ_{if}. In this work, B_{if} is given by the transition probabilities of allowed GT transitions, which are dominant,

$$B_{if}(GT) = \frac{1}{2J_i+1} \left(\frac{g_A}{g_V} \right)^2_{\rm eff} \langle f || \sum_k \sigma^k t^k_{\pm} || i \rangle^2 , \tag{2}$$

where $(g_A/g_V)_{\rm eff} = 0.74 (g_A/g_V)_{\rm bare}$ is an effective quenched value.

The theoretical formalism used to calculate the GT strengths is based on the QRPA. The quasiparticle basis corresponds to a deformed selfconsistent Skyrme Hartree-Fock calculation with the density-dependent Skyrme SLy4 force and pairing correlations treated in BCS approximation. The residual interactions are spin-isospin forces in both particle-hole and particle-particle channels. The initial and final states in the laboratory frame are expressed in terms of the intrinsic states using the Bohr-Mottelson factorization, which is a very good approximation for well deformed nuclei. The effect of these ingredients (deformation, pairing, residual forces) on the decay rates has been studied elsewhere [11, 12]. More details of the formalism can be found in Ref. [13, 14]. In particular, the parameters for residual forces used here are the same as those used in Ref. [4], where good agreement was obtained with the experimental GT strength distributions and β-decay half-lives. The success of this theoretical formalism in reproducing terrestrial decay rates supports its application to the calculation of the weak interaction rates in stellar matter.

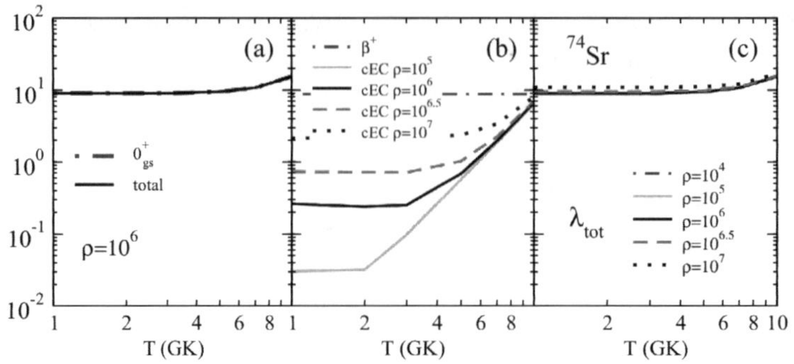

Figure 1. Weak decay rates λ (s^{-1}) for ^{74}Sr as a function of temperature for various densities.

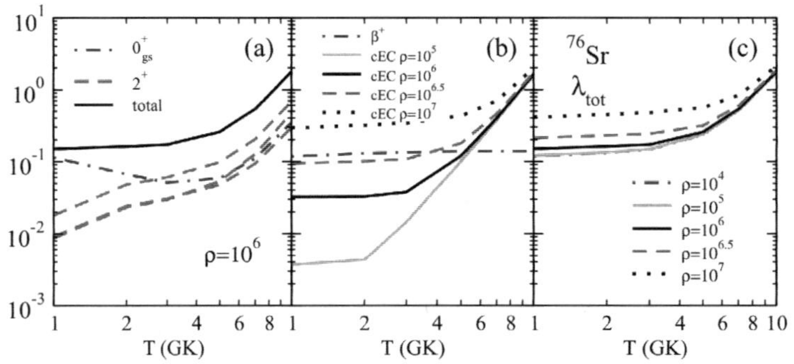

Figure 2. Same as in Fig. 1, but for ^{76}Sr.

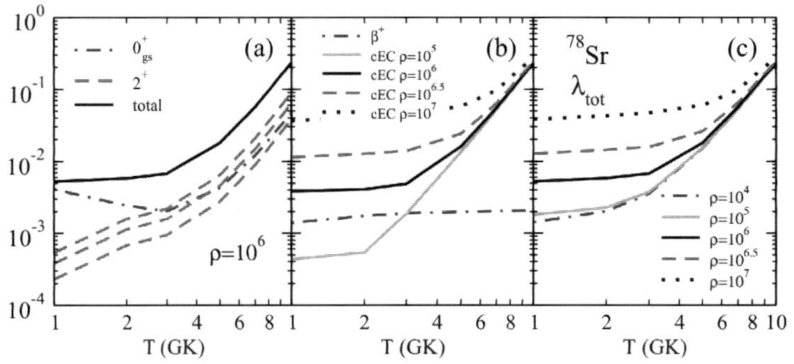

Figure 3. Same as in Fig. 1, but for ^{78}Sr.

The phase space factors in Eq. (1) are decomposed into their β^+ and cEC contributions,

$$\Phi_{if} = \Phi_{if}^{cEC} + \Phi_{if}^{\beta^+} , \tag{3}$$

where

$$\Phi_{if}^{cEC} = \int_{\omega_\ell}^{\infty} \omega\sqrt{\omega^2 - 1}(Q_{if} + \omega)^2 F(Z, \omega) \times S_e(\omega)\left[1 - S_\nu(Q_{if} + \omega)\right] d\omega , \tag{4}$$

Journal of Physics: Conference Series **366** (2012) 012039 doi:10.1088/1742-6596/366/1/012039

for continuum electron capture, and

$$\Phi_{if}^{\beta^+} = \int_1^{Q_{if}} \omega\sqrt{\omega^2-1}(Q_{if}-\omega)^2 F(-Z+1,\omega) \times [1-S_p(\omega)]\,[1-S_\nu(Q_{if}-\omega)]\, d\omega\,, \tag{5}$$

for positron emission. In these expressions ω is the total energy of the electron (positron) in units of $m_e c^2$ and $F(Z,\omega)$ is the Fermi function [15] that takes into account the distortion of the β-particle wave function due to the Coulomb interaction.

$$Q_{if} = \frac{1}{m_e c^2}\left(M_p - M_d + E_i - E_f\right) \tag{6}$$

is the total energy available in the decay in units of $m_e c^2$. It is written in terms of the nuclear masses of parent (M_p) and daughter (M_d) nuclei and their excitation energies E_i and E_f, respectively. In Eq. (4), the lower integration limit is given by $\omega_\ell = 1$ if $Q_{if} > -1$, or $\omega_\ell = |Q_{if}|$ if $Q_{if} < -1$. S_e, S_p, and S_ν, are the electron, positron, and neutrino distribution functions, respectively. Its presence inhibits or enhances the phase space available. In rp scenarios the commonly accepted assumptions state that $S_\nu = S_p = 0$.

The electron distribution is described by a Fermi-Dirac distribution

$$S_e = \frac{1}{\exp\left[\left(\omega-\mu_e\right)/(kT)\right]+1}\,, \tag{7}$$

assuming that nuclei at these temperatures are fully ionized and the electrons are not bound to nuclei. The chemical potentials μ_e are determined from ρ and T.

3. Results and discussion

Figs. 1, 2, and 3 show the decay rates versus T for ^{74}Sr, ^{76}Sr, and ^{78}Sr, respectively. In panel (a) one can see the decomposition of the total rate into their contributions from the decay of the ground state $0^+_{\rm gs}$ and from the decay of the excited 2^+ states. Panel (b) contains the decomposition of the rates into their β^+ and cEC components evaluated at different densities. Panel (c) shows the total rates at various densities.

The results decomposed into their contributions from various parent states (a) show that the contributions of the decays from excited states increase with T, as they become more and more thermally populated. In general they do not represent significant contributions to the total rates and can be neglected in most cases [6, 7]. However, in those cases where the excitation energy of the 2^+ excited state is very low, as in the cases discussed here for Sr isotopes, the contributions of the low-lying excited states compete with those of the ground state already at temperatures in the range of rp process.

Concerning the competition between β^+ and cEC rates (b) one can clearly distinguish between the three isotopes. The most exotic isotope (Fig. 1) shows a clear dominance of the β^+ rates over the cEC ones that can be neglected except at very high densities beyond rp-process conditions. On the other hand, the opposite is true with respect to the more stable isotopes (Fig. 3), where the β^+ rates are completely negligible. The origin of these features can be understood from the behavior of the phase space factors as a function of the available energy Q_{if}. In the case of the $N=Z$ WP nuclei (Fig. 2), there is a competition between β^+ and cEC rates that depends on the nucleus, on the temperature, and on the density and that must be analyzed case by case.

Finally, the total rates in (c) are a consequence of the competition between β^+ and cEC rates mentioned above. Since the β^+ decay rate is independent of the density and depends on T only through the contributions from excited parent states, the total rates are practically constant for

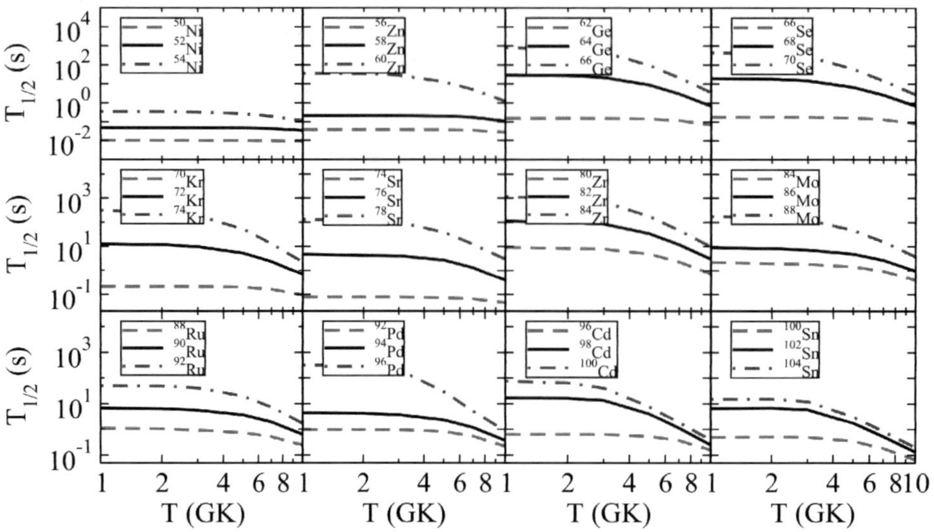

Figure 4. Half-lives as a function of T at a density $\rho Y_e = 10^6$ mol cm^{-3}.

the most exotic isotope in Fig. 1, only modulated by the small contribution from cEC. But in general this is not true for WP nuclei. Fig. 4 contains the half-lives ($T_{1/2} = \ln 2/\lambda$), as a function of T at a fixed $\rho Y_e = 10^6$ mol cm^{-3} for different isotopes. The decrease of $T_{1/2}$ as T increases is apparent.

Summarizing, we have analyzed the relevant ingredients to describe the decay rates at the peak rp-process conditions of ρ and T. We have studied the contributions to the decay rates coming from excited states in the parent nucleus, which are populated as T raises. It is found that they start to play a role above T =1-2 GK. The effect of the cEC rates is enhanced as T and ρ increase and they compete with the β^+ decay rates in the WP nuclei at rp peak conditions.

Acknowledgments
This work was partially supported by MICINN (Spain) under contract Nos. FIS2008-01301 and FIS2011-23565, as well as by Consolider-Ingenio 2010 Program CPAN CSD2007-00042.

References

[1] Schatz H *et al.* 1998 *Phys. Rep.* **294** 167
[2] Poirier E *et al.* 2004 *Phys. Rev.* C **69** 034307
[3] Nácher E *et al.* 2004 *Phys. Rev. Lett.* **92** 232501
[4] Sarriguren P 2009 *Phys. Rev.* C **79** 044315
[5] Sarriguren P *et al.* 2005 *Eur. Phys. J.* A **24** 193
[6] Sarriguren P 2009 *Phys. Lett.* B **680** 438
[7] Sarriguren P 2011 *Phys. Rev.* C **83** 025801
[8] Fuller G M, Fowler W A and Newman M J 1980 *Astrophys. J. Suppl. S.* **42** 447; 1982 *Astrophys. J.* **252** 715; 1982 *Astrophys. J. Suppl. S.* **48** 279; 1985 *Astrophys. J.* **293** 1
[9] Langanke K and Martinez-Pinedo G, 2000 *Nucl. Phys.* A **673** 481; 2003 *Rev. Mod. Phys.* **75** 819
[10] Nabi J U and Klapdor-Kleingrothaus H V 1999 *At. Data Nucl. Data Tables* **71** 149; 2004 **88** 237
[11] Sarriguren P *et al.* 1999 *Nucl. Phys.* A **658** 13
[12] Sarriguren P *et al.* 2001 *Nucl. Phys.* A **691** 631
[13] Muto K *et al.* 1992 *Z. Phys.* A **341** 407
[14] Sarriguren P *et al.* 1998 *Nucl. Phys.* A **635** 55
[15] Gove N B and Martin M J 1971 *Nucl. Data Tables* **10** 205

Journal of Physics: Conference Series **366** (2012) 012040 doi:10.1088/1742-6596/366/1/012040

Experimental evidences for low-lying octupole isovector states

M Scheck

Institut für Kernphysik, TU Darmstadt, D-64289 Darmstadt, Germany

E-mail: mscheck@ikp.tu-darmstadt.de

Abstract. An evaluation of data obtained with the $(n, n'\gamma)$ technique revealed in near spherical nuclei strong M1 components in transitions connecting higher-lying 3^-_i states with the 3^-_1 state. The $\langle 3^-_1|M1|3^-_i\rangle$ matrix elements of these transitions are in the order of 1 μ_N. In the quadrupole sector strong $\langle 2^+_1|M1|2^+_{ms}\rangle$ matrix elements serve as fingerprints for the identification of low-lying 2^+_{ms} isovector states, which are commonly referred to as mixed-symmetry states. The ratio of the two matrix elements scales for most candidates to unity indicating similar physics. Additional evidence for the assignment of the 3^-_i states as low-lying octupole isovector states is provided by particle-scattering experiments, e.g. (e, e') or (p, p'). The octupole states under investigation exhibit enhanced scattering cross sections. This observation is in accordance with the assignment of a one-phonon nature to these octupole excitations.

1. Introduction

Low-lying isovector states, such as the 1^+ Scissors mode in deformed nuclei [1, 2] or the 2^+_{ms} state [3, 4] in near spherical nuclei, continue to have impact on the understanding of low-energy nuclear structure physics. The importance of these kind of nuclear excitations is based on the fact, that their observation yields direct experimental information about the strength of the isovector (proton-neutron) part of the nuclear interaction. The existence of isovector states is a feature of the nucleus as a two-component quantum system. From a more general perspective, these class of excitations can be seen as the anti-symmetric coupling of proton and neutron subsystems. Similar excitations have been observed in other two-component quantum systems, as for example quantum dots [2].

Low-lying quadrupole isovector, in the language of the IBA [5] so-called mixed-symmetry, states are well established. Currently, their fine details are subject to experimental and theoretical work. Therefore, nowadays, the theoretical description often uses microscopic models, such as the shell model [6] (e.g. see ref. [7]) or the Quasiparticle-Phonon Model (QPM) [8] (see ref. [9] and references therein). However, the sdf-IBA-2 predicts low-lying isovector excitations for the octupole degree of freedom [10] as well. Within the $U(1) \otimes U(5) \otimes U(7)$ limit of the spf-IBA-2 the experimental signature of these octupole mixed-symmetry states is a strong M1 transition to the first excited 3^-_1 state:

$$B(M1, 3^-_{ms} \rightarrow 3^-_1) = \frac{9}{\pi}(g'_\pi - g'_\nu)^2 \frac{N_\pi N_\nu}{N}. \qquad (1)$$

Thus, the strength of the M1 transition connecting the two 3^- states only depends of the proton (π) or neutron (ν) g-factors, $g'_\rho(\rho = \pi \text{ or } \nu)$, the number of neutron and proton bosons, N_ρ, and

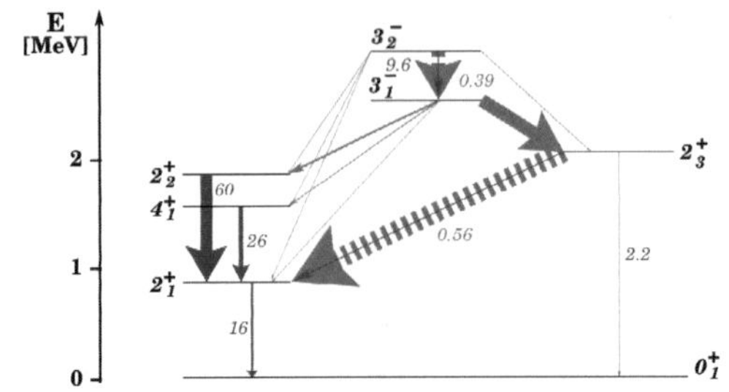

Figure 1. Partial level scheme of ^{94}Mo: E2 transitions are marked with blue arrows, E1 transitions with green arrows and M1 transitions with dashed, red arrows. The thickness of the arrows is for each multipolarity proportional to the strength of the transition. The numbers give the B($\pi L, J^\pi \to (J^\pi)'$) strength in units of W.u. (E2) and μ_N^2 (M1). Data are taken from [11].

the total number of bosons, $N = N_\nu + N_\pi$. However, the prediction of the excitation energy, where the isovector-octupole states are expected is vaguely given as above 3.5 MeV.

As long as particle-hole excitations over a shell closure can be excluded, the microscopic structure of octupole excitations is dominated by particle-hole excitations including the unique parity intruder subshell. The latter is a necessary condition for a negative parity excitation. Therefore, in the low-energy regime octupole excitations have, compared to quadrupole states, the advantage that less microscopic configurations are available. This lower density of two-quasiparticle excitations near the Fermi surface results in much simpler wavefunctions. Nuclei, for which the Fermi levels of protons as well as neutrons are situated between the intruding unique-parity subshell and the subshell three units lower in total (j) and orbital (l) angular momentum, exhibit enhanced effects associated with the octupole degree of freedom [12]. The main signatures for enhanced octupole collectivity are a low-lying first excited 3_1^- state and a high B(E3, $0_{gs}^+ \to 3_1^-$) value. Near spherical nuclei in such mass regions (Z or $N \approx 34, 56, 88, 134$) are favorable cases to search for octupole-isovector excitations.

2. Evaluated data and discussion

An evaluation of data sets obtained in inelastic neutron scattering (INS) experiments [13] revealed for several nuclei near the N=50, Z=50, and N=82 shell closures higher-lying 3^- states, which decay via strong M1 transitions to the first excited 3_1^- state.

The INS technique represents one of the most powerful experimental tools for the investigation of the low-energy level scheme of stable nuclei [14]. This spectroscopic complete tool allows for the determination of transition energies, branching ratios, and γ-ray angular distributions. Consequently, it is possible to deduce level energies, level spins and multipole-mixing ratios. The use of mono-energetic neutrons, as it is successfully done for decades at the University of Kentucky, additionally allows the application of the Doppler-shift attenuation method (DSAM) [15] to determine level lifetimes and, therefore, in combination with the other observables also absolute transition probabilities.

The limitations of INS are given by the maximum angular momentum transfer of six units and by the excitation energy range, from which on the level density becomes too high. The latter results in a lower excitation cross section for the individual states, what results in low γ-ray yields and in crowded spectra with a large number of overlapping peaks. This drawback led to considerable uncertainties in the data sets discussed in this contribution.

States that were identified by having a strong M1 component in their $3_i^- \to 3_1^-$ transition to be a candidate for an isovector octupole excitation are listed in table I of ref. [13]. Uncertainties

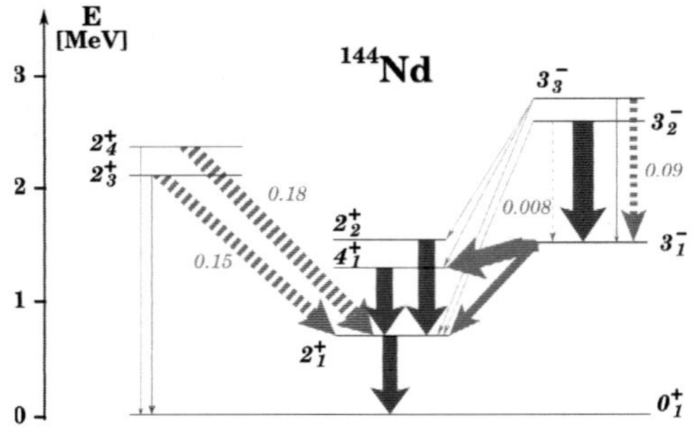

Figure 2. Partial level scheme of ^{144}Nd: E2 transitions are marked with blue arrows, E1 transitions with green arrows and M1 transitions with dashed, red arrows. The thickness of the arrows is for each multipolarity proportional to the strength of the transition. The numbers give the $B(M1, J^\pi \to (J^\pi)')$ strength in units of μ_N^2. Data are taken from [16].

in the evaluated data sets, due to the above mentioned reasons, are discussed extensively in the given reference. Two examples of partial level schemes are shown in figures 1 and 2. While for ^{94}Mo an additional E2 component exhausting 50 % of the $B(E2, 2_1^+ \to 0_{gs}^+)$ strength indicates a mixing of the octupole isovector excitation with the quadrupole-octupole coupled 3^- two-phonon state, the octupole isovector candidate (3_3^- level) in ^{144}Nd exhibits only a small mixing with the quadrupole-octupole coupled 3_2^- state at 2605 keV. However, as later will be discussed, the candidate in ^{92}Zr given in ref. [13] is not the state at 3039.8 keV.

For the two Cd isotopes, where candidates for isovector octupole excitations were assigned, it should be mentioned, that an assignment of octupole isovector states based on strong M1 transitions is not as trivial as it seems. Recent measurements of the $B(E2, 0_{gs}^+ \to 2_1^+)$ strength in light Sn isotopes (e.g. see ref. [17]) provided evidence for proton contributions in the wavefunctions of their 2_1^+ states. This observed weakening of the Z=50 shell closure allows for the contribution of two-quasiparticle components containing the $g_{7/2}$ subshell in the wavefunction of excited 3^- states. The decay of such a 3^- state to the first excited 3_1^- state with components containing the $g_{9/2}$ subshell in the wavefunctions results in strong spin-flip M1 transitions.

The ratio of the M1 matrix element connecting the 3_{IV}^- octupole isovector candidate and the first excited 3_1^- state and the M1 matrix element connecting the 2_{ms}^+ quadrupole mixed symmetry state and the first excited 2_1^+ state scale for almost all candidates to unity. The comparable strengths of the matrix elements indicate similar physics.

Besides the first evidence provided by the strong M1 components in the $3_i^- \to 3_1^-$ transitions, an evaluation of data sets obtained in light-particle scattering experiments, such as (p, p') and (e, e'), provides additional evidence. These data sets exhibit enhanced cross sections for the direct excitation of the candidates. As the isovector octupole excitations are one-phonon states a relatively high excitation probability can be expected. Two exemplary spectra obtained with the (p, p') reaction measured using 200 MeV protons at iThemba Labs (RSA) are shown in figure 3. Indeed, the excitation probability for the candidate in ^{94}Mo is, next to the 3_1^- symmetric octupole phonon, the only excited 3^- state with a noteworthy cross section. The ratio of the maximum scatting cross sections measured in this experiment for the excitation of the first 3_1^- state and the candidate for the isovector excitation are presented in table 1. Additionally, the ratio of the 2_{ms}^+ quadrupole mixed-symmetry state and the first excited 2_1^+ state is given. Furthermore, the ratios of the excitation probabilities for ^{144}Nd, as obtained in (p, p'), (d, d') [18] and (e, e') experiments [19], are presented. Indeed, the enhanced excitation strength from the ground state indicates an one-phonon nature for the proposed candidates. For ^{92}Zr the results of the (p, p')

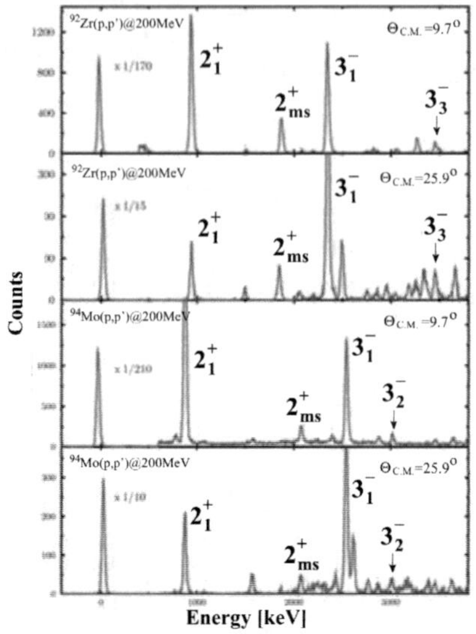

Figure 3. Spectra of ^{92}Zr and ^{94}Mo as measured in the (p, p') reaction using 200 MeV protons at iThemba labs (RSA). Figure is taken from [20].

measurement qualify a state at 3450 keV as isovector octupole excitation, not as stated in [13] the state at 3039.8 keV. In the $(n, n'\gamma)$ reaction [21] at 3452 keV a level with a transition to the 3_1^- state was observed. However, for two of the four transition depopulating this level to states with a firmly assigned positive parity a multipole-mixing ratio with values deviating from zero was measured. The latter indicates a positive parity of the state observed in the $(n, n'\gamma)$ experiment. Nevertheless, a scenario with two degenerated levels, one with positive and one with negative parity, is possible. Certainly, further experimental investigations are necessary to clarify the situation.

Table 1. Ratios of selected quantities from light-particle scattering experiments. The same ratios as presented for the first excited 3^- state and the 3^-_{IV} octupole isovector candidate are given for the first 2^+ state and the firmly identified 2^+ mixed-symmetry state(s). For ^{144}Nd the summed B(E2) strength for the two fragments of the mixed-symmetry state were considered.

Nucleus	Method	Octupole	Quadrupole	Reference
^{92}Zr	(p, p')	$\frac{\sigma(3_1^-)_{max}}{\sigma(3_{IV}^-)_{max}} = 8.7$	$\frac{\sigma(2_1^+)_{max}}{\sigma(2_{IV}^+)_{max}} = 4.0$	[20]
^{94}Mo	(p, p')	$\frac{\sigma(3_1^-)_{max}}{\sigma(3_{IV}^-)_{max}} = 20.5$	$\frac{\sigma(2_1^+)_{max}}{\sigma(2_{ms}^+)_{max}} = 13.2$	[20]
^{144}Nd	(p, p'), $((d, d')$	$\frac{B(E3,0^+\to3_1^-)}{B(E3,0^+\to3_{IV}^-)} = 4.7(7)$	$\frac{B(E2,0^+\to2_1^+)}{B(E2,0^+\to2_{ms}^+)} = 9.4(15)$	[19]
	(e, e')	$\frac{B(E3,0^+\to3_1^-)}{B(E3,0^+\to3_{IV}^-)} = 5.8(12)$	$\frac{B(E2,0^+\to2_1^+)}{B(E2,0^+\to2_{ms}^+)} = 8.1(10)$	[18]

3. Summary

Data sets of near spherical nuclei, obtained with the $(n, n'\gamma)$ technique, reveal strong M1 components in $3_i^- \rightarrow 3_1^-$ transitions. For several of the 3_i^- states light-particle scattering experiments exhibit enhanced excitation cross sections from the ground state. The two experimental features are with respect to the first excited 3^- state in the order of magnitude as observed for the combination of 2_{ms}^+ quadrupole mixed-symmetry state and the 2_1^+ quadrupole phonon. The possibility of assigning the 3_i^- states as candidates for isovector octupole excitations was discussed.

4. Acknowledgments

Financial support from the Deutsche Forschungsgemeinschaft under Grant SFB 634 is gratefully acknowledged. Furthermore, the author appreciated stimulating discussions with P. A. Butler, C. Fransen, K. Heyde, N. Pietralla, G. Rainovski, V. Werner, and S. W. Yates. The author is indebted to the courtesy of C. Walz allowing the use of his data prior to publication.

References

[1] Bohle D et al. 1984 *Phys. Lett.* **137B** 27
[2] Heyde K, von Neumann Cosel P and Richter A 2010 *Rev. Mod. Phys.* **82** 2365
[3] Iachello F 1984 *Phys. Rev. Lett.* **53** 1427
[4] Pietralla N von Brentano P and Lisetzky A F 2008 *Prog. Part. Nucl. Phys.* **60** 225
[5] Iachello F and Arima A 2006 *The interacting boson model* (Cambridge University Press)
[6] Heyde K L G 1994 *The nuclear shell model* (Berlin Heidelberg Springer Verlag)
[7] Holt J D et al. 2007 *Phys. Rev.* C **76** 034325
[8] Soloviev V G 1992 *Theory of Atomic Nuclei Quasiparticles and Phonons* (London: IOP Publishing Ltd)
[9] Lo Iudice N, Stoyanov Ch and Pietralla N 2009 *Phys. Rev.* C **80** 024311
[10] Smirnova N A et al. 2000 *Nucl. Phys.* A **678** 235
[11] Fransen C et al. 2003 *Phys. Rev.* C **67** 024307
[12] Butler P A and Nazarewicz W 1996 *Rev. Mod. Phys.* **68** 349
[13] Scheck M et al. 2010 *Phys. Rev.* C **81** 064305
[14] Garrett P E, Warr N and Yates S W 2000 *J. Res. Natl. Inst. Stand. Technol.* **105** 141
[15] Belgya T, Molnàr G and Yates S W 1996 *Nucl. Phys.* A **607** 43
[16] Hicks S et al. 1998 *Phys. Rev.* C **98** 2264
[17] Ekström A at al. 2008 *Phys. Rev. Lett.* **101** 012502
[18] Perrino R et al. 1993 *Nucl. Phys.* A **561** 343
[19] Pignanelli M et al. 1993 *Nucl. Phys.* A **559** 1
[20] Walz C 2010 *Master Thesis, TU Darmstadt* unpublished
[21] Fransen C et al. 2005 *Phys. Rev.* C **71** 054304

Journal of Physics: Conference Series **366** (2012) 012041 doi:10.1088/1742-6596/366/1/012041

Massive Majorana neutrinos in nuclear processes

Fedor Šimkovic[1,2]

[1]Laboratory of Theoretical Physics, JINR, RU-141980 Dubna, Moscow region, Russia
[2]Department of Nuclear Physics and Biophysics, Comenius University, Mlynská dolina F1, SK-84215 Bratislava, Slovakia

E-mail: simkovic@fmph.uniba.sk

Abstract. The observation of neutrino oscillations has opened a new excited era in neutrino physics and represents a big step forward in our knowledge of neutrino properties. The observed small neutrino masses have profound implications for our understanding of the Universe and are now a major focus in astro, particle and nuclear physics and in cosmology. Of crucial importance is the problem of the nature of neutrinos (Dirac or Majorana). In this context the experimental and theoretical study of the neutrinoless double-beta decay and double-electron capture is at the forefront.

1. Introduction

Neutrinos are an enigma, billions of them come streaming out of the Sun's core and some neutrinos are created by high-energy cosmic rays from deep space striking air molecules in the Earth's upper atmosphere. Terrestrial neutrinos are produced at nuclear reactors, accelerators and due to natural radioactivity.

Neutrinos are electrically neutral, and are able to pass through ordinary matter almost undisturbed. This makes neutrinos extremely difficult to detect and to learn more about their basic properties. The recent neutrino oscillation experiments have convinced us that neutrinos are massive particles. However, the problem of absolute scale of neutrino masses, CP violation in neutrino sector, magnetic moments and the nature of neutrinos (Dirac or Majorana), possible existence of sterile neutrinos is still waiting for a solution.

Precise determinations of the neutrino properties will shed light on the breaking of the SM symmetries associated with total baryon and lepton number and lepton flavors. Moreover, neutrinos are potentially a window (through the so-called see-saw mechanism, where the smallness of neutrino masses is explained by the interplay of weak-scale Dirac masses with much larger Majorana masses) to very high mass scales, directly inaccessible to foreseeable colliders.

At present the attention of physics community is concentrated on challenging problem, finding whether neutrinos are indeed Majorana particles (i.e. identical to its own antiparticle) as many particle models suggest or Dirac particles (i.e. is different from its antiparticle).

2. Neutrinoless double-beta decay

The total lepton number (LN) violating neutrinoless double beta decay ($0\nu\beta\beta$-decay) [1],

$$(A, Z) \to (A, Z+2) + e^- + e^-, \tag{1}$$

Table 1. Averaged $0\nu\beta\beta$ nuclear matrix elements $\langle M^{0\nu\beta\beta}\rangle$ and their variance σ (in parentheses) evaluated in the RQRPA and QRPA [7]. $M^{2\nu}_{GT}$ denotes the $2\nu\beta\beta$-decay nuclear matrix element deduced from $T^{2\nu-exp}_{1/2}$. In column 5 the $0\nu\beta\beta$ half-lives evaluated with the RQRPA average nuclear matrix element and for assumed $m_{\beta\beta} = 50$ meV are shown. For ^{136}Xe there are two entries: the upper two use the upper limit of the 2ν matrix element while the lower two use the ultimate limit, vanishing 2ν matrix element. $g_A = 1.25$ is assumed.

Nuclear transition	$M^{2\nu}_{GT}$ [MeV^{-1}]	$\langle M^{0\nu\beta\beta}\rangle$ RQRPA	QRPA	$T^{0\nu\beta\beta}_{1/2}$ ($\langle m_{\beta\beta}\rangle = 50$ meV) [y]
^{76}Ge $\rightarrow$ ^{76}Se	0.15 ± 0.006	5.44(0.23)	6.32(0.32)	$4.46^{0.52}_{-0.44} \times 10^{26}$
^{82}Se $\rightarrow$ ^{82}Kr	0.10 ± 0.009	4.86(0.20)	5.65(0.27)	$1.26^{0.17}_{-0.14} \times 10^{26}$
^{96}Zr $\rightarrow$ ^{96}Mo	$0.11^{+0.03}_{-0.06}$	2.01(0.20)	2.09(0.03)	$3.49^{2.48}_{-0.93} \times 10^{26}$
^{100}Mo $\rightarrow$ ^{100}Ru	0.22 ± 0.01	4.28(0.28)	5.25(0.31)	$9.99^{1.64}_{-1.32} \times 10^{25}$
^{116}Cd $\rightarrow$ ^{116}Sn	0.12 ± 0.006	3.41(0.24)	3.99(0.15)	$1.45^{0.26}_{-0.21} \times 10^{26}$
^{128}Te $\rightarrow$ ^{128}Xe	0.034 ± 0.012	4.82(0.15)	5.49(0.16)	$2.04^{0.23}_{-0.20} \times 10^{27}$
^{130}Te $\rightarrow$ ^{130}Xe	$0.036^{+0.03}_{-0.009}$	4.40(0.13)	4.92(0.12)	$9.70^{1.05}_{-1.65} \times 10^{25}$
^{136}Xe $\rightarrow$ ^{136}Ba	0.030	2.89(0.17)	3.11(0.13)	$2.11^{0.27}_{-0.23} \times 10^{26}$
	0	2.53(0.17)	2.73(0.13)	$2.76^{0.40}_{-0.33} \times 10^{26}$

is the most powerful tool to clarify if the neutrino is a Dirac or a Majorana particle. From the experimental point of view LN violation in $0\nu\beta\beta$-decay is observed through the appearance of two electrons in the final state with no missing energy.

The inverse value of the $0\nu\beta\beta$-decay half-life for a given isotope (A,Z) is a product of the effective mass of Majorana neutrinos $m_{\beta\beta}$, the known phase-space factor $G_{0\nu}(Q_{\beta\beta}, Z)$ (depending on nuclear charge Z and the energy release $Q_{\beta\beta}$ of the reaction) and the nuclear matrix element $M^{0\nu}$, which depends on the nuclear structure of the particular isotope under study [1]:

$$(T^{0\nu\beta\beta}_{1/2})^{-1} = G^{0\nu}(Q_{\beta\beta}, Z)\, |M^{0\nu\beta\beta}|^2\, |m_{\beta\beta}|^2, \tag{2}$$

where nuclear matrix element $M^{0\nu\beta\beta}$ consists of Fermi, Gamow-Teller and tensor parts as

$$M^{0\nu\beta\beta} = -\frac{M_F}{g_A^2} + M_{GT} + M_T. \tag{3}$$

Experimentally, half-lives are measured or constrained, and the effective Majorana neutrino mass $m_{\beta\beta}$,

$$m_{\beta\beta} = U^2_{e1}m_1 + U^2_{e2}m_2 + U^2_{e3}m_3, \tag{4}$$

is the ultimate goal. Here, U_{ei} and m_i (i=1,2,3) are elements of Pontecorvo-Maki-Nakagawa-Sakata (PMNS) neutrino mixing matrix [2, 3] and masses of neutrinos, respectively.

From the most precise experiments on the search for $0\nu\beta\beta$-decay the following bounds were inferred: $|m_{\beta\beta}| < (0.20 - 0.32)\ eV$ (^{76}Ge [4]), $(0.30 - 0.71)\ eV$ (^{130}Te [5]) and $(0.50 - 0.96)\ eV$ (^{130}Mo [6]). Here, the $0\nu\beta\beta$-decay NMEs of calculated with Brueckner two-nucleon short-range correlations are used [7]. A new generation of the $0\nu\beta\beta$-decay experiments, in particular GERDA/MAJORANA(^{76}Ge), SuperNEMO(^{82}Se), CUORE(^{130}Te), MOON(^{100}Mo), COBRA(^{116}Cd), LUCIFER(^{82}Se, ^{116}Cd), EXO(^{136}Xe), Kamland-ZEN(^{136}Xe), SNO+(^{150}Nd) and other experiments, hope to probe $m_{\beta\beta}$ down to 10-50 meV [1], what is the value predicted by making assumption about the inverted hierarchy of neutrino masses. These experiments with about up to 1 ton of radioactive isotope would require 5-10 years of measurements.

Journal of Physics: Conference Series **366** (2012) 012041 doi:10.1088/1742-6596/366/1/012041

Table 2. The calculated 0νECEC half-lives of ^{152}Gd for $m_{\beta\beta}$ = 50 meV. The second and third columns show the quantum numbers of the electron holes. Here, n is the principal quantum number, j is the total angular momentum, and l is the orbital momentum. The column four shows the radiative widths of the excited electron shells. The column five shows the mass difference of the initial and final atoms. The columns six and seven present the minimum and maximum half-lives of the 0νECEC transitions obtained within the spherical QRPA ($M^{0\nu ECEC}$ = 7.6). The last two columns show the minimum and maximum half-lives calculated by assuming the nuclear matrix element of the deformed QRPA ($M^{0\nu ECEC}$ = 3.2).

					spherical QRPA		deformed QRPA	
Nucl.	n 2j l	n 2j l	Γ_{ab}	Δ	$T^{min}_{1/2}$	$T^{max}_{1/2}$	$T^{min}_{1/2}$	$T^{max}_{1/2}$
	a	b	[keV]	[eV]	[y]	[y]	[y]	[y]
^{152}Gd	110	210	23	-0.8 ± 0.2	8.5×10^{27}	8.7×10^{28}	4.7×10^{28}	4.8×10^{29}
	110	211	23	-1.3 ± 0.2	7.6×10^{30}	2.0×10^{31}	4.2×10^{31}	1.1×10^{32}
	110	310	32	-7.1 ± 0.2	1.7×10^{31}	2.0×10^{31}	9.4×10^{31}	1.1×10^{32}

For $m_{\beta\beta}$ equal to 50 meV the half-lives for double β-decaying nuclei of interest are presented in Table 1. The associated NMEs were calculated in the framework of the Quasiparticle Random Phase Approximation (QRPA) and renormalized QRPA by assuming Brueckner short-range correlations and the weak coupling constant of free nucleon $g_A = 1.25$ [7]. We see that half-lives are of 10^{26} years and more.

3. Resonant neutrinoless double-electron capture

There has been an increased theoretical and experimental interest [8, 9, 10, 11, 12, 13, 14, 15, 16] to another LN violating process, namely to the neutrinoless double-electron capture (0νECEC) [17]:

$$(A, Z) + e_b^- + e_b^- \rightarrow (A, Z-2)^{**}. \tag{5}$$

Here, the two asterisks denote the possibility of leaving the system in an excited nuclear and/or atomic state, the latter being characterized by two vacancies in the electron shell of the otherwise neutral atom.

The process of the $0\nu ECEC$ has been revisited for those cases where the two participating atoms are nearly degenerate in mass [10]. The reaction (5) could in principle be detected by monitoring the X-rays or Auger-electrons emitted from excited electron shell of the atom and the electromagnetic decay of the excited nucleus (in case of a non-ground-state transition). However, the main question remains, which atomic systems are favorable for the detection of the 0νECEC, what would mean that neutrino is a Majorana particle.

A new theoretical approach developed by Šimkovic and Krivoruchenko [9] and Krivoruchenko et al. [10] allowed a unified description of the oscillations of stable and quasi-stationary atoms, which take place with violation of the total lepton number conservation and are followed by de-excitation with emission of photons.

The 0νECEC transition rate near the resonance is of Breit-Wigner form [9, 10, 17],

$$\Gamma^{0\nu\mathrm{ECEC}}_{ab} = \frac{|V_{ab}|^2}{\Delta^2 + \frac{1}{4}\Gamma^2_{ab}}\Gamma_{ab}. \tag{6}$$

The degeneracy parameter Δ can be expressed as $\Delta = Q - B_{ab} - E_\gamma$. Q stands for a difference between the initial and final atomic masses in ground states and E_γ is an excitation energy of

Journal of Physics: Conference Series **366** (2012) 012041 doi:10.1088/1742-6596/366/1/012041

the daughter nucleus. $B_{ab} = E_a + E_b + E_C$ is the energy of two electron holes, whose quantum numbers (n, j, l) are denoted by indices a and b and E_C is the interaction energy of the two holes. The binding energies of single electron holes $E_{a,b}$ are known with accuracy with few eV [18]. The width of the excited final atom with the electron holes is given by

$$\Gamma_{ab} = \Gamma_a + \Gamma_b + \Gamma^*. \tag{7}$$

Here, $\Gamma_{a,b}$ is one-hole atomic width and Γ^* is the de-excitation width of daughter nucleus, which can be neglected. Numerical values of Γ_{ab} are about up to few tens eV [19].

For favorable cases of a capture of $s_{1/2}$ and $p_{1/2}$ electrons the explicit form of lepton number violating amplitude associated with nuclear transitions $0^+ \to J^\pi = 0^{\pm 1}, 1^{\pm 1}$ can be find in [10]. After factorization the electron shell structure and nuclear matrix element and by considering the $0^+ \to 0^+$ nuclear transition we have

$$V_{ab} = \frac{1}{4\pi} G_\beta^2 m_{\beta\beta} \frac{g_A^2}{R} < F_{ab} > M^{0\nu ECEC}. \tag{8}$$

Here, $< F_{ab} >$ is a combination of averaged upper and lower bispinor components of the atomic electron wave functions [10] and $M^{0\nu ECEC}$ is the nuclear matrix element. We note that by neglecting the lower bispinor components $M^{0\nu ECEC}$ takes the form of the $0\nu\beta\beta$-decay NME for ground state to ground state transition after replacing isospin operators τ^- by τ^+.

Based on the most recent data and realistic evaluation of the decay half-lives, a complete list of the most perspective isotopes for which the 0νECEC capture may have the resonance enhancement was provided for further experimental study [10]. Some isotopes such as ^{156}Gd were found to admit several closely lying resonance levels. Assuming an effective mass for the Majorana neutrino of 50 meV and appropriate value of nuclear matrix element, some half-lives were found to be as low as 10^{25} years in the unitary limit.

A detailed calculation of the 0νECEC of ^{152}Gd associated with the ground-state to ground-state nuclear transitions was performed in [20]. Improved measurements of Q-value for this transitions with accuracy of about 100 eV [15] was considered. The nuclear matrix element of $^{152}\mathrm{Gd} \to {}^{152}\mathrm{Sm}$ transition was calculated within spherical QRPA.

However, nuclei participating in this 0νECEC ground state to ground state nuclear transition are deformed. The deformation parameter β can be extracted from values on measured E2 probability ($Q_p = \sqrt{16\pi B(E2)/5e^2}$, the sign cannot be extracted). Within the deformed QRPA [21, 22, 23] nuclear matrix element of the $^{152}\mathrm{Gd} \to {}^{152}\mathrm{Sm}$ transition becomes by factor 2.3 smaller when compared with the spherical case. From Table 2 we see that the effect of nuclear deformation has a strong impact on the 0νECEC half-life of ^{152}Gd. Currently, it is the smallest known half-life among known 0νECEC transitions, which is 2-3 orders of magnitude longer than the half-life of $0\nu\beta\beta$ decay of ^{76}Ge (see Table 1).

Precise measurements of Q-values between the initial and final atomic states [15, 24, 25, 26], additional spectroscopic information on the excited nuclear states (energy, spin and parity) and reliable calculation of corresponding NMEs are highly required to improve predictions of half-lives of the resonant 0νECEC. It is expected that the accuracy of 10 eV in the measurement of atomic masses will be achievable in the near future.

4. Conclusions

The search for the $0\nu\beta\beta$-decay represents the new frontiers of neutrino physics, allowing in principle to fix the neutrino mass scale, the neutrino nature and possible CP violation effects. Many next generation $0\nu\beta\beta$-decay experiments are in preparation or under consideration. In the case nature prefers inverted hierarchy of neutrino masses they have a chance to observe this rare process. It would prove that the total LN is not a conserved quantity and that neutrinos are massive Majorana fermions.

Journal of Physics: Conference Series **366** (2012) 012041 doi:10.1088/1742-6596/366/1/012041

It is reasonable to hope that the search for the 0νECEC of atoms, which are sufficiently long lived to conduct a practical experiment, may also established the Majorana nature of neutrinos. This possibility is considered as alternative and complementary to searches for the $0\nu\beta\beta$-decay.

To interpret the data from LN violating nuclear processes accurately better understanding of the nuclear structure effects important for the description of the corresponding nuclear matrix elements is needed. In this connection it is crucial to develop theoretical methods capable to evaluate reliably the nuclear matrix elements, and to realistically asses their uncertainties. Nuclear matrix elements are as important as data.

References

[1] F.T. Avignone, S.R. Elliott and J. Engel, Rev. Mod. Phys. **80**, 481 (2008).
[2] B. Pontecorvo, J. Exptl. Theoret. Phys. **33**, 549 (1957). [Sov. Phys. JETP **6**, 429 (1958)]; J. Exptl. Theoret. Phys. **34**, 247 (1958) [Sov. Phys. JETP **7**, 172 (1958)].
[3] Z. Maki, M. Nakagawa, and S. Sakata, Prog. Theor. Phys. **28**, 870 (1962).
[4] L. Baudis *et al.* (Heidelberg-Moscow Collaboration), Phys. Rev. Lett. **83**, 41 (1999).
[5] E. Andreotti et al. (CUORE Collaboration), arXiv:1012.3266; M.Sisti (CUORE Collaboration), J. Phys. Conf. Ser. **203**, 012069 (2010).
[6] R. Arnold et al. (NEMO Collaboration), Phys. Rev. Lett. **95**, 182302 (2005).
[7] F. Šimkovic, A. Faessler, H. Müther, V. Rodin, M. Stauf, Phys. Rev. C **79**, 055501 (2009).
[8] Z. Sujkowski and S. Wycech, Phys. Rev. C **70**, 052501 (2004).
[9] F. Šimkovic, M.I. Krivoruchenko, Phys. Part. Nucl. Lett. **6**, 298 (2009).
[10] M.I. Krivoruchenko, F. Šimkovic, D. Frekers, A. Faessler, Nucl. Phys. A **859**, 140 (2011).
[11] A.S. Barabash, Ph. Hubert, A. Nachab, and V. Umatov, Nucl. Phys. A **785**, 371 (2007).
[12] A.S. Barabash, Ph. Hubert, A. Nachab, S.I. Konovalov, I.A. Vanyushin, and V.I. Umatov, Nucl. Phys. A **807**, 269 (2008).
[13] P. Belli *et al.*, Nucl. Phys. A **842**, 101 (2009).
[14] N.I. Rukhadze *et al.*, Nucl. Phys. A **852**, 197 (2011).
[15] S. Eliseev, et al., Phys. Rev. Lett. **106**, 052504 (2011).
[16] D. Frekers et al., Nucl. Phys. A **860**, 1 (2011).
[17] J. Bernabéu, A. De Rujula, and C. Jarlskog, Nucl. Phys. B **223**, 15 (1983).
[18] F.B. Larkins, At. Data and Nucl. Data Tables **20**, 313 (1977).
[19] J.L. Campbell and T. Papp, At. Data and Nuclear Data Tables **77**, 1 (2001).
[20] F. Šimkovic, M.I. Krivoruchenko, Phys. Part. Nucl. Lett. **6**, 298 (2009).
[21] M. Yousuf Saleh, V. Rodin, A. Faessler, F. Šimkovic, Phys. Rev. C **79**, 014314 (2009).
[22] Dongliang Fang, A. Faessler, V. Rodin, M. Yousuf Saleh, F. Šimkovic, Phys. Rev. C **81**, 037303 (2010).
[23] Dong-Liang Fang, A. Faessler, V. Rodin, F. Šimkovic, Phys. Rev. C **82**, 051301 (2010).
[24] S. Eliseev, et al., Phys. Rev. C **83**, 038501 (2011).
[25] S. Eliseev, et al., Phys. Rev. C **84**, 012501 (2011).
[26] M. Goncharov, et al., Phys. Rev. C **84**, 028501 (2011).

Journal of Physics: Conference Series **366** (2012) 012042 doi:10.1088/1742-6596/366/1/012042

Simple structures in complex nuclei versus complex structures in simple nuclei: a nuclear moments perspective

Andrew E. Stuchbery

Department of Nuclear Physics, Research School of Physics and Engineering, The Australian National University, Canberra, ACT 0200, Australia

E-mail: andrew.stuchbery@anu.edu.au

Abstract. Patterns in g-factor systematics across the nuclear landscape and the pathway to collectivity are explored for heavy nuclei ($A > 90$). The emergence of simple collective structures in complex nuclei with many valence nucleons is viewed from the perspective of the magnetic moments. Examples of the collective complexities in 'simple' nuclei with few valence nucleons, that are revealed by magnetic moment measurements and may be pre-cursors for the development of collectivity in more complex nuclei, are also discussed.

1. Introduction

The quest to understand the microscopic origins of 'simple' collective structures in 'complex' nuclei is ongoing. Insights into the evolution of nuclear structure as the number of valence nucleons changes can be gained from trends in observables such as $R_{42} = E_x(4_1^+)/E_x(2_1^+)$ (see for example [1] and references therein). In this paper trends in the g factors for 2_1^+ states in heavy even-even nuclei will be reviewed and the path from spherical to deformed nuclei considered from the point of view of the magnetic moments. Nuclear collectivity will then be considered in relation to the complexities of the low-excitation levels in nominally 'simple' spherical nuclei wherein the lowest states may contain collective components, which are absent in a shell-model description.

2. Simple patterns in complex nuclei: magnetic moments perspective

The trends in $g(2_1^+)$ values for medium to heavy nuclei have been discussed previously, however it is timely to reconsider these systematics in the light of theoretical advances (see below) and the availability of new data, which either increase the span of the data set (e.g. [2]) or significantly improve the precision (e.g. [3]). Discussion here will be limited to two regions. Data for the rare-earth region ($50 \leq Z < 82$ and $82 \leq N < 126$) are displayed in figure 1 as a function of the valence proton fraction, N_p/N_t, where N_p is the number of valence protons or proton holes relative to the nearest magic number, and N_t is the total number of valence nucleons (protons plus neutrons) counted in the same way. Color has been used to indicate the R_{42} value associated with each measured g factor. The data for $N_p/N_t < 0.7$ cluster around a straight line, which can be interpreted in terms of the proton-neutron Interaction Boson Model expression $g = g_\nu + (g_\pi - g_\nu)(N_p/N_t)$. Although the boson g factors from the fit, $g_\nu = 0.23$ (neutrons)

Published under licence by IOP Publishing Ltd

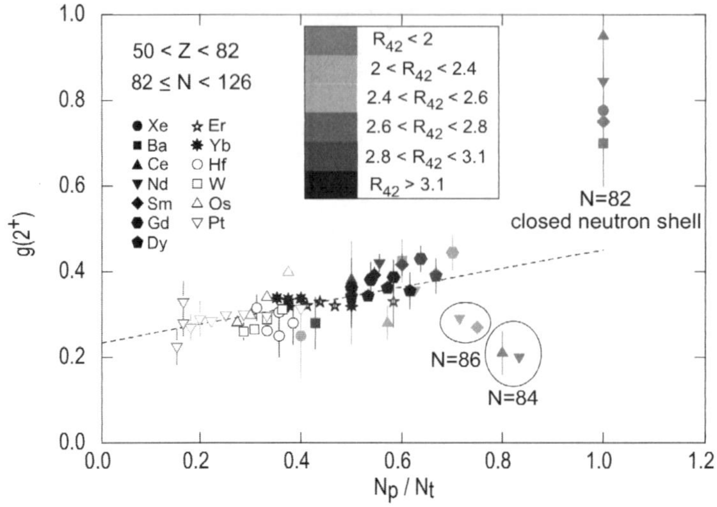

Figure 1. $g(2_1^+)$ as a function of the valence proton fraction for nuclei with $50 \leq Z < 82$ and $82 \leq N < 126$. The broken line indicates the trend for $N_p/N_t < 0.7$. Data points (from [5]) are colored to indicate R_{42}.

and $g_\pi = 0.45$ (protons), depart significantly from the nominal $g_\nu = 0$ and $g_\pi = 1$, the trend for an increase in $g(2^+)$ as the shell closures at $N = 82$ and $N = 126$ are approached is correct; in contrast, this experimental feature of the rare-earth region is contrary to the Z/A estimate, even when corrections for pairing are included [4].

It is known from observables such as R_{42} that heavy nuclei make a rapid transition from spherical to deformed behavior. The rapidity of this transition is evident in the g factors as well. Aside from the semimagic $N = 82$ isotones, the nuclei that show strong departures from the collective trend line all have $R_{42} < 2.4$, and 4 or fewer valence neutrons. Turning to figure 2 and the $A \sim 100$ region ($40 \leq Z < 50$ and $50 \leq N < 82$), the g factors departing strongly from the trend for collective nuclides again have $R_{42} < 2.4$. In both regions, $g(2^+) \sim 1$ for the semimagic nuclei, the specific value being determined by the particular proton configuration. Then, with the addition of two neutrons, $g(2^+)$ falls well below the collective trend.

This phenomenon was noted and discussed about a decade ago [6]. An important observation is that the coupling between protons and neutrons in these nuclei with few valence nucleons is rather weak. For example, the level spectrum of ${}^{144}_{60}Nd_{84}$ can be obtained by superimposing the level schemes of ${}^{142}_{60}Nd_{82}$ (the 4-proton excitation) and ${}^{148}_{64}Gd_{84}$ (the 2-neutron excitation; ${}^{142}_{64}Gd_{82}$ is approximately a closed-shell nucleus). It was suggested that (i) coupling between the valence protons and neutrons is relatively weak because the protons and neutrons are in different shells, and (ii) neutron excitations are then favored in the lowest 2^+ state because their residual interactions are more attractive than those between protons. Considerable progress has been made since then. The shell model calculations of Holt *et al.* [7] for the N=52 isotones, shown in the insert of figure 2, give a good description of the data and illuminate the underlying nuclear structure phenomena. The weak coupling of the proton and neutron excitations leads to a 'configurational isospin polarization', which means that the 2_1^+ and 2_2^+ states in ^{92}Zr, for example, do not form the fully-symmetric and mixed-symmetry states of the proton-neutron interacting boson model. The consequent difference in their g factors has been observed [8].

Research into the g factors of nuclei near closed shells, the formation of mixed symmetry

Journal of Physics: Conference Series **366** (2012) 012042 doi:10.1088/1742-6596/366/1/012042

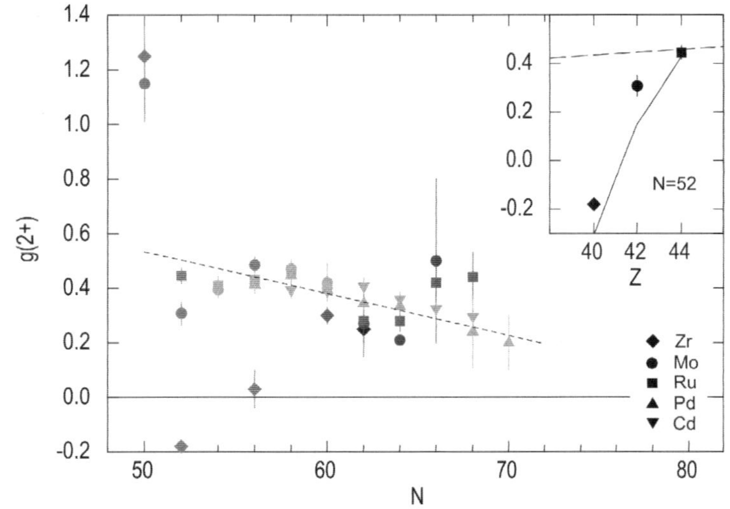

Figure 2. $g(2_1^+)$ as a function of neutron number for nuclei with $40 \leq Z < 50$ and $50 \leq N < 82$. Data are summarized in [3]. The broken line indicates the trend for $56 \leq N \leq 70$. See figure 1 for color key. The inset shows $g(2_1^+)$ for the $N = 52$ isotones. The dashed line is Z/A; the solid line is the shell model calculation of Holt *et al.* [7].

states, and the on-set of collectivity, is on-going. In the $A \sim 80$ region the $N = 48$ isotones $^{84}_{36}$Kr and $^{86}_{38}$Sr follow the pattern of $g(2_1^+) \ll Z/A$. On the other hand $^{132}_{54}\mathrm{Te}_{80}$, which has two valence protons and two neutron holes, has $g \sim Z/A$ [9, 10]. This case will be discussed elsewhere [11].

Whereas the behavior of $g(2_1^+)$ follows a similar path to collectivity in the $A \sim 100$ and rare earth regions, the trends differ once collectivity sets in. In the $A \sim 100$ nuclei the g factors are *not* well correlated with N_p/N_t. Instead they decrease steadily as N increases beyond mid-shell at $N = 66$. Why the difference between these two regions? It must stem from the sensitivity of the g factors to the underlying single-particle composition of the quadrupole collectivity.

The question has been illuminated by microscopic calculations using the tidal-wave model of Frauendorf and collaborators [3]. This model, which describes the yrast states of transitional and deformed nuclei by means of the self-consistent cranking model, allows the calculation of the magnetic moment directly from the nucleonic currents. It was found that the decrease in $g(2_1^+)$ along the isotope chains in the $A \sim 100$ region is primarily due to an increasing angular momentum contribution from the $h_{11/2}$ neutrons, which can begin even below $N = 64$, as illustrated for the Mo isotopes in figure 3. The mechanism is akin to the strong increase of angular momentum carried by neutrons in deformed nuclei, caused by the rotational alignment of the $i_{13/2}$ and $j_{15/2}$ neutrons, which reduces the g factors of high-spin states below Z/A.

It is therefore proposed that the difference in the g-factor systematics observed for the $N = 50 - 82$ shell compared to the $N = 82 - 126$ shell stems from the difference in the location of the high-spin intruder orbital. As seen in figure 3, the neutron $h_{11/2}$ orbit is near the top ($N = 82$ end) of the $N = 50 - 82$ shell. In contrast, the $i_{13/2}$ orbit is much nearer to the middle of the $N = 82 - 126$ shell. A study of the data in figure 1 in the tidal-wave model is needed. The aim would be to provide a microscopic basis for the trends observed, including the trend of rising g factors towards the $N = 126$ end of the shell.

Journal of Physics: Conference Series **366** (2012) 012042 doi:10.1088/1742-6596/366/1/012042

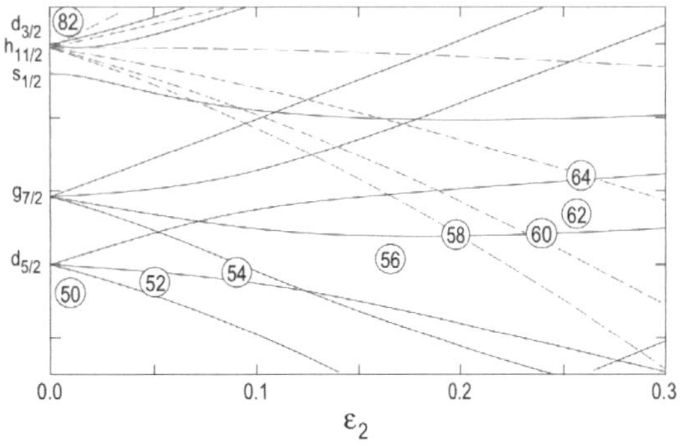

Figure 3. Nilsson diagram for neutrons in Mo isotopes, which illustrates the intrusion of the $h_{11/2}$ shell among the positive parity orbits and the consequent impact of $h_{11/2}$ orbit on the g factors in this region [3]. Neutron numbers are indicated at the calculated deformations.

3. Moments and quadrupole collectivity in neutron-rich S and Ar isotopes

New regions of quadrupole collectivity are being explored in neutron-rich nuclei. The sulfur and argon isotopes between $N = 20$ and $N = 28$ show collective features that can be explored through shell model calculations [12–15]. For example, shell model calculations with the SDPF-NR interaction [13, 14] are very successful in describing the $E2$ transition rate and $g(2^+)$ in ^{40}S. The experimental $B(E2)$ suggests prolate deformation, and the shell model calculations give $Q(2_1^+)$ and $B(E2)$ values consistent with the same deformation in the rotor model. The g factor, however, is not near Z/A, but near zero in both theory and experiment [16, 17]. The microscopic reason is that intrinsic-spin contributions from the $f_{7/2}$ neutrons cancel out the positive contribution from the orbital proton current. According to the shell-model calculations, the collectivity stems from the motion of relatively few nucleons. Rather than picture the quadrupole collectivity here in terms of the rotation of a deformed fluid, it is therefore better to understand it in terms of the symmetries of the underlying shell structure: quasi-$SU(3)$ for the $\nu f_{7/2} - p_{3/2}$ orbits and pseudo-$SU(3)$ for the $\pi d_{3/2} - s_{1/2}$ orbits [13].

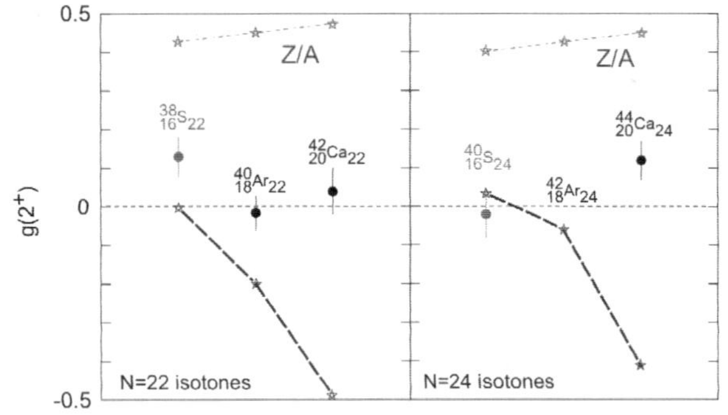

Figure 4. Comparison of experimental [16–19] and theoretical g factors in the SDPF-NR shell model (dashed blue line).

Despite these successes of the SDPF-NR shell model in the case of ^{40}S, the theoretical g factors for the neighboring $N = 22$ isotones ^{38}S and ^{40}Ar underestimate the experimental values significantly, as shown in figure 4. There is a more dramatic disagreement between theory and experiment for ^{42}Ca$_{22}$ and ^{44}Ca$_{24}$. In addition to the experimental data shown in figure 4, the trends have been confirmed in our new $g(2_1^+)$ measurements on the neutron-rich isotopes ^{42,44,46}Ar performed at NSCL. These experimental g factors suggest that, along with the symmetries in the shell-model space that cause quadrupole collectivity, there are also contributions from collective core-excitations that are not explicitly included in the shell-model space. Indeed, it has been known since the 1960s that the 2_1^+ states in the Ca isotopes are not pure $\nu f_{7/2}^n$ configurations, but in ^{42,44}Ca they are a strong mixture of shell model and deformed, multiparticle-multihole configurations [20]. This mixing makes the $g(2_1^+)$ values positive in ^{42,44}Ca. In contrast, the negative g factor in ^{46}Ca is closer to the shell model value [19], which reflects the fact that the deformed contribution decreases towards $N = 28$.

There is a puzzling discrepancy between theory and experiment [12, 15] for $B(E2; 0_1^+ \rightarrow 2_1^+)$ in the $N = 28$ nucleus ^{46}Ar, where shell model calculations with several alternative interactions in the SDPF model space all overestimate the $E2$ strength. One possibility is that the experimental $B(E2)$ values based on intermediate energy Coulomb excitation are not correct (see [21]). However another possibility, should the Coulomb-excitation data prove correct, is that the discrepancy may stem from the fact that the shell-model interactions have been tuned to fit the energy spectra in a limited basis space, without explicitly including the coupling of the shell-model configurations to the deformed core excitations. The shell model interactions might implicitly include the effect of the core excitations that occur near $N = 20$, but not account for the fact that they diminish toward $N = 28$.

4. Conclusion

Patterns in $g(2_1^+)$ systematics for heavy nuclei have been examined and found to be sensitive to the underlying single-particle structure. Near closed shells the magnetic moments show pronounced changes that depend on the strength of the coupling between protons and neutrons, which is known to be important in the development of 'simple' collective excitations in 'complex' nuclei. Studies of magnetic moments can therefore provide a sensitive probe of the pathway to collectivity in atomic nuclei.

From another perspective, magnetic moment measurements on 'simple' nuclei near closed shells can sometimes show that collective components are already present in the nuclear wavefunctions. It is a matter for further research to investigate the ways in which these collective 'complexities' may be precursors of the 'simple' collective structures that emerge in more 'complex' nuclei.

Acknowledgments

The author is grateful to many colleagues who have contributed to the work discussed here. This research was supported in part by the ARC Discovery Scheme, Grant No. DP0773273.

References

[1] Cejnar P, Jolie J and Casten R F 2010 *Rev. Mod. Phys.* **82** 2155
[2] Berant Z *et al* 2009 *Phys. Rev. C* **80** 057303
[3] Chamoli S K *et al* 2011 *Phys. Rev. C* **83** 054318
[4] Stuchbery A E 1995 *Nucl. Phys.* **A 589** 222
[5] Stone N J 2011 **INDC(NDS)-0594** http://www-nds.iaea.org/publications/indc/indc-nds-0594/
[6] Stuchbery A E 2001 *Nuclear Physics A* **682** 470
[7] Holt J D, Pietralla N, Holt J W, Kuo T T S and Rainovski G 2007 *Phys. Rev. C* **76** 034325
[8] Werner V *et al* 2008 *Phys. Rev. C* **78** 031301
[9] Stone N J *et al* 2005 *Phys. Rev. Lett.* **94** 192501
[10] Stuchbery A E and Stone N J 2007 *Phys. Rev. C* **76** 034307
[11] Danchev M *et al* 2011 these proceedings and to be published
[12] Scheit H *et al* 1996 *Phys. Rev. Lett.* **77** 3967
[13] Retamosa J, Caurier E, Nowacki F and Poves A 1997 *Phys. Rev. C* **55** 1266
[14] Caurier E, Martínez-Pinedo G, Nowacki F, Poves A and Zuker A P 2005 *Rev. Mod. Phys.* **77** 427
[15] Robinson S J Q, Sharon Y Y and Zamick L 2009 *Phys. Rev. C* **79** 054322
[16] Davies A D *et al* 2006 *Phys. Rev. Lett.* **96** 112503
[17] Stuchbery A E *et al* 2006 *Phys. Rev. C* **74** 054307
[18] Stefanova E A *et al* 2005 *Phys. Rev. C* **72** 014309
[19] Taylor M *et al* 2005 *Physics Letters B* **605** 265
[20] Gerace W and Green A 1967 *Nuclear Physics A* **93** 110
[21] Mengoni D *et al* 2010 *Phys. Rev. C* **82** 024308

Journal of Physics: Conference Series **366** (2012) 012043 doi:10.1088/1742-6596/366/1/012043

Investigations of Skin Nuclei in a Density Functional Approach

N. Tsoneva[a,b], **H. Lenske**[a]

[a] Institut für Theoretische Physik, Universität Gießen, D-35392 Gießen, Germany
[b]Institute for Nuclear Research and Nuclear Energy, 1784 Sofia, Bulgaria

E-mail: tsoneva@theo.physik.uni-giessen.de

Abstract.
A theoretical method incorporating the density functional theory and quasiparticle-phonon model is applied for systematic study of low-energy excitations of different multipolarities related to neutron or proton skins in stable and exotic nuclei. Furthermore, the unique character of the observed excitations identified with pygmy dipole or pygmy quadrupole resonances is confirmed in quasiparticle-random-phase-approximation plus multiphonon calculations. The achievements of the multiphonon approach for description of pygmy and other strengths are discussed.

1. Introduction

The observation of skin related dipole excitations in stable and unstable nuclei with neutron excess [1, 2, 3, 4] reveals new insight into the dynamical properties of nuclear matter at low densities. An enhanced dipole strength, located close to the particle emission threshold was related to a new dipole mode named pygmy dipole resonance (PDR). Presently, the nature of the PDR is a subject of outstanding discussions. Nevertheless the most common theoretical interpretation associated it with oscillations of a weakly bound neutrons (for $N/Z>1$) or protons (for $N/Z\approx1$) with respect to an isospin symmetric core [3, 4, 5, 6, 7, 8, 9].

A question, coming up immediately in this connection, is to what extent the presence of a neutron or proton skin will affect excitations of other multipolarities and parities and *vice versa*. Promising candidates are low-energy 1^+, 2^+, 3^- states ect., especially those in excess of the spectral distributions known from stable nuclei. Recently, quadrupole response functions were investigated, theoretically in neutron-rich Sn nuclei. A close connection of low-energy 2^+ excitations and nuclear skins was found. These quadrupole states were related to pygmy quadrupole resonance (PQR) [10].

In recent experimental studies of $(\alpha, \alpha'\gamma)$ reactions the isospin character of the PDR in ^{124}Sn was investigated [6].

2. Theoretical Model

The model Hamiltonian resembles in structure the standard QPM model [11] but in detail differs in the physical content in important aspects as discussed in Ref. [8, 9]. Consequently the nuclear ground state is obtained in a fully microscopic Hartree-Fock-Bogolubov (HFB) approach[12] where dynamical effects beyond mean-field are taken into account. That goal is achieved by using the energy density functional theory (EDFT)[9].

Journal of Physics: Conference Series **366** (2012) 012043 doi:10.1088/1742-6596/366/1/012043

The nuclear excitations are described in terms of quasiparticle-random-phase- approximation (QRPA) phonons [11]:

$$Q^{+}_{\lambda\mu i} = \frac{1}{2}\sum_{jj'} \left(\psi^{\lambda i}_{jj'} A^{+}_{\lambda\mu}(jj') - \varphi^{\lambda i}_{jj'} \widetilde{A}_{\lambda\mu}(jj') \right), \tag{1}$$

where $j \equiv (nljm\tau)$ is a single-particle proton or neutron state; $A^{+}_{\lambda\mu}$ and $\widetilde{A}_{\lambda\mu}$ are time-forward and time-backward operators, coupling two-quasiparticle creation or annihilation operators to a total angular momentum λ with projection μ by means of the Clebsch-Gordan coefficients $C^{\lambda\mu}_{jmj'm'} = \langle jmj'm'|\lambda\mu\rangle$. The excitation energies of the phonons and the time-forward and time-backward amplitudes $\psi^{\lambda i}_{j_1 j_2}$ and $\varphi^{\lambda i}_{j_1 j_2}$ in Eq. (1) are determined by solving QRPA equations [11].

Furthermore, the QPM provides a microscopic approach to multiconfiguration mixing [11]. For spherical even-even nuclei the model Hamiltonian is diagonalized by using of an orthonormal set of wave functions constructed from one-, two- and three-phonon configurations [13]:

$$\Psi_\nu(JM) = \Bigg\{ \sum_i R_i(J\nu) Q^{+}_{JMi} + \sum_{\substack{\lambda_1 i_1 \\ \lambda_2 i_2}} P^{\lambda_1 i_1}_{\lambda_2 i_2}(J\nu) \left[Q^{+}_{\lambda_1\mu_1 i_1} \times Q^{+}_{\lambda_2\mu_2 i_2} \right]_{JM} \tag{2}$$

$$+ \sum_{\substack{\lambda_1 i_1 \lambda_2 i_2 \\ \lambda_3 i_3 I}} T^{\lambda_1 i_1 \lambda_2 i_2 I}_{\lambda_3 i_3}(J\nu) \left[\left[Q^{+}_{\lambda_1\mu_1 i_1} \otimes Q^{+}_{\lambda_2\mu_2 i_2} \right]_{IK} \otimes Q^{+}_{\lambda_3\mu_3 i_3} \right]_{JM} \Bigg\} \Psi_0,$$

where R, P and T are unknown amplitudes, and ν labels the number of the excited states.

In the procedure the exact commutation relations including the internal fermionic structure of the phonon operators are implemented. It is applied also for transition operators and transition matrix elements of multiphonon states [14]. This is a necessary condition in order to account for the Pauli principle.

3. Discussion

From QRPA calculations in Sn isotopes with $N/Z = 1.2 \div 1.68$, we observe a sequence of low-lying 1^- excitations with energies $E^* = 5.5 - 8$ MeV. These states are related to oscillations of the least bound neutrons (the proton contribution is less than 1%) [8, 9]. We find similar results also in N=50, 82 isotones [3, 4] and in the newly calculated ^{86}Kr nucleus from N=50 chain. In the lightest $^{100-104}$Sn isotopes, the lowest dipole QRPA states, located at $E^* = 8.1 - 8.3$ MeV, are dominated by proton excitations involving quasibound proton states [9]. Electromagnetic breaking of isospin symmetry is the main reason for the persisting of low-energy dipole strength close to ^{100}Sn. The observed sequences of low-energy neutron or proton dipole states are related to neutron or proton PDR, respectively.

In this connection, more detailed analysis of the dipole excitations is achieved by investigating proton and neutron transition densities in various energy and mass regions [4, 9]. The unique behavior of the dipole transition densities in the PDR region is found as a common feature of the lowest-lying dipole states in Z=50, N=50, 82 nuclei [3, 4, 9] making it meaningful to identify the PDR as a distinct and genuine excitation different from the giant dipole resonance (GDR). Furthermore, the change of the character of the transition densities from neutron to proton one is observed for lowest-mass tin isotopes thus confirming the transition of a neutron PDR to a proton PDR [9]. From QRPA calculations of total PDR strengths it is found to be closely related to the neutron skin thickness [3, 4, 5, 8, 9]. Namely, the PDR strength increases with the increase of N/Z ratio for both isotopic and isotonic chains. This result is further justified

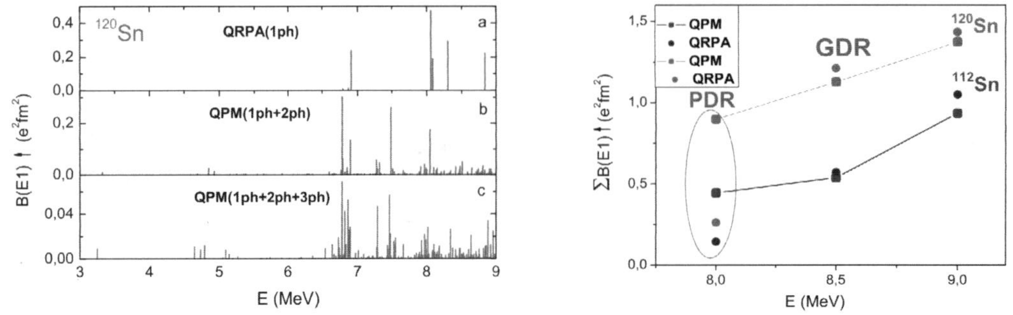

Figure 1. (left) Calculations of electric dipole strength up to E*= 9 MeV in ^{120}Sn: a/QRPA, b/ two-phonon QPM and c/ three-phonon QPM;
(right) Comparison of QRPA and three-phonon QPM B(E1) strengths summed up to E*= 8 MeV (PDR region), 8.5 MeV (GDR region) and 9 MeV (GDR region), respectively.

in our multiphonon QPM calculations presented in this paper. As an example case we consider the comparison between QRPA and QPM calculations of summed B(E1) strengths at different energies in ^{112,120}Sn which is presented in Fig.1(right). In addition, calculations of the excited QRPA 1^- states and their fragmentation pattern obtained within two- and three-phonon QPM are shown in Fig.1(left)a,b,c, respectively.

According to analysis of QRPA transition densities and state vectors structure (see for details in Ref. [9]), the summed B(E1) strengths in the energy region E*≤8 MeV refers to almost pure PDR strengths while for the regions E*≤8.5 MeV and E*≤9 MeV the GDR contribution is dominant. The comparison of QRPA and two- and three-phonon QPM excited 1^- states indicates that for the PDR region the coupling of QRPA PDR and GDR phonons to multiphonon states is very important. The result is a shift of E1 strength towards lower energies which is well visible in Fig.1(left)a,b,c and from Fig.1(right). Consequently in the energy interval $0 < E < 8$ MeV the QPM calculations give about twice as much total summed E1-strength than the QRPA calculations, as shown in Fig.1 (right). Differently, with the increase of the excitation energy toward the GDR the QRPA and QPM summed B(E1) transition strengths are of comparable size which suggests the prevalence of $p - h$ excitations of GDR type. Thus, our QPM calculations of ^{112}Sn (N/Z=1.24) and ^{120}Sn (N/Z=1.4) confirm experimental and QRPA observations of increasing PDR strength with the increase of N/Z ratios [3]. The comparison of ^{112}Sn and ^{120}Sn nuclei (see in Fig.1(right)) shows that the relative difference between the E1 strengths calculated for certain energy regions, both for the QRPA and QPM cases, reduces with the increase of the excitation energies toward GDR. This behavior is expected because of the weak dependence of GDR on N/Z ratios in tin isotopes.

From theoretical investigations of 2^+ excitations in Sn nuclei we find a clustering of quadrupole states, at low-excitation energies $E* = 2 - 4MeV$ which structure is dominated by neutron two-quasiparticle (2QP) excitations from the valence shells. The most important proton contribution in all isotopes is due to the $[1g_{9/2}2d_{5/2}]_\pi$ 2QP component, which, however, never exceeds 5% [10]. This dependence is illustrated for the $[2_2^+]_{QRPA}$ states in Fig.2(left). The figure reflects the change of the proton binding energy ϵ_b of the $g_{9/2}$ level when approaching the $N = Z$ limit [10].

Journal of Physics: Conference Series **366** (2012) 012043 doi:10.1088/1742-6596/366/1/012043

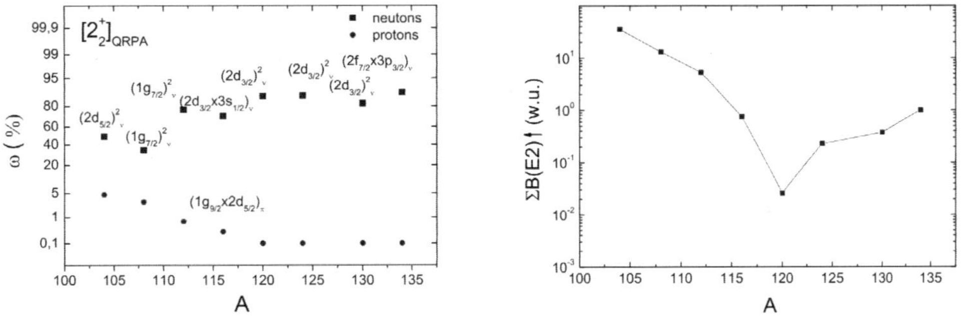

Figure 2. (left) QRPA structure of the $[2_2^+]_{QRPA}$ states calculated in Sn isotopes. The main neutron and proton 2QP components with a state vector probability $\omega_{j_1j_2}(\lambda i) = |\psi_{j_1j_2}^{\lambda i}|^2 - |\varphi_{j_1j_2}^{\lambda\mu i}|^2$ larger than 0.1% are presented. (right) QRPA results for B(E2)↑ transition probabilities summed over the $[2_2^+]_{QRPA}$ and $[2_3^+]_{QRPA}$ states in Sn isotopes.

In the neutron sector, the contributions follow closely the evolution of the shell structure as seen from Fig.2(left). In most cases the $[2_2^+]_{QRPA}$ state vectors are dominated by re-scattering contributions related to re-orientation of the s.p. angular momenta [10].

Theoretical results of summed B(E2) strength distributions in $^{104\div134}$Sn isotopes are presented in Fig.2(right). A sizable increase of B(E2) strength at $E_x \approx$ 2-4 MeV is observed with increasing of N/Z ratio [10] associated with the mass dependence of the neutron 2QP transition matrix elements and decreasing neutron binding energy ϵ_b. For $N/Z < 1.4$ the B(E2) transitions follow the increasing contribution of the protons, which are coupled directly by their physical charge to the electromagnetic field. Correspondingly, the connection of the low-energy 2^+ states with a PQR is clearly seen in the analysis of transition densities [10]. Results from QRPA proton and neutron quadrupole transition densities of the PQR mode in ^{124}Sn are presented in Fig.3(left). Strong neutron oscillations at the nuclear surface play a dominant role in the PQR energy range which is in agreement with the observations from [10]. Similarly to the PDR, a transition from a neutron PQR to a proton PQR in ^{104}Sn is observed for the mass region where the neutron skin reverses into a proton skin [10].

Investigations of B(E2) and B(M1) transition rates of low-lying 2^+ states in ^{124}Sn in QPM presentation indicate differences from known collective states and scissors modes. The results on B(E2) transition rates are compared to experimental data in ^{124}Sn in Fig. 3(right). A good agreement of the data with the calculated total B(E2) value related to PQR is obtained.

QPM calculations of low-energy E1 and M1 excitations in ^{138}Ba [5] and ^{90}Zr nucleus confirm unambiguously the predominantly electric character of the observed low-energy dipole strength in these nuclei. Furthermore, our recent QPM studies of the fine structure of M1 strength in ^{90}Zr lead to interesting and new observations of the origin of the magnetic excitations as a composition of spin-flip and orbital modes of sizable sizes where the multiphonon counterparts play an important role. The results are of significant importance for the analysis of strength functions used in the (n, γ) reactions data.

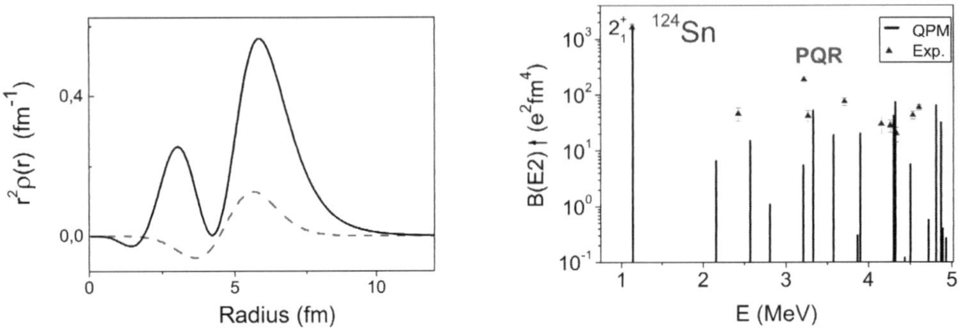

Figure 3. (left) QRPA calculations of quadrupole transition densities related to the main proton and neutron components in the state vectors of $[2_i^+]_{QRPA}$ states with excitation energies $E* = 2 - 5MeV$; (right) QPM calculations of B(E2) transitions of low-energy 2^+ excited states in ^{124}Sn in comparison with experimental data [15, 16, 17].

4. Conclusions

In conclusion, systematic investigations of low-energy dipole and other multipole excitations in Z=50 and N=50, 82 nuclei reveal new modes of excitations related to PDR and PQR. A close correlation of PDR B(E1) and PQR B(E2) transition strengths related to N/Z ratios and thicknesses of a neutron or proton skin is found. The unique character of both modes is further confirmed by analysis of related dipole and quadrupole transition densities. A sizable increase of low-energy E1 strength with respect to the pure QRPA PDR strength is obtained from multiphonon QPM calculations in ^{112}Sn and ^{120}Sn. The result is related to contribution of multiphonon states and of the coupling to the GDR. As a general behavior we find an increase of the PDR with the neutron number.

The knowledge of the fine structure of the low-energy dipole strength by means of its multiphonon presentation is found essential for description of M1 resonances, dipole response functions and related corresponding neutron-capture cross sections relevant for nucleosynthesis.

References

[1] A. Zilges *et al.*, *Phys. Lett. B* **542**, 43 (2002).
[2] P. Adrich et al., *Phys. Rev. Lett.* **95**, 132501 (2005).
[3] S. Volz et al., *Nucl. Phys. A* **779**, 1 (2006).
[4] R. Schwengner et al., *Phys. Rev. C* **78**, 064314 (2008) and refs. therein.
[5] A. Tonchev et al., *Phys. Rev. Lett.* **104**, 072501 (2010).
[6] J. Endres et al., *Phys. Rev. Lett.* **105**, 212503 (2010).
[7] N. Paar et al., *Rep. Prog. Phys.* **70**, 691 (2007) and refs therein.
[8] N. Tsoneva, H. Lenske, Ch. Stoyanov, *Phys. Lett. B* **586**, 213 (2004).
[9] N. Tsoneva, H. Lenske, *Phys. Rev. C* **77**, 024321 (2008) and refs. therein.
[10] N. Tsoneva, H. Lenske, *Phys. Lett. B* **695** 174 (2011).
[11] V.G. Soloviev, *Theory of complex nuclei* (Oxford: Pergamon Press, 1976).
[12] F. Hofmann and H. Lenske, *Phys. Rev. C* **57**, 2281 (1998).
[13] M. Grinberg, Ch. Stoyanov, *Nucl. Phys. A* **573**, 231 (1994).
[14] V.Yu. Ponomarev, Ch. Stoyanov, N. Tsoneva, M. Grinberg, *Nucl. Phys. A* **635** 470 (1998).
[15] M. Spieker, J. Enders, A. Zilges, private communication;
[16] J. Bryssincket al., *Phys. Rev. C* **61** 024309 (2000).
[17] J. Katakura, Z. Wu, *Nucl. Data Scheets* **109** 1668-1692 (2008).

Journal of Physics: Conference Series **366** (2012) 012044 doi:10.1088/1742-6596/366/1/012044

Perspectives for the gamma–ray spectroscopy at LNL: the GALILEO project

Calin A. Ur

Istituto Nazionale di Fisica Nucleare, Sezione di Padova, 35131 Padova, Italy

E-mail: `calin.ur@pd.infn.it`

Abstract. GALILEO is a new 4π high–resolution γ–ray array in which GASP tapered detectors and Gammapool Cluster detectors will be used together. The array will be located at the Laboratori Nazionali di Legnaro and it will make use of the stable beams provided by the Tandem–ALPI–PIAVE accelerator complex and, in the future, of the exotic radioactive ion beams provided by SPES. The project requires the transformation of the original EUROBALL 7–capsules cluster detectors into triple cluster detectors and several R&D activities are ongoing for the development of a triple cluster cryostat and for building the corresponding anti–Compton shields. The development of the front–end, digital sampling, preprocessing and readout electronics is taking advantage of the most recent advances obtained within the AGATA project. A strong physics case was made for the building of the GALILEO array based on Letters of Intent submitted by several research groups from all over the world. In the present paper a brief summary of the status and the perspectives of the GALILEO project is given.

1. Introduction

A new gamma-ray array is proposed to be built and installed at the Laboratori Nazionali di Legnaro (LNL). The original idea was to reshape all the EUROBALL 7-capsules cluster detectors into triple cluster detectors and build with them a full 4π high efficiency gamma-ray array. Arrays of non-segmented Ge detectors require the use of anti-Compton shields in order to provide good quality gamma-ray coincidence data. The cost estimate for building the anti-Compton shields showed that the project would be highly demanding in terms of funding. While this project can be pursued in the future within a wide European collaboration, it was decided to build a Phase I array that makes use of part of the EUROBALL cluster detectors and the GASP detectors. This phase is very important as it includes R&D activities for the development of the triple cluster detectors and their anti-Compton shields, activities that constitute the basis for building the larger array. The proposed configuration of the array offers the possibility to have in short time a competitive device at a reasonable cost. The physics case of the project is supported by a worldwide collaboration of research groups active in the field of gamma-ray and particle spectroscopy whose interest was expressed in Letters of Intent proposing the investigation of isospin symmetry in isobar multiplets or mirror nuclei, critical point symmetries, excitation of exotic collective modes, astrophysical interest cross sections, high–lying isomers acting as waiting points for the s–processes. The proposed experiments require the use of stable beams provided by the Tandem–ALPI–PIAVE accelerator complex of LNL and, in perspective, of the SPES radioactive beams. The new array will benefit of the continuous improvement of the heavy ions beams delivered by the accelerator complex Tandem-ALPI-PIAVE; a lot of effort is put into developing new beams with the PIAVE ECR source and to increase the intensity and energy of the delivered beams. Several options for the coupling of the array with ancillary detectors for the detection of light charged particles, neutrons or heavy nuclei were considered as solutions to improve the arrays sensitivity.

2. Physics case

Nuclear structure research is evolving towards extreme regions of spin and excitation energies and neutron-proton ratio. Gamma-ray spectroscopy, as the main tool to investigate the structure of nuclei, has to respond to the increasing experimental difficulties. During the last two decades, LNL has hosted several gamma-ray arrays such as GASP [1], EUROBALL [2] and CLARA [3], producing valuable physics results and providing the whole nuclear physics community with excellent and forefront research devices. Presently, the AGATA Demonstrator [4] is installed and running experiments at LNL. The power of the GALILEO array to address new and challenging problems of the nuclear matter stays in a combination of the high gamma-ray detection efficiency with a new highly performant data taking electronics allowing for higher event rates and with the use of state-of-the-art ancillary detectors such as charged-particle and neutron arrays, plunger and recoil detectors, to improve the sensitivity. The physics case of GALILEO is addressing hot topics of the present worldwide nuclear structure research such as: 1) the study of N Z nuclei that will provide the most relevant information about the properties of the neutron-proton pairing interaction and its evolution with the spin and excitation energy; 2) the study of isospin symmetry as a tool to understand fundamental questions such as the charge-symmetry and charge-independence of the nuclear two-body interaction; 3) the study of neutron-rich nuclei near major shell closures is of considerable interest due to the influence of the neutron excess on the nuclear potential that may lead to the formation of new shell closures and new regions of deformation; 4) accurate mapping of the proton-dripline location through the study of exotic proton and alpha decays of the nuclear states in very neutron deficient nuclei; 5) study of the connection between molecular-like nuclear states, such as superdeformed alpha-cluster states, and nuclear resonances through their particle and gamma-ray decays; 6) study of the gamma-ray decay of the GDR to investigate the isospin symmetry and its breaking by Coulomb interaction in heavy N=Z nuclei, at finite temperature; 7) investigation of the warm rotation regime in nuclei will provide crucial information to understand the transition of the nuclear structure from regular near the ground state to chaotic in the resonance region, and it will also test the robustness of the shell structure with temperature; 8) identification of the critical point symmetries corresponding to the nuclear shape- phase transition, such as the E(5) on the transition from spherical to deformed γ-unstable shape and the X(5) on the transition from spherical to axially deformed shape, requires thorough knowledge of the whole nuclear region around the predicted cases of such symmetries; 9) study of particular cases of triple shape coexistence in neutrondeficient nuclei to understand the underlying nucleon configurations favoring such peculiar phenomena; 10) determination of the n-capture cross section on the sprocess branching points through the surrogate nuclear reaction method will be used to test the predictions of s-process calculations.

3. The GALILEO array

Ten triple clusters can be obtained from five original EUROBALL cluster detectors. Thorough Monte Carlo simulations were performed with a GEANT4 code derived from the AGATA simulation package [5]. Calculations with ten triple clusters indicate as the most powerful configuration when they are placed at 90° 24 cm from the target position. The forward and backward angles with respect to the beam direction are filled with GASP tapered HPGe detectors placed 22.5 cm from the target position and distributed in five rings ranging from 21° to 159°. The original anti–Compton shields will be used for the GASP detectors. The cluster detectors will be surrounded with new anti-Compton shields built of the single crystals recovered from the original cluster shields. An artistic view of the array from the backside is shown in Fig. 1. Simulations give for this configuration a total photopeak efficiency of 8% and a peak-to-total ratio of 50

Figure 1. Artistic view of the GALILEO array. One can see the 90° ring of ten triple cluster detectors and the compact shell of 15 HPGe GASP detectors at the backward angles.

3.1. The triple cluster detectors

While the GASP detectors can be used as they are, in the case of the triple cluster detectors a completely new development is required for building a dedicated cryostat to hold the three capsules. The end–cap is proposed to be realized in carbon fiber. Carbon fiber is as resistant as aluminum from the mechanical point of view. Vacuum tightness was tested with a prototype made for a GASP detector end-cap. The Zeolites container is made such to ensure easy access, mounting and dismounting when reconditioning is needed. The cold finger was designed such to minimize the heat transfer between the cold part and the outside environment and the transmission of mechanical vibration from dewar to capsules. A new concept dewar is also developed. The whole ensemble is optimized for a better thermal insulation, less heat transfer, lower LN2 consumption and better mechanical rigidity.

The performance of the array at high fold gamma-ray coincidence is strongly dependent on the peak-to-total ratio. In case of a low value the coincidence is dominated by the Compton scattered events. To increase the peak-to-total ratio one needs to surround the Ge detectors with anti-Compton shields made of BGO scintillator crystals. In the case of the GASP detectors their original shields can be used but for the ten triple cluster detectors one needs to develop new shields. The possibility to build the new anti-Compton shields out of the single crystals from the original EUROBALL cluster shields is under investigation. The proposed geometry for these shields would leave empty spaces corresponding to the corners of the triple cluster. This lack can be overcome by doing a shared Compton suppression with the neighboring shields standing

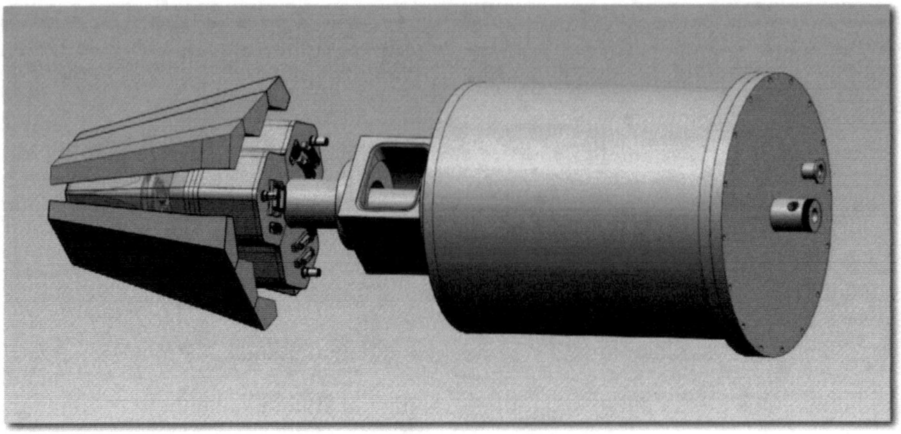

Figure 2. Drawing of the complete ensemble of a triple cluster detector and its anti-Compton shield.

in front of the openings. The complete ensemble of triple cluster detector and anti-Compton shield is shown in Fig. 2.

3.2. The holding structure
A new mechanical frame was designed and is currently under construction for the GALILEO array. The new design was specifically made to allow easy access to the reaction chamber and special care was taken for the integration of different ancillary detectors. The frame consists of a modular structure as shown in Fig. 3. There is a central ring for holding the 10 triple cluster detectors with their anti-Compton shields. The backward part of the frame will hold 15 GASP-type detectors. On the forward part of the frame other 15 GASP-type detectors will be placed in the same geometry as in the backward direction but divided into two structures: one close to the central ring, holding 10 detectors and another one holding 5 detectors. This makes possible to dismount completely the 5 most forward detectors to allow for the insertion of ancillary detectors such as neutron detectors or recoil filter detectors. A preliminary design of the holding structure is shown in Fig. 3.

3.3. Electronics
A common preamplifier is developed for both the GASP-type detectors and the triple cluster detectors, based on the AGATA SingleCore-type preamplifier [6]. The main characteristics of the preamplifier are: differential output, compatibility with different types of FET (BF862 or IF1320), variable threshold fast reset circuit, output energy range of up to 10 MeV.

A new electronics for readout and preprocessing of the digital data is under development in collaboration with the AGATA project. The experienced gained until now in the handling of the digital signals and the advances in the electronic components allowed for the development of a very compact PCI express based solution. The new solution has the advatage of a lower power consumption, no need of dedicated cooled racks, use of all the services already implemented in the host workstations, such as network connection or CPU processing, in a transparent and cost effective way. The boards are based on the use of the XILINX VIRTEX-6 FPGA.

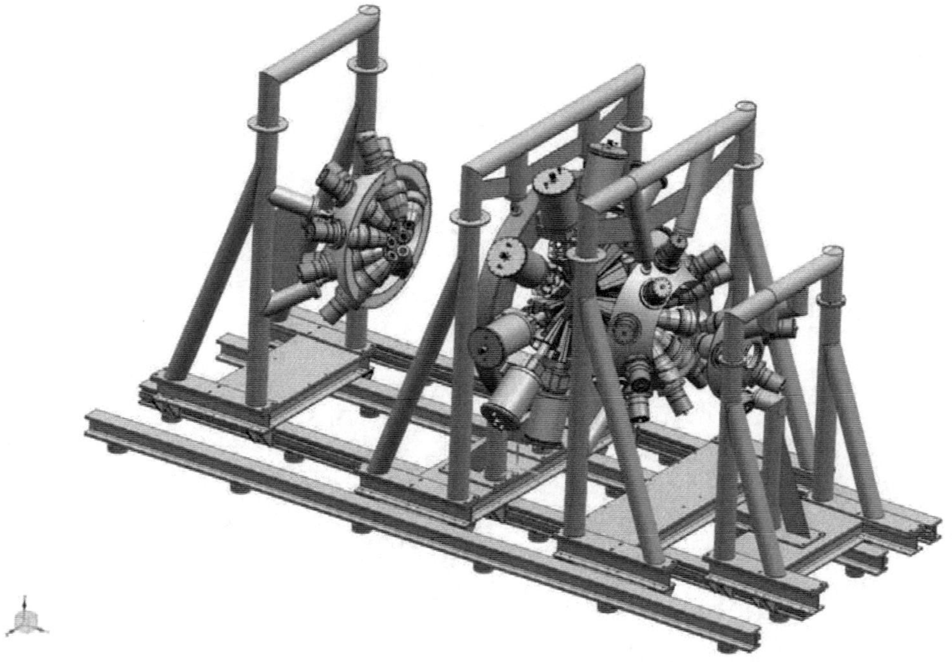

Figure 3. Preliminary layout of the GALILEO frame. The four different structures of the frame are illustrated. They will move on rails to give access to the reaction chamber.

4. Conclusions

The development of the gamma–ray array is important for the extension of the nuclear structure studies towards exotic nuclei populated with small cross–sections. Recent developments in the detector and electronics technologies allow for new designs. The present paper reported on the status of the GALILEO project. The new array will be dedicated to gamma–ray spectroscopy studies with stable and exotic beams at the Laboratori Nazionali di Legnaro. The project involves the development of a triple cluster detector with a new conception cryostat and of a new front–end and read–out electronics in collaboration with the AGATA project.

Acknowledgments

The author gratefully acknowledges the substantial contribution of C.Fanin for the mechanical design, E.Farnea for the GEANT4 simulations, M.Bellato, D.Bortolato, A.Pullia and R.Isocrate for the electronics development, D.Rosso, P.Cocconi and L.Costa for the cryogenic and vacuum systems development. He also acknowledges the support of the mechanical workshops of INFN Padova and Laboratori Nazionali di Legnaro for the work done.

References

[1] Bazzaco D 1992 *Proc. Workshop on Large γ–ray Detector Arrays* (Chalk River Canada) AECL-10613, 376
[2] Simpson J 1997 *Z. Phys.* A **357** 139
[3] Gadea A *et al.* 2004 *Eur. Phys. J.* A **20** 193
[4] Wiens A *et al.* 2010 *Nucl. Instr. and Meth.* A **618** 223
[5] Farnea E 2010 *Nucl. Instr. and Meth.* A **621** 331
[6] Pascovici G, Pullia A, Zocca F, Bruyneel B, Bazzacco D 2008 *WSEAS Transactions on Circuits and Systems* **7** 470

Journal of Physics: Conference Series **366** (2012) 012045 doi:10.1088/1742-6596/366/1/012045

Study of the Order-to-Chaos transition in ^{174}W with the AGATA-Demonstrator

V Vandone[1], S Leoni[1], D Assanelli[1], S Bottoni[1], A Bracco[1], F Camera[1], F C L Crespi[1], A Giaz[1], R Nicolini[1], L Pellegri[1], G Benzoni[2], N Blasi[2], C Boiano[2], S Brambilla[2], B Million[2], O Wieland[2], A Corsi[3], D Bortolato[4], E Calore[4], A Gottardo[4], D R Napoli[4], E Sahin[4], J J Valiente Dobon[4], D Bazzacco[5], M Bellato[5], E Farnea[5], S Lunardi[5], D Mengoni[5], C Michelagnoli[5], D Montanari[5], F Recchia[5], C A Ur[5], A Gadea[6], T Hüyük[6], N Cieplicka[7], A Maj[7], M Kmiecik[7], A Atac[8], S Akkoyun[8], A Kaskas[8], J Nyberg[9], P A Söderström[9]
and the AGATA collaboration

[1]Universitá degli Studi di Milano e INFN sezione di Milano via Celoria 16, 20133, Milano, Italy
[2]INFN sezione di Milano via Celoria 16, 20133, Milano, Italy
[3]CEA, Centre de Saclay, IRFU/Service de Physique Nuclaire, F-91191 Gif-sur-Yvette, France
[4]INFN Laboratori Nazionali di Legnaro, Italy
[5]Universitá di Padova e INFN, sezione di Padova, Padova, Italy
[6]IFIC, Valencia, Spain
[7]The Niewodniczanski Institute of Nuclear Physics, PAN, Krakow, Poland
[8]Department of Physics, Faculty of Science, Ankara University, Ankara, Turkey
[9]Department of Physics and Astronomy, Uppsala University, Sweden

E-mail: `valeria.vandone@mi.infn.it`, `silvia.leoni@mi.infn.it`

Abstract. The transition between order and chaos is studied in the warm rotating nucleus ^{174}W by γ-spectroscopy, focusing on the conservation of selection rules of the K quantum number with the excitation energy, where K is the projection of the total angular momentum on the symmetry axis. The ^{174}W nucleus was populated by the fusion-evaporation reaction of ^{50}Ti (at 217 MeV) on a ^{128}Te backed target. The measurement was performed in July 2010 at Legnaro National Laboratories of INFN using the AGATA Demonstrator HPGe-array coupled to an array of 27 BaF_2 scintillators, named Helena.
The data analysis concentrates on γ-γ coincidence matrices selecting the γ-decay flow populating low-K and high-K structures. By a statistical fluctuation analysis the total number of low-K and high-K bands can be evaluated as a function of excitation energy. Comparisons with cranked shell model calculations at finite temperature are used to extract information on the onset of the chaotic regime as a function of excitation energy.

1. Introduction

The transition between order and chaos in a quantum mechanical system is a very fascinating subject that is currently investigated in different research fields, ranging from nuclear to atomic, molecular and solid-state physics.
Warm rotating nuclei produced at high spin by heavy-ion fusion-evaporation reactions give

the possibility to investigate this topic in nuclear physics. In fact, close to the yrast line (at temperature T$\sim$0), the nucleus behaves as an ordered system with well-defined quantum numbers and associated selection rules. Moving toward higher excitation energy, energy levels start to interact giving rise to a gradual loss of selection rules. As a consequence, quantum numbers lose their meaning with temperature. The extreme regime is the chaotic region of the compound nucleus, where only energy, spin and parity are well defined for each energy level.
The weakening of the selection rules with temperature can be studied focusing on the K-quantum number (i.e. the projection of the total angular momentum on the symmetry axis). Close to the yrast line K is a well-defined quantum number, and low-K and high-K γ-decay flows are well-distinguished. As the internal energy increases, a statistical K-mixing process takes place, due to the high density of states. Therefore, by studying the γ-decay flow in coincidence with low-K and high-K states in the large region between the cold regime close to the yrast line and the chaotic compound nucleus levels, information on the order-to-chaos transition can be obtained [1].
The nucleus ^{163}Er is the only investigated case [2], giving indication of a gradual transition between order and chaos at around 2.5 MeV of excitation energy. Therefore, additional experimental investigations are nedeed and an interesting case is represented by the ^{174}W nucleus [3], here discussed. In particular, ^{174}W is characterized by the existence of low-K ($\sim$3-4) and high-K ($\sim$8-12) rotational bands extending up to spin 39$\hbar$, with two high-K bands (K=8 and K=12) built on isomeric states, with lifetimes larger than 120 ns.

Preliminary results from an experiment performed in July 2010 at Legnaro National Laboratory of INFN, using the first phase of the Advanced GAmma Tracking Array (AGATA) [4], coupled to a multiplicity filter of 27 BaF_2 scintillators (the Helena array), will be presented in this contribution. The experimental data will be compared to Cranked Shell Model calculations at finite temperature for the specific nucleus ^{174}W [5].

2. Experimental details

The ^{174}W nucleus has been populated by the fusion-evaporation reaction of ^{50}Ti (at 217 MeV) on a ^{128}Te target (1 mg/cm^2 thick, backed by 50 mg/cm^2 of ^{nat}Pb). The reaction is expected to reach spins as high as $\sim$60$\hbar$ and energy of $\sim$4 MeV above the yrast line, therefore providing a good population of the warm rotational regime.
The experimental setup consisted of four triple clusters of the AGATA HPGe-array (i.e. the AGATA Demonstrator) placed at 16 cm from the target (with an absolute efficiency of $\sim$5% at 1.3 MeV), coupled with an array of 27 BaF_2 scintillator detectors, covering $\sim$25% of the total solid angle.
To probe the gradual weakening of selection rules of the K-quantum number, γ-γ matrices selecting low-K and high-K structures have been constructed and analyzed by statistical fluctuation techniques [6]. This allows to estimate the number of low-K and high-K bands and their correlations, as a function of excitation energy. In particular, the selection of low-K and high-K transitions will require the use of both prompt and delayed γ coincidences with respect to a time reference given by the Helena array. Furthermore, the scintillator array helps focusing on the high-multiplicity and high-energy part of the γ-cascades, where the full transition into the chaotic regime is expected to take place.

3. Preliminary results

The experiment was performed with an average current of 1 pnA (in order to prevent target damage), requiring as trigger conditions either four-fold events in AGATA or three-fold events in AGATA in coincidence with at least one event in Helena, since the analysis is concentrated on γ-γ matrices.
The first part of the data analysis has been focused on the presorting of the data: all detectors

Journal of Physics: Conference Series **366** (2012) 012045 doi:10.1088/1742-6596/366/1/012045

have been carefully calibrated in a very wide energy-range: an AmBe(Ni) source, which provides calibration lines up to 9 MeV, has been used for both AGATA and Helena energy-calibrations; in addition 3 other sources were used (i.e. ^{137}Cs, ^{60}Co and ^{88}Y) to have low-energy calibration point for the scintillator spectra. Time-difference spectra between the 12 AGATA detectors have been constructed and carefully aligned, resulting in a time resolution of the order of 20 ns. A tracking algorithm has been applied to the calibrated Ge energies in order to reduce the Compton scattering and improve the peak-to-background ratio of the AGATA array. In figure 1 we show a comparison between a γ-ray spectrum constructed using only the core electrode signals (raw data) and the energy spectrum obtained after applying the tracking procedure [4].

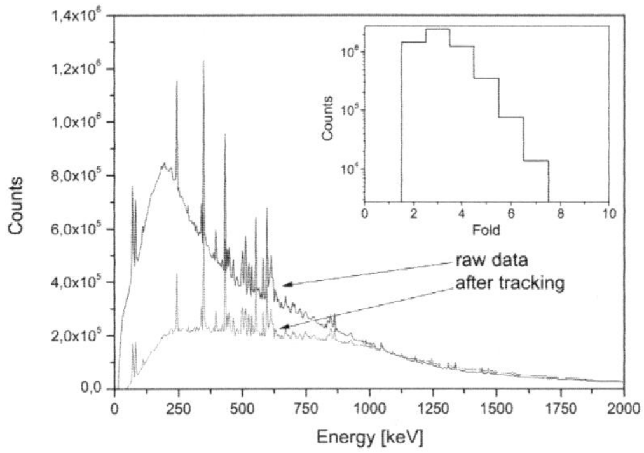

Figure 1. Comparison between γ-ray spectra before and after the γ-ray tracking procedure. The strong peaks correspond to the γ-transitions of the yrast line of ^{174}W. The inset shows the Ge fold distribution.

One can notice the large reduction of low-energy background events, resulting in a significant improvement of the peak-to-background ratio. The inset of figure 1 shows the Ge fold distribution, which is found to be peaked around 3, an essential requirement for the data analysis based on γ-γ matrices.

The Helena array has been mainly used as a multiplicity filter. To enhance the selectivity to the ^{174}W nucleus and to focus on the high-multiplicity and high-energy part of the γ-cascades (where the transition into the chaotic regime is expected to take place) the fold and sum-energy measured in Helena have been used.

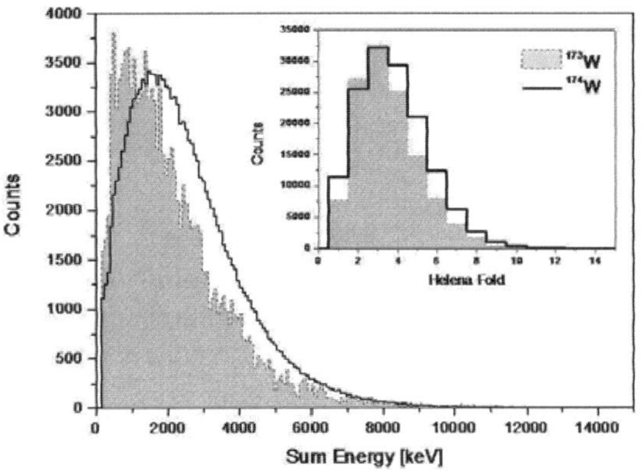

Figure 2. Comparison between the sum-energy spectra obtained for ^{174}W (black line) and ^{173}W (grey filled) nuclei. The inset shows the Helena fold distribution for the two nuclei.

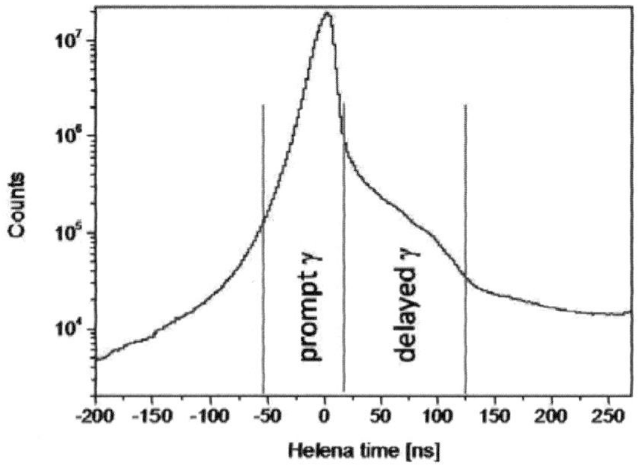

Figure 3. BaF_2 scintillators time spectrum. The red lines denote the limits for the prompt and delayed time-gate, used in the subsequent data-analysis.

Figure 2 shows the sum-energy and fold distributions of the Helena array, in coincidence with

low spin transitions of ^{174}W and ^{173}W. These are the two main evaporation channels open in the reaction, with relative intensity of the order of 40% and 50%, respectively. As expected, the sum-energy and fold distributions associated to ^{174}W are peaked at higher values, indicating the possibility to favor the selection of this type of events by requiring an high-fold and high-sum energy gating condition. Based on these considerations and since the sum-energy information is not so selective, we have chosen as a discrimination condition for the further analysis, fold$_{Helena} \geq 4$. The time information of the Helena array has also been used, the time-spectra of each BaF_2 detector (constructed with respect to the trigger signal) have been aligned to each other. Figure 3 shows the Helena time spectrum, evidencing the regions of prompt and delayed γ-events, the latter ones being mainly associated to high-K transitions built on isomeric states.
Since the intensity of γ-rays belonging to high-K states is very weak, they can not be used as a gate in energy to build γ-γ matrices, because the selectivity would be too poor. An alternative way to proceed is selecting these high-K bands through the long lived isomeric gammas, associated to their decay. This possibility has been confirmed by the following: i) by gating on the delayed part of the time spectrum, the isomeric γ transitions associated to the decay of the high-K bands show an increased intensity, while contributions from prompt γ-rays are largely suppressed (see figure 4); ii) when gating on the energy of isomeric γ-transitions the corresponding time spectrum shows an enhanced bump in the delayed region (as shown in the inset of figure 4) as compared to the time-spectrum gated on energy of prompt γ-rays.

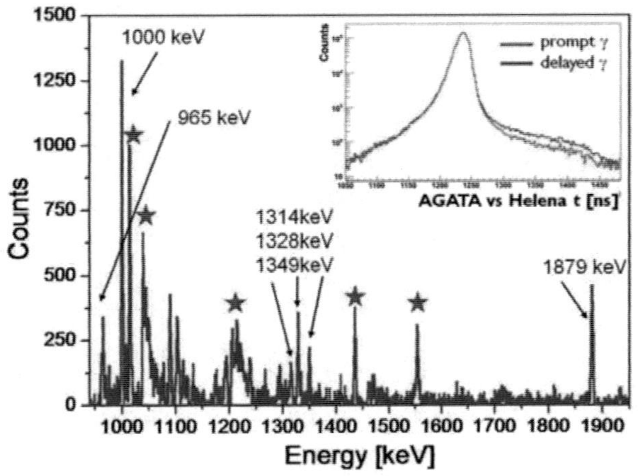

Figure 4. Energy spectrum obtained by gating on delayed transitions: isomeric γ-rays are indicated by arrows; red stars denote peaks due to neutrons and environmental background. The inset shows the time-spectra obtained by gating on energy of prompt (red line) and delayed (blue line) γ-rays.

These results clearly indicate that we can use time-gate instead of energy-gate to select the contribution of high-K bands from the total decay flux. Therefore two different γ-γ matrices have been built: the first with gate on prompt transitions in the Helena array, the second requiring delayed gammas in the scintillators. The obtained matrices show in both cases a ridge-valley structure typical of rotational nuclei: the ridges are populated by discrete bands close to the yrast line (ordered region), the valley by γ transitions from the warmer regime, where band mixing sets in (onset of chaos). Before performing any further analysis the uncorrelated background (e.g. Compton scattered γ-rays) and the discrete resolved lines have been subtracted [6]. Figure 5 shows projections at the average transition energy of 820 keV of the matrices built with gate on prompt (on the left) and delayed transitions (on the right). The corresponding ridge structure are populated by the total discrete bands and the high-K bands only, respectively.
A quantitative study of the ridge structure of both matrices can be performed by a statistical fluctuation analysis [6], providing the total number of bands of ^{174}W and the number of bands of high-K nature.

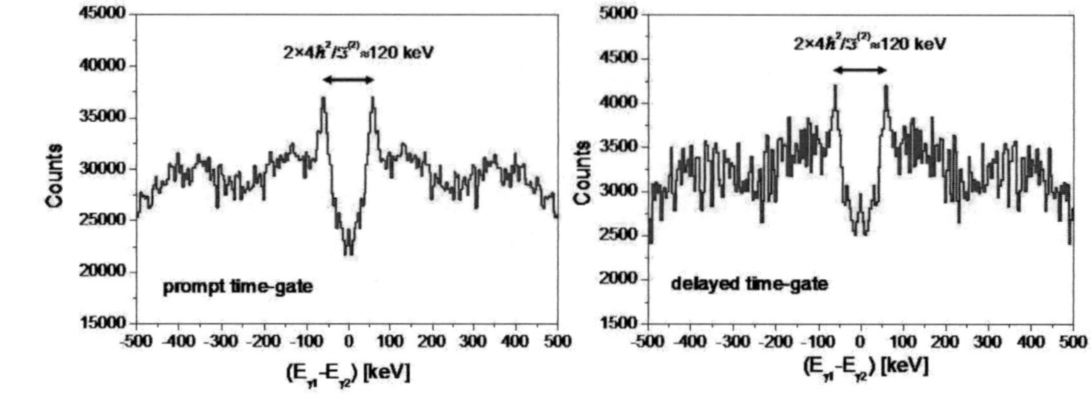

Figure 5. Projections of γ-γ matrices built with gate on prompt (on the left) and delayed (on the right) transitions at the average transition energy of 820 keV. The observed ridge structures have in both cases a separation of $2\times 4\hbar^2/\Im^{(2)} \approx 120$ keV, where $\Im^{(2)}$ is the dynamical moment of inertia of the discrete rotational bands.

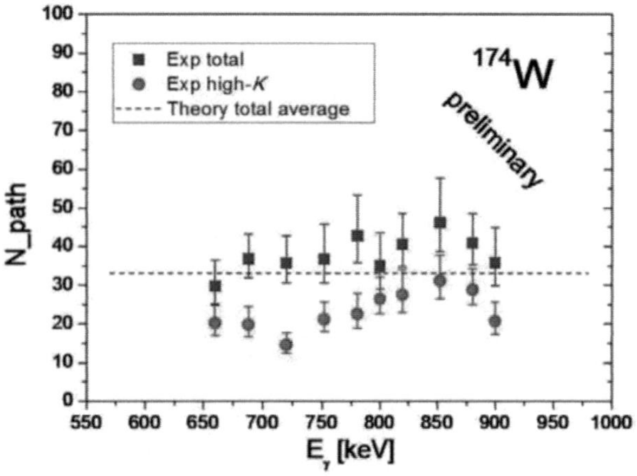

Figure 6. The number of discrete paths (regular bands) obtained by the statistical fluctuation analysis of the ridge structures of ^{174}W. Blue squares indicate the total number of bands (after correction for contaminants of ^{173}W and ^{175}W in the region $E_\gamma \approx$650-800 keV), red circles refer to discrete bands of high-K nature. The dashed line is the prediction of the Cranked Shell Model for the total number of discrete bands, averaged over the 20-50$\hbar$ spin region.

Figure 6 shows very preliminary results about the experimental number of bands for the ^{174}W nucleus, as a function of the transition energy E_γ. We find that there are ~35-40 bands in total (that is the sum of low-K and high-K contribution) (blue squares) out of which roughly half are of high-K nature (red circles). This is similar to what was previously obtained for the ^{163}Er nucleus [2]. In addition, the results for the total number of bands in ^{174}W are in quite good agreement with preliminary Cranked Shell Model calculation [5] (dashed line in figure 6).

References

[1] Bracco A et al 2002 *Rep. Prog. Phys.* **65** 299;
[2] Benzoni G et al 2005 *Phys. Lett.* B **72** 034307;
[3] Tandel S K et al 2006 *Phys. Rev.* C **73** 044306 and 2008 *Phys. Rev.* C **77** 024313;
[4] Bazzacco D 2004 *Nucl. Phys.* A **746** 248c;
[5] Matsuo M 2004 *Nucl. Phys.* A **736** 223;
[6] Døssing T et al 1996 *Phys. Rep* **268** 1;
[7] Bracco A et al 1996 *Nuc. Phys.* A **599** 83c;

Journal of Physics: Conference Series **366** (2012) 012046 doi:10.1088/1742-6596/366/1/012046

Advanced fission models in nuclear data calculations

T M Shneidman[1], **A V Andreev**[1], **M Sin**[2], **C Massimi**[3,4], **G Vannini**[3,4] **and A Ventura**[4,5]

Abstract. Transition states at the saddle points and superdeformed or hyperdeformed states in the secondary wells of multiple-humped potential barriers play an important role in low-energy fission processes. In the present work discrete collective spectra at large nuclear deformations are predicted by means of the dinuclear model and combined with the optical model for fission of the Empire-3 system of codes.The formalism is applied to the $^{233}U(n,f)$ reaction and the computed cross section compared with recent experimental results of the n_TOF Collaboration. Angular anisotopies of fission fragments are evaluated with an improved version of the scission-point model.

[1] Joint Institute for Nuclear Research, Dubna, Russian Federation
[2] Nuclear Physics Department, University of Bucharest, Romania
[3] Dipartimento di Fisica dell' Università di Bologna, Italy
[4] Istituto Nazionale di Fisica Nucleare, Sezione di Bologna, Italy
[5] ENEA, Centro Ricerche Ezio Clementel, Bologna, Italy

E-mail: alberto.ventura@enea.it

1. Introduction

Accurate neutron cross sections are of great importance to the design of innovative systems for nuclear energy production and nuclear waste incineration, such as accelerator driven systems and Generation IV fast reactors, whence the need for high-precision measurements and evaluations of neutron data based on reliable reaction models and codes.

In order to perform accurate measurements of radiative capture and fission cross sections, an innovative neutron time-of-flight (n_TOF)[1] facility has been constructed at CERN, Geneva. Neutrons are produced by 20 GeV/c protons from the CERN Proton Synchrotron impinging on a lead block surrounded by a water layer acting as coolant and moderator of the neutron spectrum, which spans several orders of magnitude, from thermal energy to approximately 1 GeV.

As far as fission is concerned, during the first experimental campaign (2002-2004), the n_TOF Collaboration measured (n,f) cross sections of nuclides relevant to the $Th-U$ fuel cycle (^{232}Th, $^{233,234,236}U$) and of minor actinides commonly appearing in the spent fuel of present-day reactors (^{237}Np, $^{241,243}Am$, ^{245}Cm). In the second experimental campaign, started in 2008, measurements are planned or under way for fission cross sections of $^{240,242}Pu$ and angular distributions of fission fragments (^{232}Th, ^{234}U, ^{237}Np).

The main purpose of the present project is to establish a set of advanced fission models and computer codes suited to the theoretical interpretation of the n_TOF measurements, and, as an useful by-product, to the evaluation of neutron cross sections of the mentioned nuclides in the energy range of interest to fission reactors. The preliminary calculations described in the

Published under licence by IOP Publishing Ltd

Journal of Physics: Conference Series **366** (2012) 012046 doi:10.1088/1742-6596/366/1/012046

present work are focused on the $^{233}U(n, f)$ cross section, measured by the n_TOF Collaboration from thermal energy up to 20 MeV[2],[3].

2. Models, codes and results

The present section is dedicated to a concise description of the system of codes used in calculating fission cross sections, to the models adopted in determining part of the fission input and to the comparison of calculated results with experimental data.

2.1. The fission channel of the Empire-3 system

Empire-3, used in cross section calculations, is a modular system of nuclear reaction codes, based on advanced nuclear models and designed for calculations over a broad range of incident energies and incident particles (neutrons, protons, photons, light and heavy ions)[4].

The code accounts for the major nuclear reaction mechanisms, including direct, pre-equilibrium and compound nucleus ones, by means of a generalized optical model, multistep-direct and multistep-compound models, exciton model with cluster emission and statistical Hauser-Feshbach model.

With particular reference to the fission channel, Empire-3 can describe transmission through multi-humped barriers parametrized analytically or defined numerically. In the present work, use is made of the former option: the barrier as a function of a deformation parameter, β, along the fission path, is parametrized by N smoothly joined parabolas

$$V_i(\beta) = E_{fi} + \frac{(-1)^i}{2}\mu(\hbar\omega_i)^2(\beta - \beta_i)^2. \qquad (i = 1, N) \tag{1}$$

Here, odd values of index i refer to humps, even values to wells, $\mu \approx 0.054A^{5/3}$ MeV^{-1} is a semiempirical inertial mass parameter, while heights E_{fi} and curvatures $\hbar\omega_i$ can either be fitted on theoretical values or adjusted on experimental data. In principle, N is an arbitrarily high odd number, but in the present work calculations are performed for $N = 3$ (two humps) and $N = 5$(three humps).

On the humps and in the wells, the fissioning nucleus has a discrete and a continuous spectrum. The discrete part consists of rotational bands built on collective (vibrational) states or on non-collective (few quasi-particle) states, each state being labelled with its angular momentum, J, projection of J on the nuclear symmetry axis, K and parity π

$$E_n(J, K, \pi) = E_{fi} + \epsilon_n(K, \pi) + \frac{\hbar^2}{2I_n}\left[J(J+1) - K^2\right] \quad . \tag{2}$$

The collective band-head energies, $\epsilon_n(K, \pi)$ and moments of inertia, I_n, are theoretically evaluated within the framework of the dinuclear model, as discussed in the next subsection.

Above the lowest-lying non-collective excitation, a two-quasiparticle state in the case of an even-even compound nucleus, like ^{234}U, the discrete spectrum is replaced with a continuum, *i. e.* a non-collective level density computed in the frame of the generalized superfluid model, with an energy-dependent collective enhancement factor, as described in Ref.[4].

The intermediate potential wells may have an imaginary component, according to the optical model for fission, which simulates the damping of collective states in the wells themselves due to the coupling with non-collective states in the ground-state well. In this way, the transmittance trough the multiple-humped barrier, computed in JWKB approximation as described in Ref.[5], will be the sum of a direct contribution, due to the probability flux that crosses the whole barrier, and an indirect contribution, due to the flux absorbed in an intermediate well and reemitted in the fission channel.

Journal of Physics: Conference Series **366** (2012) 012046 doi:10.1088/1742-6596/366/1/012046

In the case of the $^{233}U(n,f)$ reaction, however, vibrational states inside the wells do not give a sizable contribution, because the excitation energy is always greater that the barrier. Therefore, the analysis will be focused on collective transition states on the humps.

2.2. Transition states in the dinuclear model

A theoretical estimate of the fission barrier parameters, such as heights and curvatures of the parabolic approximation given by formula (1), can be obtained by self-consistent microscopic methods (Hartree-Fock-Bogoliubov or relativistic mean field) or by macroscopic-microscopic methods (Strutinsky's shell and pairing corrections to liquid drop energies): after defining a set of collective coordinates suited to the description of nuclei at large deformations, the potential energy of the system as a function of convenient subsets of the above mentioned coordinates lends itself to the determination of fission paths.

In the present work we have used the macroscopic-microscopic method, with shell corrections evaluated by means of the two-centre shell model[6], suited to the description of a system of two clusters (prefragments) of masses A_1, A_2, charges Z_1, Z_2 in terms of mass asymmetry, $\eta = (A_1 - A_2)/(A_1 + A_2)$, charge asymmetry, $\eta_Z = (Z_1 - Z_2)/(Z_1 + Z_2)$, distance, d, of cluster centres (elongation) and cluster deformations, $\beta_i = c_i/a_i$, $(i = 1,2)$, where c_i and a_i are the major and minor semiaxis of the ith cluster, schematized as an ellipsoid.

The potential energy, U, of the system is the sum of the shell-corrected potential energies of the two clusters, their nuclear and Coulomb interaction and the rotational energy. Fission paths can be determined on the potential energy surface depending on mass asymmetry and elongation, once the other collective variables have been fixed so as to minimize U for given η and d.

In their equilibrium condition, the two ellipsoids are in a pole-to-pole configuration, with an elongation slightly greater than the sum $c_1 + c_2$ of the major semiaxes by a quantity of the order of $0.5 \div 1$ fm, owing to the interplay of the attractive nuclear and repulsive Coulomb interactions.

The model predicts for the compound nucleus ^{234}U a three-humped fission barrier with peak heights decreasing with increasing elongation. At each point on the fission path, the nuclear wave function is a superposition of different dinuclear configurations; the dominant one is $^{230}Th +^4 He$ at normal deformation, $^{212}Po +^{22} O$ at the first hump, $^{194}Os +^{40} S$ at the second hump, $^{128}Sn +^{106} Mo$ at the third hump and $^{132}Te +^{102} Zr$ at the scission point. Charge and mass multipole moments and moments of inertia can be computed in the frame of the dinuclear model as described in Ref. [7].

In principle, the collective spectrum of a dinuclear system is obtained by diagonalizing a Hamiltonian depending on 15 collective coordinates, mass asymmetry, elongation, 3 Euler angles describing rotations of the system as a whole, 3 + 3 Euler angles describing rotations of the two clusters, 2 + 2 Bohr coordinates describing internal excitations of the two clusters, on the assumption that their deformations are quadrupolar. In practice, not all degrees of freedom have the same importance and several simplifying assumptions can be made.

If one is interested in the lowest-lying collective bands of both parities of a system with stable octupole deformation, like a compound nucleus at the saddle points and in the intermediate wells of a fission barrier, use can be made of the bending approximation[8], where the relevant collective motions are basically three: exchange of nucleons between the two clusters, rotations of the dinuclear system as a whole and oscillations of the heavy cluster around its equilibrium position inside the system. In fact, the frequencies of oscillation of the light deformed cluster around the pole-to-pole orientation of the system are much higher than those of the heavy cluster and give rise to collective states at higher energy than the first few-quasiparticle states. The bending Hamiltonian can be diagonalized analytically and the resulting spectrum consists of rotational bands of type (2), with band-head energies that do not depend on parity

$$\epsilon_n(K,\pi) = \hbar\omega_b(2n + |K| + 1) \quad . \tag{3}$$

In other words, when $K \neq 0$, states of the same angular momentum J and opposite parity are degenerate. As an example, Fig. 1 shows the calculated collective spectrum of ^{234}U at the second saddle point. Such a degeneracy does not appear in the spectrum at ground-state deformation, owing to the dominant contribution of the mononucleus configuration, which does not have a stable octupole configuration. There, the theoretical collective bands turn out to be in excellent agreement with the experimental ones.

Since we are interested in reproducing the measured fission cross section of ^{233}U up to 20 MeV, collective spectra at the saddle points and in the intermediate wells of fission barriers are computed also for the fissioning nuclei ^{233}U and ^{232}U, in order to evaluate second- and third-chance fission ($(n, n'f)$ and $(n, 2nf)$, respectively). While the spectra of transition states are kept fixed, we take the freedom to adjust barrier heights and curvatures for a better fit of experimental data. Results are shown in Fig. 2 for three-humped and two-humped barrier calculations.

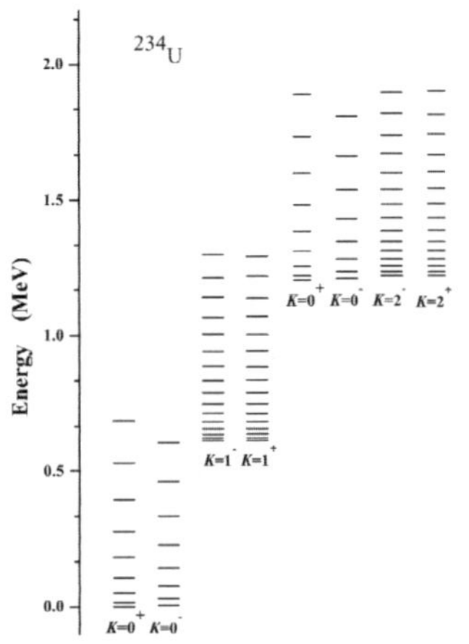

Figure 1. Collective spectrum of ^{234}U at the second saddle point.

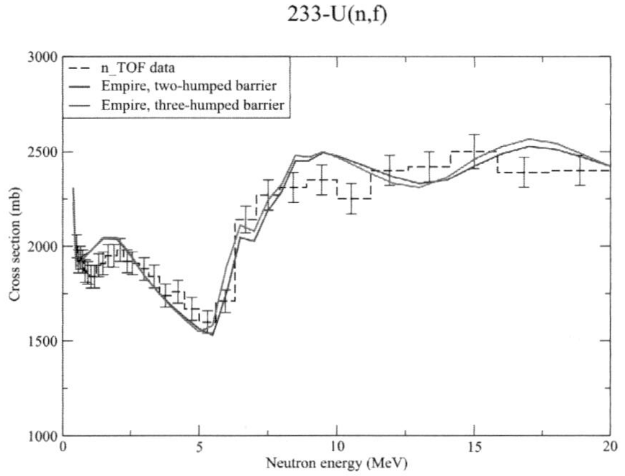

Figure 2. Calculated and experimental $^{233}U(n, f)$ cross sections. Experimental data are taken from Ref.[3]

2.3. Distributions of fission fragments in the scission-point model

The fission cross section alone is not enough to check the transition states, since it can be equally well reproduced with different sets of collective bands. Angular distributions of fission fragments are much more sensitive to the (K, π) quantum numbers of the bands than the fission cross section itself.

In order to evaluate angular distributions in a static fission model, two extreme assumptions can be made: either angular distributions depend on the (K, π) quantum numbers of states at the outer saddle point, approximately conserved in the descent from saddle to scission, or on the (K, π) states at the prescission configuration. Testing the validity of the first assumption requires modifying the statistical model routines of the Empire code and is under study, testing the second assumption is possible within the framework of the scission-point model[9].

The model assumes the existence of a well defined scission configuration, up to which strong non-adiabatic effects maintain statistical equilibrium among various degrees of freedom and beyond which there is no nuclear interaction of the fragments. It is thus possible to calculate the probabilities of different scission configurations and, consequently, to describe various fission characteristics, such as mass, charge and kinetic energy distributions of the fragments.

As for angular distributions, if the dinuclear system at the scission point has excitation energy E^*, angular momentum J, with projection M on the direction of the incident neutron beam and parity π, the angular distribution of fission fragments reads

$$W_{JM\pi}(A_1, A_2, Z_1, Z_2, E^*, \theta) = A(J)\left(\frac{2J+1}{2}\right)\sum_{nK}\exp\left[-\frac{E_{nJMK\pi}}{T(E^*)}\right]|d^J_{MK}(\theta)|^2 \quad , \qquad (4)$$

where $A(J)$ is a normalization constant, $E_{nJMK\pi}$ the eigenvalues of the dinuclear-model Hamiltonian,$T(E^*)$ the nuclear temperature and $d^J_{MK}(\theta)$ the Wigner reduced matrix elements. The angular anisotropy of fission fragments, $W(\theta = 0)/W(\theta = \pi/2)$, evaluated with an improved version[10] of the scission point model, as a function of incident neutron energy is compared with experimental data in Fig.3. The agreement between theory and experiments might be improved at low neutron energy by considering the contributions of bands built on quasiparticle states, not yet included in the model. In any case, further analysis is needed. As a perspective for

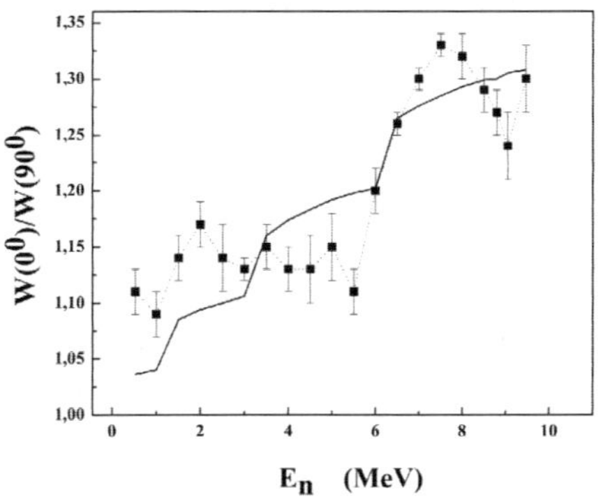

Figure 3. Calculated and experimental angular anisotropy of fission fragments *vs* incident neutron energy. Data are taken from Ref. [11].

future work, we intend to use the dinuclear model for a consistent evaluation of the collective enhancement factor of level densities at the saddle points and to extend the present analysis to the fission cross section of the adjacent nucleus ^{234}U, also recently measured by the n_TOF Collaboration[12], where collective bands in the intermediate wells play a sizable role in the under-threshold region.

References

[1] Gunsing F *et al* (the n_TOF Collaboration) 2007 *Nucl. Instr. Meth.* **B 261** 925 and references therein.
[2] Calviani M *et al* (the n_TOF Collaboration) 2009 *Phys. Rev. C* **80** 044604.
[3] Belloni F*et al* (the n_TOF Collaboration) 2011 *Eur. Phys. J.* **A 47** 2.
[4] Herman M *et al* 2007 *Nucl. Data Sheets* **108** (2007) 2655.
[5] Sin M *et al* 2006 *Phys. Rev. C* **74** 014608.
[6] Maruhn J and Greiner W 1972 *Z. Phys.* **251** 431.
[7] Shneidman TM *et al* 2000 *Nucl. Phys.* **A 671** 119.
[8] Adamian GG *et al* 2007 *Phys. At. Nuclei* **70** 1350.
[9] Wilkins BD, Steinberg EP and Chasman RR 1976 *Phys. Rev. C* **14** 1832.
[10] Andreev AV *et al* 2004 *Eur. Phys. J.* **A 22** 51.
[11] Simmons JE and Henkel RL 1960 *Phys. Rev.* **120** 198; Leachman RB and Blumberg L 1965 *Phys. Rev.* **137** B814.
[12] Paradela C *et al* (the n_TOF Collaboration) 2010 *Phys. Rev. C* **82** 034601.

XIX International School on Nuclear Physics, Neutron Physics and Applications IOP Publishing
Journal of Physics: Conference Series **366** (2012) 012047 doi:10.1088/1742-6596/366/1/012047

Measurement of Neutrons in Different Pb/U Setups Irradiated by Relativistic Protons and Deuterons by means of Activation Samples

V Wagner[1a,b], **O Svoboda**[a], **J Vrzalová**[a,b], **M Suchopár**[a,b], **B Geier**[a], **A Kugler**[a] **and M Honusek**[a]
for the collaboration Energy and Transmutation of Radioactive Waste

a) Nuclear Physics Institute of ASCR, CZ-25068 Řež, Czech Republic
b) Faculty of Nuclear Sciences and Physical Engineering of CTU, Břehová 7, CZ-110 00 Praha 1, Czech Republic

E-mail: wagner@ujf.cas.cz

Abstract. The collaboration Energy and Transmutation of Radioactive Waste uses different setups consisting of lead, uranium and graphite irradiated by relativistic protons and deuterons to study transmutation of radioactive materials by produced neutrons. Our group measured spatial distribution of neutrons by means of activation samples during the assembly irradiation by the JINR Nuclotron beams. We also present results of simulations using MCNPX code and their comparison with obtained experimental data. We use Au, Al, Bi, In and Ta foils as activation detectors, but unfortunately almost no experimental cross-section data for observed threshold (n,xn) reactions are available for higher neutron energies. Therefore we carried out series experiments devoted to determination of neutron cross-sections of various threshold reactions using different quasi-monoenergetic neutron sources.

1. The spallation neutron production and transmutation of actinides and fission products

Accelerator Driven Transmutation Systems (ADTS) based on a subcritical nuclear reactor driven by an external spallation neutron source are being investigated with increasing interest in the last three decades. The transmutations by very intensive fast and resonance neutron fields require more precise knowledge, particularly about emitted and moderated neutrons, residual nuclei and transmutations reactions. The studies of simple or more complex model set-ups simulating ADTS features are necessary. The collaboration Energy and Transmutation of Radioactive Waste is an international project that investigates neutron production by spallation reactions and their transport as well as transmutation of actinides and fission products in corresponding neutron fields.

Different set-ups are used to simulate properties of future ADTS. Lead or uranium targets of such set-ups are irradiated by proton or deuteron beams generated by the JINR Dubna accelerators. The first type of set-ups is simple lead target surrounded by paraffin or graphite moderator (GAMMA 1 and GAMMA 3). More complex set-up consists of lead target, natural uranium blanket and polyethylene biological shielding. Such set-up is named as "Energy plus Transmutation" and the systematic set of

[1] To whom any correspondence should be addressed.

experiments with irradiation by proton and deuteron beams with different energies was performed [1]. Last year, experiments with new big uranium target KVINTA were started.

The measurements of the spatial distribution of neutron field inside such set-ups represent an important part of described studies. The activation detectors of neutrons are used for determination of neutron intensities and energies. The big advantages of this type of neutron detectors are their small size and their simplicity of usage.

2. The measurement of neutron reaction cross-sections

Unfortunately, almost no experimental cross-section data for most of the observed threshold (n,xn), (n,p) and (n,α) reactions are available for used activation detector materials and higher neutron energies. This is main motivation for the cross-section measurements done by means of two quasi-monoenergetic neutron sources based on proton beam, lithium target and reaction ^{7}Li(p,n)^{7}Be. The first neutron source is based on the cyclotron at Nuclear Physics Institute (NPI) at Řež with proton energy up to 38 MeV [2] and the second one is based on the cyclotron at The Svedberg Laboratory (TSL) in Uppsala with possible proton energy up to 200 MeV [3]. We performed five cross-section measurements exploiting the neutron source at NPI in Řež and seven irradiations by means of the neutron source at TSL in Uppsala using different energies of produced neutrons from 17 up to 94 MeV.

We studied Al, Au, Bi, In, Ta, Y and I materials usually in the form of thin foil. The activation method was used for determination of cross-sections. The first results were published in [4] and [5].

2.1. The sources of uncertainties

The very detailed analysis of all possible sources of statistical and systematic uncertainties is the most important condition for obtaining good quality results. The cross-section σ is determined by means of equation:

$$\sigma = \frac{N_{yield} \cdot S \cdot A}{N_n \cdot N_A} \tag{1}$$

where S is area of the foil, A is molar weight and N_A is Avogadro number. Then first source of possible uncertainties is connected with the neutron number N_n determination. The yield of radioisotope N_{yield} produced in the studied reaction is obtained by gamma spectroscopy using equation (2):

$$N_{yield} = \frac{S_p \cdot C_{abs}(E)}{I_\gamma \cdot \varepsilon_P(E) \cdot COI \cdot C_{area}} \frac{t_{real}}{t_{live}} \frac{1}{m_{foil}} \frac{e^{(\lambda \cdot t_0)}}{1-e^{(-\lambda \cdot t_{real})}} \frac{\lambda \cdot t_{irr}}{1-e^{(-\lambda \cdot t_{irr})}} \tag{2}$$

where S_p is the gamma peak area, C_{abs} is self-absorption correction, I_γ is gamma line intensity, $\varepsilon_p(E)$ is detector efficiency, COI is correction for real coincidences, C_{area} is square-emitter correction, t_{real}/t_{live} is dead time correction, m_{foil} is the sample mass and the last two fractions represent decay during cooling and measurement and decay during irradiation. Then the second source of uncertainties is connected with accuracy of the different described gamma spectroscopy corrections and parameters.

The last source of uncertainties is connected with the neutron spectrum of the used source. Such sources are quasi-monoenergetic and their neutron spectrum contains beside the main neutron peak also a lower continuum. This background is negligible problem in the case of neutron energies which are near to the studied reaction threshold. The accurate subtraction of such background is very important if the cross-sections measurement is performed within neutron energies, which are much higher than the energy threshold.

We have prepared systematic studies of different neutron reaction cross-sections on yttrium. This monoisotopic material is very useful as a high energy neutron activation detector. It has many useable threshold (n,xn) reactions and we use such activation detectors very intensively during our ADTS studies. But the experimental information about cross-sections for almost all reactions is very scarce, see EXFOR [6].

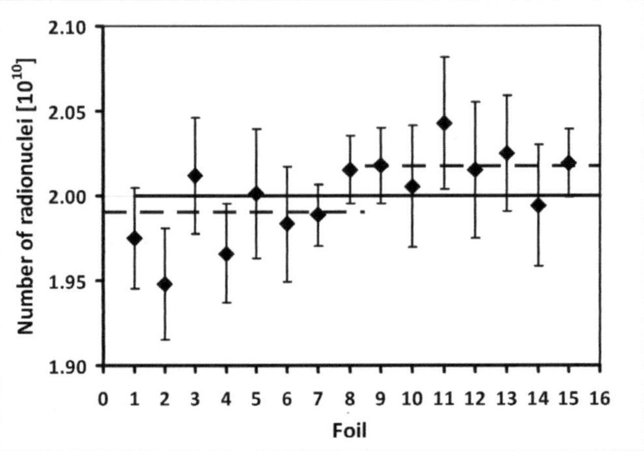

Figure 1) Number of radioactive nuclei ^{88}Y determined by means of gamma line with energy 898 keV, first eight values (side A was near to detector) and last seven values (side B was near to detector).

As part of this preparation we made methodical experiment by means of the NPI neutron source. The yttrium foil and gold foil (a common material with detailed knowledge of his properties) were irradiated by neutrons with energy 32.5 MeV obtained using proton beam with energy 35 MeV. The irradiation took 22 hours and high activities of both samples were obtained. Gamma measurements were performed for different distances of sample (15, 23, 33, 53, 70, 93 and 173 mm) to the detector and several measurements (between 6 and 15) were taken for each distance. The sample was turned to the opposite side after every measurement to analyze if there is an influence of geometry on measurements with different side nearer to detector. Magnitudes of the different gamma spectroscopy corrections are quickly decreasing with increasing distance to the detector. But it is necessary to use short distances in the case of low activity samples and to know accuracy of correction determination.

The yttrium foil was thicker and not fully homogenous. This was the reason why there was small difference between measurements with different side nearer to the detector for the nearest distances to the detector. But the difference is smaller than 1.4 % for second nearest position and within statistical error of single measurement, see figure 1. No such difference was observed in the case of thinner and homogenous gold foil. A methodical study of the uncertainties concerning the determination of other corrections was performed. The biggest influence has uncertainty of efficiency determination which is about 3 %.

2.2. The study of population of ground and isomeric ^{87}Y states

The only two threshold reactions ^{89}Y(n,2n)^{88}Y and ^{89}Y(n,3n)^{87}Y are possible for neutron energy 32,5 MeV. The first reaction has the threshold energy 11.6 MeV and influence of the neutron background is important. The amount of ^{88}Y nuclei produced by background is about 65 %. The cross-sections needed for background influence determination were calculated by means of the deterministic code TALYS [7] and the spectra of neutron source described in [2] were used. The last cross-section from EXFOR data base is for neutron energy 28 MeV. We obtained cross-section for new energy region.

The second reaction ^{89}Y(n,3n)^{87}Y is very interesting. Prior to their further decay, the nuclei are produced either in their ground state of half-life of 79.8 hours, or in the 380.79 keV isomeric state of half-life 13.38 hours, see figure 2. The isomeric state decays by gamma transition to the ground state with probability 98.4 %. The beta decay of this state is within our accuracy negligible.

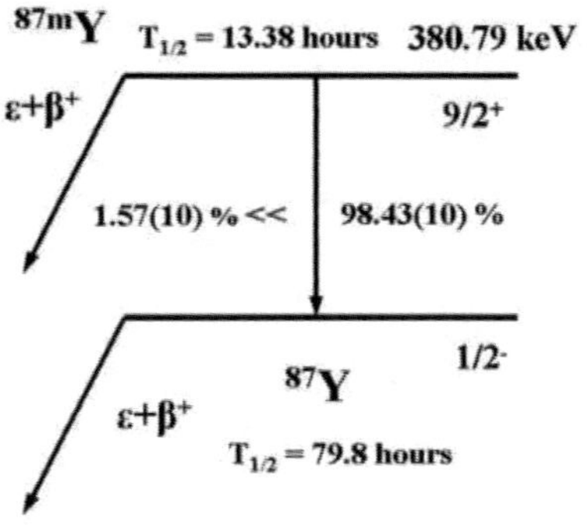

Figure 2) Decay scheme of isotope ^{87}Y and its isomeric state ^{87m}Y

We made many measurements in different times and we obtained detailed description of isomeric state decay and population and decay of ^{87}Y ground state, see figure 3. From the described data, we derived cross-sections of the ^{89}Y(n,3n)^{87}Y and ^{89}Y(n,3n)^{87m}Y reactions. Advantage is threshold of this reaction which is 21 MeV in this case. This is the reason why the amount of ^{87}Y nuclei produced by the neutron background is relatively very small - about 7%.

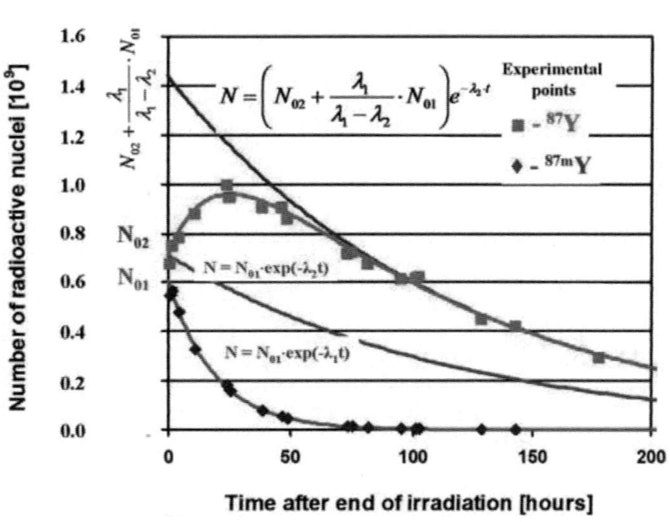

Figure 3) Decay and population of ground state and decay of isomeric state.

Number N_1 of nuclei in isomeric state is described by equation (3):

$$N_1 = N_{01}e^{-\lambda_1 \cdot t} \tag{3}$$

where N_{01} is number of nuclei in isomeric state just after end of irradiation and λ_1 is decay constant of isomeric state. Number N_2 of nuclei in ground state is described by equation (4):

$$N_2 = \left(N_{02} + \frac{\lambda_1}{\lambda_1 - \lambda_2}\cdot N_{01}\right)e^{-\lambda_2 \cdot t} - \frac{\lambda_1}{\lambda_1 - \lambda_2}\cdot N_{01}e^{-\lambda_1 \cdot t} \tag{4}$$

where N_{02} is number of nuclei in ground state just after end of irradiation and λ_2 is decay constant of ground state. The influence of decay and population of both states during irradiation is necessary to include for obtaining the cross-sections of ^{87}Y ground and isomeric states production. The cross-section of isomeric state is about 2.8times larger then the cross-section of ground state, see figure 4.

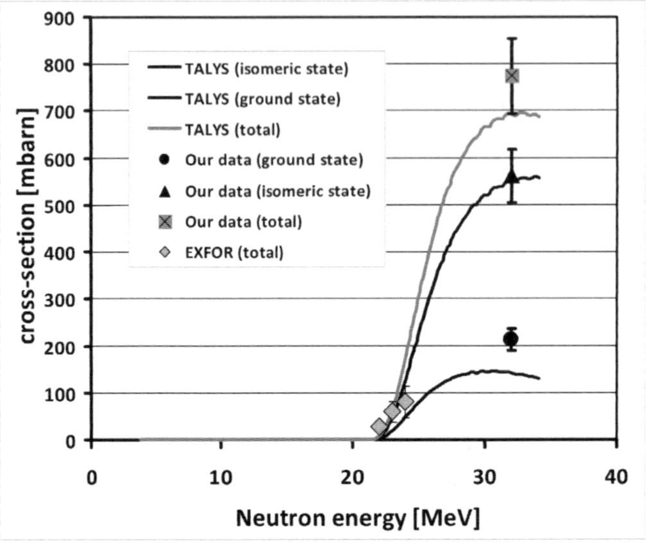

Figure 4) Cross-sections of ^{87}Y ground and isomeric states production, experimental data marked by points and TALYS calculations are described by lines. Our data are preliminary.

3. Conclusions

The activation neutron detectors represent a convenient tool for the determination of the spatial distribution of a neutron field. The measurements of cross-sections in wide range of energies are needed to increase the usability of such detectors. We have used two quasi-monoenergetic neutron sources to obtain needed cross-sections. The methodic measurement of neutron reactions on yttrium for energy 32.5 MeV tested the possibility to obtain excitation functions of such reactions. The origins and magnitudes of the uncertainties were studied in detail.

Acknowledgements

This work was supported by the EFNUDAT program, grant number CTU0808214 and by the F4E program of the Nuclear Reaction Department (NPI), F4E-2008-GRT-014.

References

[1] A. Krása et al., "Neutron production in a Pb/U-setup irradiated with 0.7-2.5 GeV protons and deuterons", Nucl. Instr. and Meth. in Nucl. Res., A615, No.1, 70-77 (2010)
[2] P. Bém et al., "The NPI cyclotron-based fast neutron facility", *Proceedings of the International Conference on Nuclear Data for Science and Technology* – ND2007, 555-558 (2007)
[3] A. V. Prokofiev et al., "The TSL Neutron Beam Facility", *Rad. Prot. Dos.,* 126, 18-22 (2007)
[4] O. Svoboda et al., *Proceedings of the International Conference on Nuclear Data for Science and Technology–ND2010*, Jeju, South Korea, 1709-1712 (2010)
[5] J. Vrzalová et al., *Proceedings of the NEMEA-6 scientific workshop on Nuclear Measurements, Evaluations and Applications*, Krakow, Poland, 259-265 (2010)
[6] Experimental Nuclear Reaction Data (EXFOR/CSISRS) [http://www.nndc.bnl.gov/exfor], 7.10.2011
[7] A. J. Koning et al., TALYS 1.0., *Proceedings of the International Conference on Nuclear Data for Science and Technology-ND 2007*, 2007, 211-214

Lifetime and magnetic moment measurements

V. Werner[1], G. Ilie[1,2], D. Radeck[1,3], T. Ahn[1], C.W. Beausang[4], L. Bettermann[1,3], R.J. Casperson[1], R. Chevrier[1,5], N. Cooper[1], T.C. Bonniwell[4], A. Heinz[1,6], E. Holland[1], D. McCarthy[1,7], B. Pauerstein[4], M.K. Smith[1], J.R. Terry[1,8], E. Williams[1]

[1]Wright Nuclear Structure Laboratory, Yale University, New Haven, Connecticut 06520, USA
[2]National Institute for Physics and Nuclear Engineering, P.O. Box MG-6, Bucharest-Măgurele, Romania
[3]Institut für Kernphysik, Universität Zu Köln, D-5037 Köln, Germany
[4]Department of Physics, University of Richmond, Richmond, Virginia 23173, USA
[5]Department of Physics, University of Caen Basse-Normandie, 14000, France
[6]Chalmers University of Technology, Department of Fundamental Physics, 41296 Goteborg, Sweden
[7]Department of Physics, University of Surrey, Guilford, GU27XH, United Kingdom
[8]Los Alamos National Laboratory, Los Alamos, NM, USA

E-mail: volker.werner@yale.edu

Abstract. The g-Plunger technique has been introduced, which allows for the simultaneous measurement of lifetimes and absolute values of g-factors using a plunger device in inverse kinematics. Experiments on the ^{104,106,108}Pd isotopes using Coulomb excitation resulted in g-factors in agreement with literature data. Lifetimes of 2_1^+ states have in part significantly changed in this work.

1. Introduction

Plunger devices for lifetime measurements in combination with the Recoil Distance Doppler Shift (RDDS) method have been used since many years [1]. Great advances were achieved through the introduction of the Differential Decay Curve Method (DDCM) [2], and modern Cologne-type plunger devices became a standard, which ensure a fixed distance through capacitive feedback [3], such as the ("new") Yale Plunger Device (NYPD) [4]. With the application of many methods to inverse kinematics as a consequence of the increased use of rare isotope beams, also the effects of nuclear deorientation in plunger experiments, as discussed, e.g., in [5, 6], needs to be re-addressed. In contrast to normal kinematics reactions, in inverse kinematics, due to the high charge states reached using fast heavy beams, deorientation effects become important in the regime of few-picosecond lifetimes. Whereas deorientation leads to the necessity of corrections in RDDS analyzes [10, 11], we show that it can be effectively used to obtain relative magnetic moments for excited nuclear states, applying techniques developed in normal and inverse kinematics (see, e.g., Refs. [7, 8, 9]).

We chose ^{106}Pd and neighboring Pd isotopes to test the technique to measure excited state lifetimes and relative g-factors in parallel, henceforth referred to as g-Plunger technique. The g-factor of the first excited 2^+ state, $g(2_1^+)$ in ^{106}Pd, is well established [12] and serves as a reference point for much other g-factor data in this mass region [13]. In the following, the method to fix parameters of the hyperfine interaction and use them to obtain g-factors will be outlined. Preliminary results on $^{104-106}$Pd, for lifetimes and g-factors of their 2_1^+ states will be given.

2. The g-Plunger technique

2.1. Experiment

We chose the Coulomb excitation mechanism in order to directly populate only the first excited 2^+ state of even-even nuclei, in this case $^{104-108}$Pd. At a sufficiently low beam energy, feeding from higher-lying states is negligible, tremendously simplifying the RDDS analysis. A scheme of the plunger setup is shown in Fig. 1. The heavy-ion beams at energies of 324 MeV, 330 MeV, and 336 MeV, respectively, were produced by the ESTU tandem accelerator at the Wright Nuclear Structure Laboratory (WNSL) at Yale University. The beams were Coulomb-excited on an about 0.6 mg/cm^2 thick ^{24}Mg target, and subsequently stopped in a 15.7 mg/cm^2 thick Cu foil. The distance between the foils within NYPD was varied between electrical contact to about 2 mm. Coulomb-excited Pd nuclei decayed either in flight in between the two foils, or after stopping in the Cu foil. Hence, two components of the decay were visible in the spectra, Doppler shifted and unshifted. γ-rays were detected by anti-Compton shielded detectors of the YRAST-Ball array [14] in the SPEEDY setup [15], 3 placed at 138.5°, 4 at 41.5°, and an additional 2 unshielded clover detectors at 90° relative to the beam axis. The forward scattered Mg nuclei passed through the Cu foil and were detected in a Si detector behind the stopper. The requirement of a coincidence between the Si detector and any HPGe detector along with random-background subtraction ensured excitation on the target and not on the stopper foil, and removed all contributions from room background.

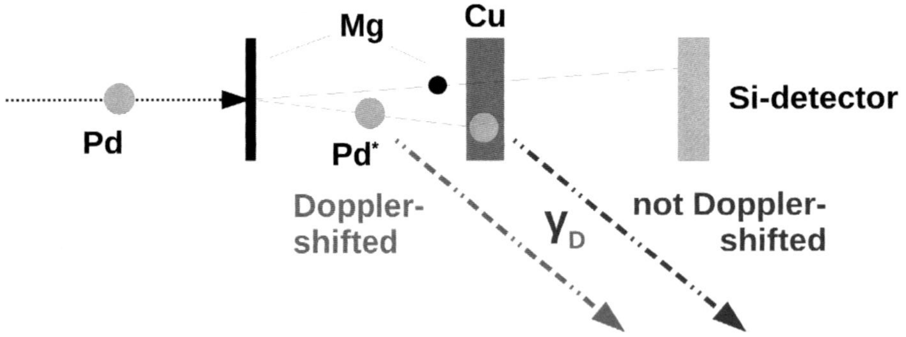

Figure 1. Schematic of the setup inside the plunger target chamber.

2.2. Lifetime analysis

The data was analyzed by fitting the in-flight decay probability

$$P_{\text{exp}}(d) = \frac{I^{SH}(d)}{I^{SH}(d) + I^{USH}(d)}\,, \tag{1}$$

obtained from fitting the Doppler shifted and unshifted peaks, after applying corrections to the measured peak areas. These corrections were due to solid angles of the HPGe detectors, introducing-factors Q_k [16] into the angular distribution function of the γ-rays, as well the solid angle of the particle detector, which was taken into account in calculations using a standard Coulomb excitation code [17]. A significant correction due to the large velocities (4% c) is from the Lorentz boost, which changes the effective detection solid angles for Doppler-shifted γ-rays (laboratory system versus nuclear inertial system) following

$$\frac{d\Omega}{d\Omega'} = \left(1 + \frac{E_\gamma - E_{\gamma,0}}{E_{\gamma,0}}\right)^2 . \tag{2}$$

Another major correction is due to the deorientation effect, which is discussed in the following sub-section. Since Clover detectors, consisting of four individual crystals, can be split into two halfs with different angles relative to the beam axis, we obtained data for a total of six angles (rings). For each ring we fitted $P_{\text{exp}}(d)$ with the decay function

$$P(d) = A \cdot (1 - \exp(-\lambda(d - d_0)) , \tag{3}$$

where A is a scaling parameter, and d_0 is the minimum achievable distance setting (just before electrical contact between the foils), which was independently obtained from a distance calibration, and from this fit. The lifetime of the 2_1^+ states was extracted using $\lambda = \frac{1}{v \cdot \tau}$.

For the first excited state of ^{106}Pd, a preliminary lifetime result is $\tau(2_1^+, {}^{106}\text{Pd}) = 19.87(14)$ ps, where the error is purely statistical. We currently assume another 1-2% systematic error. We note, that the result before deorientation correction is about 17.6 ps, making this correction the most important one in the lifetime analysis. It is due to the fact that nuclei decaying from their 2_1^+ state in the stopper foil, passed through the entire vacuum gap between the foils, in which they lost their alignment due to the hyperfine interaction. In contrast, nuclei decaying in-flight only passed through a fraction of the vacuum distance in their excited state, hence, the angular distribution of the de-exciting γ-rays is less attenuated.

2.3. Nuclear deorientation

After Coulomb excitation, nuclei leaving the target are aligned with respect to the beam axis, hence, γ-rays from decays of the 2_1^+ state have a distinct angular distribution. However, after leaving the foil they are subject to hyperfine interaction, which causes a precession of the nuclear spin $\vec{J}$ around the total spin of the ion $\vec{F} = \vec{J} + \vec{I}$, where I is the atomic spin. Since the atomic configurations of the ions after passing through the target (Mg) foil are randomized, the directions of the precessions are random. Assuming that the atomic state lifetimes are longer than the deorientation process, one expects an approximately exponential attenuation of the alignment of the ensemble of nuclei, converging to a hard-core value [6, 7, 18]. This behavior is parametrized in factors

$$G_k(t) = \alpha_k + (1 - \alpha_k) \cdot \exp(-\Gamma_k t) , \tag{4}$$

which are introduced into the γ-ray angular distribution function as

$$W(t, \vartheta) = \sum_{k=0,2,4} R_k B_k Q_k G_k(t) P_k(\cos(\vartheta)) . \tag{5}$$

Herein, R_k coefficients follow the convention of Rose and Brink [19], and B_k coefficients correspond to the initial alignment ($d = t = 0$), which is calculated from the Coulomb excitation mechanism; P_k are Legendre polynomials. The G_k parameters ($k = 2, 4$) are obtained from the ratios of the experimental angular distribution parameters $A_k = R_k B_k Q_k G_k$ at each distance over the unattenuated values as obtained from the Coulomb excitation formalism,

$$G_k(t) = \frac{A_{k,\text{exp}}(t)}{A_{k,\text{coulex}}(t = 0)} . \tag{6}$$

A fit of Eq. (4) to the experimental G_k coefficients then gives the hard-core parameters α_k, as well as the time constant of the exponential, Γ_k. A sample fit is shown in fig. 2. The parameters

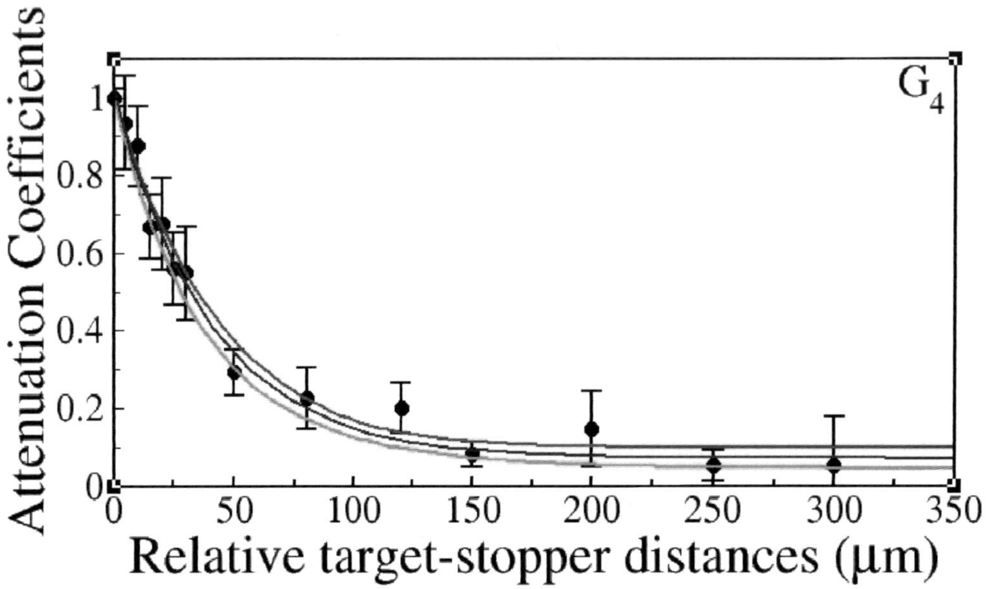

Figure 2. Sample deorientation fit from the ^{106}Pd data, showing the attenuation of the 4th order term in the angular distribution function (5).

Γ_k depend on the g-factor of the excited state,

$$\Gamma_k = \frac{|g|}{C_k}\,, \tag{7}$$

where C_k are parameters of the hyperfine interaction. We used the known g-factor $g(2_1^+) = +0.398(21)$ for ^{106}Pd to extract the parameters C_k. Since a change of the hyperfine interaction is not expected for neighborng Pd isotopes after the same reaction, hence, charge state distribution, C_k can be assumed to be the same for all Pd isotopes. For the data on ^{104,108}Pd, we hence left $|g|$ as a free parameter and obtained g-factors for the 2_1^+ states of both isotopes. Preliminary results are given in Table 1, compared to literature values. We note that the statistics in the ^{104}Pd case was low, leading to the larger error bar. Also, only absolute values of g-factors can be obtained in the g-Plunger technique.

3. Discussion

The g-factors obtained in this work agree within error with literature values. Hence, the g-Plunger technique offers a technique alternative to the commonly used transient field technique (see, e.g. Ref [20] for the measurement of magnetic moments of short-lived excited states. The advantage of the method over RIV is the possibility of a time dependent measurement. Whereas in RIV g-factors and lifetimes of excited states in multiple isotopes of an isotopic chain must be known to parametrize the hyperfine interaction, the same can be obtained in one isotope for which only the excited state g-factor needs to be known. Lifetimes and g-factors can be measured simultaneously using a plunger device. The g-Plunger technique only gives absolute values of g-factors, akin to RIV. Measuring the sign of a g-factor needs the application of an external field, such as in the transient field technique.

Table 1. Preliminary results: absolute values of the g-factors of the 2_1^+ states in ^{104,106,108}Pd. The value marked with an asterisk* was used as a reference point. Errors for values from the present work are statistical only.

	^{104}Pd	^{106}Pd	^{108}Pd
$\lvert g(2_1^+)\rvert$			
this work:	0.52(10)	0.40(2)*	0.32(5)
NNDC:	0.46(4)	0.40(2)	0.36(3)
$\tau(2_1^+$			
this work:	15.64(18) ps	19.87(14) ps	39.05(67) ps
NNDC:	14.9(9) ps	17.6(6) ps	34.6(18) ps

The lifetime values obtained in this work are systematically higher than literature values. The difference to literature values was in all cases introduced when applying the deorientation correction to the data. Problematic is especially the change in $\tau(2_1^+)$ of ^{106}Pd. If the literature lifetime value is changed to our present value, the g-factor of the state will have to be reanalyzed. This would potentially lead to a change of all g-factors that have been measured relative to the $g(2_1^+)$ factor in ^{106}Pd. We are currently reanalyzing the data in order to exclude any systematic errors in the analysis procedures, and are analyzing data sets on other isotopic chains. One recent experiment on ^{98}Ru using the g-Plunger technique has just been submitted for publication [21]. No significant change in the 2_1^+ lifetime has been found in that case.

Acknowledgments

The authors thank N. Benczer-Koller, J. Billowes, G. Kumbartzki, P. Petkov, K.-H. Speidel, and A. Stuchbery for many fruitful discussions. This work was supported by U.S. DOE under grant nos. DE-FG02-19ER-10609, DE-FG52-06NA26206 and DE-FG02-05ER41379. D.R. thanks the German Academic Exchange Service (DAAD) for financial support.

References

[1] Schwarzschild A Z and Warburton E K 1968 *Ann. Rev. Nucl. Sci.* **18** 265
[2] Dewald A, Harissopulos S and von Brentano P 1989 *Z. Phys.* A **334** 163
[3] Alexander T K and Bell A 1969 *Nucl. Instr. M.* **81** 22
[4] Krücken R 2000 *J. Res. Natl. Inst. Stand. Technol.* **105** 53
[5] Sturm R J and Guidry M W, *Nucl. Instr. Meth.* A **138** 345
[6] Petkov P *et al.* 1995 *Nucl. Phys.* A **589** 341
[7] G. Goldring 1982 *Heavy ion collision* vol. 3, ed Bock R (Amsterdam: North-Holland Publishing Company)
[8] Billowes J 1986 *Hyp. Int.* **30** 265
[9] Stone N J *et al.* 2005 *Phys. Rev. Lett.* **94** 192501
[10] Terry J R *et al.* 2009 *AIP Conf. Proc.* **1090** 337
[11] Rother W *et al.* 2011 *Nucl. Instr. Meth.* A **654** 196
[12] Raghavan P 1989 *At. Nucl. Data Tables* **42** 189
[13] Gürdal G *et al.* 2010 *Phys. Rev.* C **82** 064301
[14] Beausang C W *et al.* 2000 *Nucl. Instr. Meth.* A **452** 431
[15] Krücken R 2001 *AIP Conf. Proc.* **576** 319
[16] Barrette J, Lamoureux G and Monaro S 1970 *Nucl. Instr. Meth.* A **93** 1
[17] Winther A and De Boer J 1965 *Cal. Inst. Tech., Tech. Rep.*
[18] Stuchbery A E and Stone N J 2007 *Phys. Rev.* C **76** 034307
[19] Rose H J and Brink D M 1967 *Rev. Mod. Phys.* **39** 306
[20] Shu N K B *et al.* 1980 *Phys. Rev.* C **21** 1828
[21] Radeck D *et al.* 2011 *Phys. Rev.* C *submitted*

Journal of Physics: Conference Series **366** (2012) 012049 doi:10.1088/1742-6596/366/1/012049

Correlations in nuclear interactions between E_{CM} /u and unexplained experimental observables

W Westmeier[1], R Brandt[1], S Tyutyunnikov[2]

[1] Kernchemie, FB Chemie, Philipps-Universität, D-35043 Marburg, Germany
[2] JINR, 141980 Dubna near Moscow, Russian Federation
Corresponding author: Westmeier@Westmeier.com

Abstract: A new concept is introduced for the classification of "unresolved problems" in the understanding of interactions in thick targets irradiated with relativistic ions: The centre-of-mass energy per nucleon of a hypothetical compound nucleus from a primary interaction, E_{CM}/u, is calculated and correlated with several experimental observations in thick target irradiations. One observes in various reactions of relativistic primary ions with thick targets that there appears to be a threshold for reactions leading to "unresolved problems" which lies around **E_{CM} /u ~ 150 MeV** where E is the kinetic energy of the beam. A thick target is defined so that products from the first nuclear interaction will make secondary nuclear interactions. All "unresolved problems" are exclusively observed above this threshold, whereas below this threshold no "unresolved problems" are found.
Another (the same?) threshold exists at a nuclear temperature of 158±3 MeV as a threshold for massive pion production in nuclear interactions. Hagedorn had proposed this threshold decades ago and it is known as the **Hagedorn limit**. In this paper we will only mention, but not elaborate on Hagedorn's theoretical concept any further. Some considerations will be presented and further studies in this field will be suggested.

1. Introduction

Spallation mass-yield curves in nuclear interactions with thin targets were systematically studied in many nuclear chemistry laboratories for decades around the world. These observed spallation mass-yield curves strictly obey well-known concepts of "limiting fragmentation" and "factorisation" (see section 2.2) and are thus well understood within current theoretical models. This applies for nuclear reaction studies induced by ions from E_{total} < 1 GeV and is extending up to 80 GeV ^{40}Ar irradiations. Limited studies extend up to proton induced reactions with $E_{kinetic}$ = 300 GeV (see Ref. [1] for details).

Several articles have recently appeared describing "unresolved problems" in the study of nuclear interactions in thick targets induced by relativistic ions and their secondary reaction products [1, 2]. Product yield distributions in thick copper targets from irradiations with 72 GeV ^{40}Ar (at the LBNL, Berkeley), 44 GeV ^{12}C (at the JINR, Dubna), and 48 GeV ^{4}He (at CERN, Geneva) [3] cannot be understood with well-established theoretical concepts, thus constituting "unresolved problems". Moreover, exceedingly large neutron emission during the irradiation of thick copper, lead and uranium targets with high energy heavy ion beams having E_{total}>30 GeV have been observed in several laboratories; an exceedingly large neutron multiplicity is also considered to be an "unresolved problem".

Journal of Physics: Conference Series **366** (2012) 012049 doi:10.1088/1742-6596/366/1/012049

All attempts to characterise unresolved problems in thick-target nuclear reactions since about 1954 [4] have borne no fruit; the problem being that there are no defined ion energy, projectile mass, and target mass where these unresolved problems systematically occur.

In this paper the following approach will be introduced:
One calculates on a purely hypothetical basis the centre-of-mass energy E_{CM} per nucleon in the entrance channel of the nuclear interaction. This entrance channel is defined by the kinetic energy E_P of the primary ion (projectile) with mass A_P and the target mass A_T. The value of E_{CM} /u in units of MeV is calculated as:

$$\mathbf{E_{CM} /u = E_P * A_T / (A_P + A_T) / (A_P + A_T)} \qquad \mathbf{(1)}$$

Experimental phenomena are produced in thick targets by primary ions (primaries) up to the end of their range and, in addition, by secondary fragments (secondaries) making nuclear interactions in the thick target. The relative importance of nuclear reactions in thick targets due to secondaries compared to primaries increases with the thickness of the target. One correlates the value E_{CM}/u - which might be taken as the hypothetical average excitation energy of each nucleon in the entrance channel of the reaction - with experimentally observed phenomena.
Some correlations are presented in Section 2 for increasing E_{CM}/u. Obviously any observed correlation between E_{CM}/u of the entrance channel and interactions of secondary fragments in thick targets will not explain the reason for unresolved problems, but one does find a systematic dependence. In section 3 we will present some considerations which may be helpful as a start to understanding the observed order presented in Section 2. Section 4 contains our conclusions on the subject and new experiments are suggested which may help to shed light onto this rather old and complex set of unresolved problems.

References

[1] Brandt R et al. 2008 *Physics of Elementary Particles and Nuclei* **39(2)** 507
reprinted in *Physics of Particles and Nuclei* **39(2)** 259

[2] Hashemi-Nezhad S R et al. 2011 *Physics Research International*, Vol. 2011, Article ID 128429, doi:10.1155/2011/128429

[3] Dersch G 1986 *PhD Dissertation* Philipps Universität Marburg unpublished

[4] Rössle E and Schopper E 1954 *Zeitschrift für Naturforschung* **9** 837

As the length of this contribution exceeds space limitation we present only Abstract and Introduction. The full paper can be requested at:

Dr. W. Westmeier
Möllner Weg 32
D-35085 Ebsdorfergrund
(Germany)
Telephone: +49 6424 923000
Telefacsimile: +49 6424 923002
Mail: Westmeier@Westmeier.com

XIX International School on Nuclear Physics, Neutron Physics and Applications IOP Publishing
Journal of Physics: Conference Series **366** (2012) 012050 doi:10.1088/1742-6596/366/1/012050

Electromagnetic transition strengths in ^{155}Dy

M S Yavahchova[1], P Petkov[1,2], A Dewald[2], O Möller[2], B Saha[2], A Fitzler[2], K Jessen[2], D Tonev[1,4], N Gutev[1] T Klug[2], S Heinze[2], J Jolie[2], P von Brentano[2], D Bazzacco[3], C Ur[3], E Farnea[3,4], M Axiotis[4], S Lunardi[3], C Rossi-Alvarez[3], G de Angelis[4], D R Napoli[4], N Marginean[4], T Martinez [4] and M Caprio[5]

[1] Bulgarian Academy of Sciences, Institute for Nuclear Research and Nuclear Energy, 1784 Sofia, Bulgaria.
[2] Institut für Kernphysik der Universität zu Köln, 50937 Köln, Germany.
[3] Dipartimento di Fisica and INFN, Sezione di Padova, Padova, Italy.
[4] INFN, Laboratori Nazionali di Legnaro, Legnaro, Italy.
[5] Department of Physics, University of Notre Dame, USA.

Abstract.
Lifetimes of excited states in ^{155}Dy were measured by means of the Recoil Distance Doppler-shift technique in the coincidence mode. The experiment was performed at the Laboratori Nazionali di Legnaro with the GASP array and the Cologne plunger using the reaction ^{124}Sn(^{36}S,5n)^{155}Dy at a beam energy of 155 MeV. The Differential decay-curve method was applied for the lifetime determination. The measured transition probabilities in ^{155}Dy and the energy spectrum are compared to the predictions of the Particle plus rotor model. The comparison indicates slightly different quadrupole deformations characterizing the low-lying one-quasineutron bands which may point to a shape coexistence.

1. Introduction

Critical point symmetries are a fundamental problem in the study of nuclear structure. The X(5) symmetry[1] characterizes the critical point of the phase transition between spherical and axially deformed shape in atomic nuclei. The first example of this symmetry has been found in ^{152}Sm by Casten and Zamfir[2]. Later, other examples of X(5) nuclei have been found at N=90. The effect of the shape transition on the single particle degrees of freedom is best studied in odd-A nuclei. The mass region A$\sim$150 is characterized by a rapid shape transition. The nucleus ^{155}Dy lies in this region. The level-scheme of ^{155}Dy is well known[3, 4] but information on the collectivity was known only at high spins[5]. The first aim of the present work was to measure electromagnetic transition strengths at low and intermediate spins in the one-quasineutron bands. The second aim was to perform particle plus triaxial rotor model (PTRM) calculations for comparison with the data in order to get information on the quadrupole deformation (ϵ, γ) of the bands studied. To determine lifetimes in these bands we performed Recoil Distance Doppler-shift (RDDS)(cf. e.g. Ref[6] and references therein) measurements.

2. Experiment

To populate excited states in ^{155}Dy we used the reaction ^{124}Sn(^{36}S,5n)^{155}Dy at a beam energy of 155 MeV. The beam was supplied by the XTU tandem of the Laboratori Nazionali di Legnaro,

Italy. The target and stopper foils were mounted in the Cologne coincidence plunger[7]. The target-to-stopper distance was changed by moving the target holder. The target consisted of 0.9 mg/cm^2 tin, enriched to 97.7% in ^{124}Sn, which was evaporated onto a 1.8 mg/cm^2 ^{181}Ta foil serving as a backing and facing the beam. A 12.0 mg/cm^2 Au foil was used to stop the recoils leaving the target with a mean velocity v of 1.83(2)% of the velocity of light c. The deexciting γ-rays were registered with the GASP array[8]. This array consists of 40 germanium detectors, grouped into seven rings. The rings of interest for our RDDS experiment are these where significant Doppler shifts are observed. Namely, these are ring 0 (mean angle with respect to the beam axis of 34.6°), ring 1 (59.4°), ring 5 (120.6°) and ring 6 (145.4°). They consist of six detectors each. Spectra for the analysis of the RDDS data were obtained by setting a gate from above on the shifted component of a transition feeding directly the level of interest. In Fig.1, an example of spectra at different distances is shown to illustrate the quality of the data.

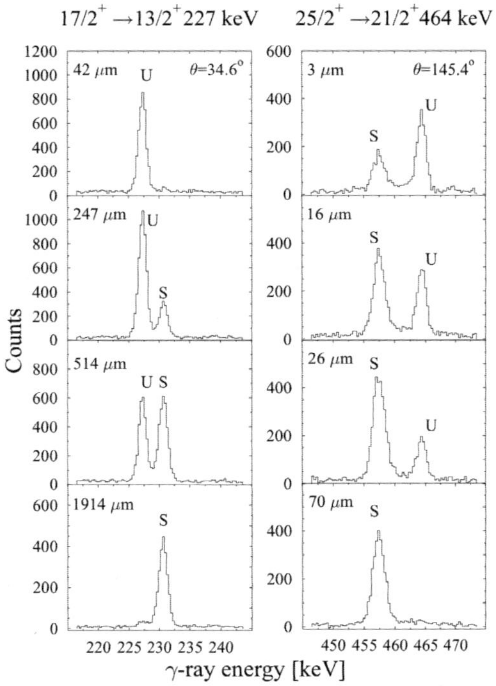

Figure 1. Gated γ -ray spectra of ^{155}Dy taken at four different distances. The spectra are measured at the indicated distances with the detectors positioned at 34.6° (ring 0) and 145.4° (ring 6) with respect to the beam axis.

3. Data analysis and results

For the data analysis the Differential decay curve method (DDCM)[9, 10] was employed. According to this method, at each target-to-stopper distance x, the lifetime τ(x) of the level of interest is calculated as

$$\tau(x) = \frac{\{B_s, A_u\}}{\frac{d}{dx}\{B_s, A_s\}} \frac{1}{v} \tag{1}$$

Here v is the mean velocity of the recoiling nuclei. The quantities in braces are the number of coincident events corresponding to detection of the Doppler-shifted (s) or the unshifted (u) components of the γ -ray transitions. $\{B_s, A_u\}$ is the area of the unshifted peak of the transition A while the derivative $\frac{d}{dx}\{B_s, A_s\}$ is the derivative of the area of the shifted component. The gate is set on the shifted component of a directly feeding transition B. Using Eq.1, a set of lifetime values (the τ-curve) is obtained which should naturally lie on a straight line when plotted versus the distance x. Deviations from such behavior point to the presence of systematics errors in the analysis. Gating from above eliminates the unobserved feeders and gating only on the shifted component cancels out completely the effect of nuclear deorientation.

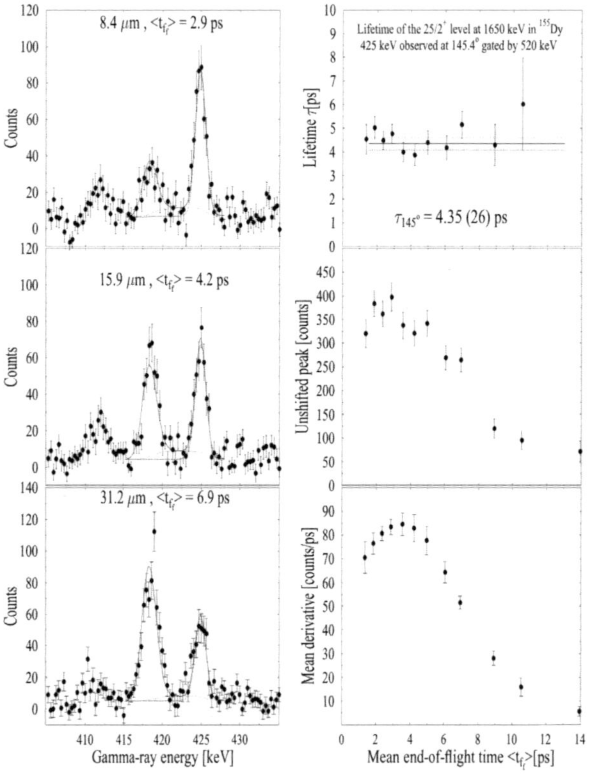

Figure 2. Example of a lifetime determinationfor the $25/2^+$ level using gated spectra measured with the detectors positioned at 145.4°.The left part of the figure presents fits[11] of the line-shape of the 425 keV γ -ray transition measured at different distances with the detectors of ring 6. The gate is set on the top of the shifted component of the 520 keV transition, which feeds directly the level of interest. The right part illustrates the lifetime determination according to Eq.(1). The τ-curve derived for the I^{π}=25/2$^+$ level in ^{155}Dy is displayed on top of the figure. The mean value τ=4.35 (26) ps is obtained as an average of the data points shown.

For the investigation of a given level we have set gates on the shifted component of a directly feeding transition in the γ-γ matrices corresponding to all rings where appreciable Doppler-shifts are observed. The spectra for every gated ring were summed up to increase the statistics. The resulting spectrum was analyzed. There are two forward and two backward rings, which means that four values for the lifetime have been determined independently.

As a result from our analysis 16 lifetimes of exited states in ^{155}Dy were determined for the first time. The relative uncertainties of the lifetimes are of the order of maximum 10%. By gating from above the problem of the unknown feeding is eliminated which ensures a high degree of reliability for the results. More details on the lifetime analysis and the results will be given in a forthcomming paper [12]. The reduced transition probabilities $B(\sigma\lambda)$ deduced from the lifetime data are presented in Fig.3 in units of e^2b^λ for electric transitions and $\mu_N b^{\lambda-1}$ for magnetic ones.

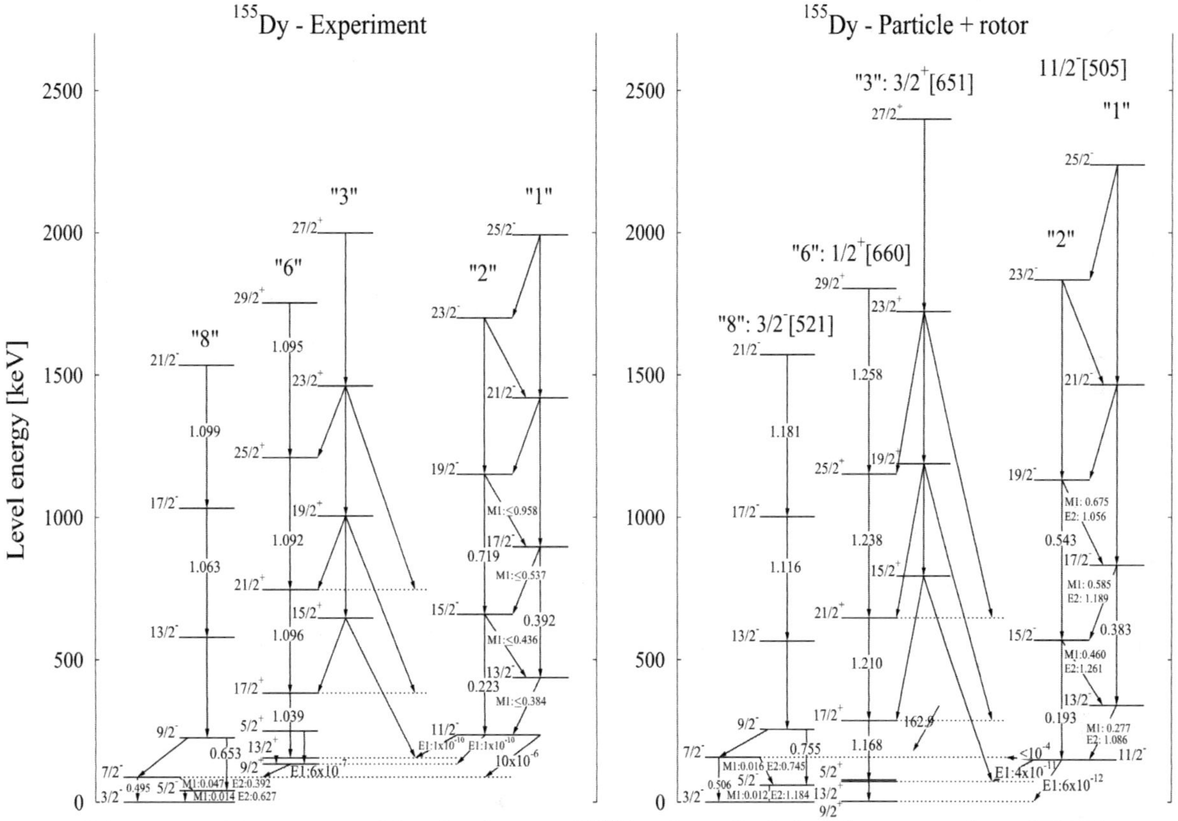

Figure 3. Experimental level scheme of ^{155}Dy from Refs.[3, 4] compared to PTRM calculations. The Nilsson configurations of the theoretical bands are indicated. See also text.

4. Discussion

To describe the reduced transition probabilities B($\sigma\lambda$) and the level scheme in ^{155}Dy we performed particle plus triaxial rotor model (PTRM) calculations within the approach of Ref.[13]. For this purpose, we used the codes GAMPN, ASYRMO PROBAMO and E1PROBAM[14] supplied by the authors of the model. First, the single particle orbitals for a fixed quadrupole deformation (ϵ,γ) are found as a solution of the Hamiltonian problem with the Modified Harmonic Oscillator potential. Then, a standard BCS calculation is performed to take into account pairing and a set of orbitals close to the Fermi surface (in our case 15) is chosen which participate in the diagonalization of the Particle-Rotor Hamiltonian in the strongly coupled basis. At the last stage, reduced transition probabilities are calculated with the resulting wave functions. Different values of the deformation parameters ϵ and γ were used to provide best PTRM fit of the level energies and the transition strengths. A value of the quadrupole deformation parameter ϵ=0.24 and a value of the asymmetry shape parameter γ=0° were found to be optimal. The results of the calculations for the $B(\mathrm{E2})$ transition strengths show a very good agreement with the experimental ones. There are some differences between the theoretically determined values in the bands and the experimental data. Some of the transition probabilities are bigger and other are smaller. This suggests that the different bands are characterized by a different quadrupole deformation in shape-transitional ^{155}Dy.

5. Summary and conclusions

Using the Recoil Distance Doppler-shift method we have determined 16 lifetimes in ^{155}Dy. For the data analysis the Differential decay-curve method was used. The experimentally determined excitation energies are in a good agreement with the particle plus rotor model. There is a good agreement also between the experimentally determined and theoretically calculated within the particle plus rotor model electromagnetic transition strengths $B(\mathrm{E2})$. The differences between the theoretically determined values in the bands and the experimental data suggest that the different bands are characterized by a different quadrupole deformation in ^{155}Dy.

References

[1] Iachello F 2001 *Phys. Rev. Lett.* **87** 052502
[2] Casten R F and Zamfir N V 2001 *Phys. Rev. Lett.* **87** 052503
[3] Reich C W 2005 *Nuclear Data Sheets* **104** 1
[4] Vlastou R *et al.* 1994 *Nucl. Phys.* A **580** 133
[5] Emling H *et al.* 1989 *Phys. Lett.* B **217** 33
[6] Alexander T K and Forster J S 1978 *Adv. Nucl. Phys.* **10** 197
[7] Dewald A *et al.* 1992 *Nucl. Phys.* A **545** 822
[8] Bazzacco D 1992 *in Proceding of the International Conference on Nuclear Sttructure at High Angular Momentum, Ottawa, Chalk River Report, AECL 10613* p.386
[9] Dewald A, Harissopoulos S and von Brentano P 1989 *Z. Phys.* A **334** 163
[10] Böhm G *et al. Nucl. Instr. Meth. Phys. Res.* A **329** 248
[11] Petkov P *et al. Nucl. Instr. Meth. Phys. Res.* A **431** 208
[12] Petkov P *et al. to be published*
[13] Larsson S E, Leander G and Ragnarsson I 1978 *Nucl. Phys.* A **307** 189
[14] Semmes P B 1991 *Computer manual, presented at Nuclear structure theory workshop, Oak Ridge, TN* (August 5-16, 1991)

Journal of Physics: Conference Series **366** (2012) 012051 doi:10.1088/1742-6596/366/1/012051

Studies of the ^{136}Xe+p and ^{136}Xe+^{12}C reactions at 1 GeV per nucleon with the SPALADIN setup

O Yordanov[1], J É Ducret[2], T Gorbinet[3], S Leray[3], A Boudard[3] and M D Salsac[3]

[1] Institute for Nuclear Research and Nuclear Energy, Sofia, Bulgaria
[2] CEA/IRFU/Service d'Astrophysique CELIA, Université de Bordeaux 1, F-33405 Talence
[3] IRFU/Service de Physique Nucléaire, CEA, Centre de Saclay, F-91191 Gif-sur-Yvette

E-mail: orliny@inrne.bas.bg

Abstract. Obtaining a correct description of spallation, in which a proton of energy in the GeV-range interacts with an atomic nucleus, is essential for different research domains in basic nuclear physics, advanced nuclear technologies, space- and medical research. Such a description is based on models developed within the larger framework of nuclear reactions at relativistic beam energies, where a complex interplay of different nuclear excitation and de-excitation mechanisms occurs. At the initial intranuclear cascade stage of the interaction of the target- and projectile nuclei a highly excited prefragment-nucleus is formed, whose decay towards stability could proeceed via different mechanisms such as sequential particle evaporation, asymmetric binary decay, fission in the case of fissile nuclei but also the so-called multifragmentation. Employing the inverse kinematics technique by impinging a heavy relativistic beam of ^{136}Xe on light targets of liquid hydrogen and ^{12}C, the SPALADIN experimental setup detects in coincidence neutrons and charged residues with $Z \geq 2$, which include the heavy residual nucleus and the light mass fragments emitted in the course of its formation.

1. Introduction

Based on the high-resolution FRS-spectrometer and the large ALADIN dipole magnet, an extensive experimental program for nuclear reaction studies in inverse kinematics has been pursued at GSI Darmstadt for more than a decade. The outgoing channels of proton-induced reactions with a non-fissile relativistic beam consist in their majority of one heavy residual nucleus, whose formation proceeds via the emission of protons, neutrons and light-mass charged fragments from a highly excited prefragment formed at the initial stages of the reaction. Experiments with the SPALADIN setup are aiming at the reconstruction of the reaction kinematics at relativistic beam energies by detecting the heavy reaction residues in coincidence with the emitted light- to intermediate mass charged fragments and neutrons. This experimental approach was already sucessfully implemented in the study of spallation of ^{56}Fe at 1 GeV per nucleon [1]. The formation of the heavy residue, depends on the excitation energy of the prefragment-nucleus, formed during the initial intranuclear cascade- or abrasion stage of the reaction. Although the excitation energy of the prefragment is not directly measurable, it can be estimated by detecting the particles emitted during evaporation. The coincident detection of the majority of reaction products is reducing the uncertainties in the input experimental data, which serve for the development of codes simulating the nuclear reactions at relativistic beam

Published under licence by IOP Publishing Ltd

Journal of Physics: Conference Series **366** (2012) 012051 doi:10.1088/1742-6596/366/1/012051

energies starting from their initial nucleon-nucleon collision stage till the formation of the final residue.

2. Experimental setup

The SPALADIN experiment was set up in the Cave-C experimental hall of GSI-Darmstadt. The ^{136}Xe-beam of 1 GeV per nucleon was provided by the UNILAC/SIS-18 accelerator facility. A set of plastic detectors generated the trigger signal for the data acquisition system and provided the monitoring of the correct entrance of the primary beam in the setup. The cryogenic target device was located downstream following the set of plastic detectors. Inside of the vacuum module of the target device two vessels filled with liquid hydrogen, a ^{12}C sample and an empty vessel were mounted. Placing a certain target sample in the beamline was done automatically via remote control. Following the target device a four-anode-, P10-operated gas ionization chamber was placed. This so-called forward ionization chamber was aimed primarily at the measurement of the charge of the heavy reaction residue, thus providing direct experimental information on element production yields and the coresponding element production cross sections. The forward ionization chamber was placed in front of the entrance window of the ALADIN dipole magnet. A Time-of-Flight wall was placed behind the exit aperture of the dipole magnet. The ToF-wall served as a detector for light- to intermediate-mass charged fragments. The simultaneous operation of the ToF-wall and the Forward ionization chamber made the setup sensitive for all residues with charges Z$\geq$2. Neutrons produced in the reaction were detected using the Large Area Neutron Detector (LAND). LAND was positioned along the initial beam direction, at a distance of 10 meters measured from the exit aperture of the ALADIN magnet. In addition to the forward ionization chamber, behind the ToF-wall a second TWIN-ionization chamber was mounted. Having an active volume of nearly 0.5 cubic meters, the TWIN-MUSIC detector offers high charge resolution, combined with horizontal position sensitivity, which is achieved by it's specially shaped anodes. By requesting the coincident detection of the same charge in both ionization chambers, events of secondary reactions of the heavy residue with parts of the setup were efficiently suppressed.

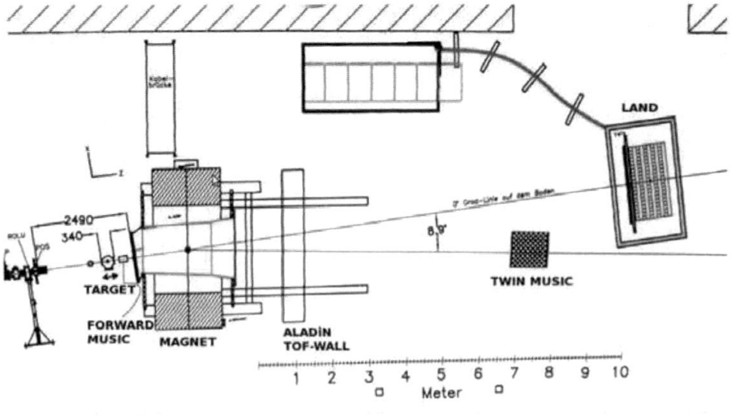

Figure 1. The SPALADIN experimental setup

Journal of Physics: Conference Series **366** (2012) 012051 doi:10.1088/1742-6596/366/1/012051

3. Experimental data

3.1. ^{136}Xe+p

The experimental data on proton induced reactions were collected by impinging the primary ^{136}Xe beam on liquid hydrogen, contained in the cryogenic module of target device. Being placed in the primary beam, the titanium target vessel has a non-negligible contribution to the total reaction rate. Therefore dedicated empty-target runs e.g. without liquid hydrogen content, were performed. After combining the two datasets collected with a filled- and an empty target vessel, histograms containing only the statistics from reactions of the primary beam with the liquid hydrogen content were obtained. The preliminary element production cross-sections measured by the forward ionization chamber show agreement with those, measured in a previous experiment with the FRS-spectrometer [2] .

The formation of a certain heavy fragment results from the amount of mass lost within the initial intranuclear cascade stage and evaporation stages of the reaction. The experimental data show, that the production of light charged residues rises if the charge of the heavy fragment is decreasing. This observation is consistent with the expectation that decreasing the mass of the heavy residues requires an increase of the excitation energies of the prefragments. This increase is manifested in the measured intensified emission of light charged residues.

The correlation between the charges of light particles and charged particle multiplicities, measured in coincidence with preselected heavy fragments, are examined too. The analysis has shown, that events of highest multiplicities obtained by the ToF-wall, involve the detection of alpha-particles only. The heaviest intermediate mass fragments, detected by the ToF-wall, are attributed mainly to events with charged particle multiplicities in the ToF-wall of one. It was found, that the charge of the intermediate mass fragment observed with the ToF-wall in an event of unit multiplicity, is comparable to the summary charge obtained in the event of highest multiplicity built up by alpha particles only. It is questionable whether the emission of an intermediate mass fragment in coincidence with a heavy residue can be described by means of evaporation. This hypothesis may be supported by the fact, that the charge of the intermediate mass fragment is related to the summary charge of the alpha particles, emitted in the event of highest multiplicity. On the other hand obtaining two heavy fragments without other charged particles being detected is an experimental sign of fission. This specific point concerning the nuclear reaction dynamics has to be addressed in further experiments.

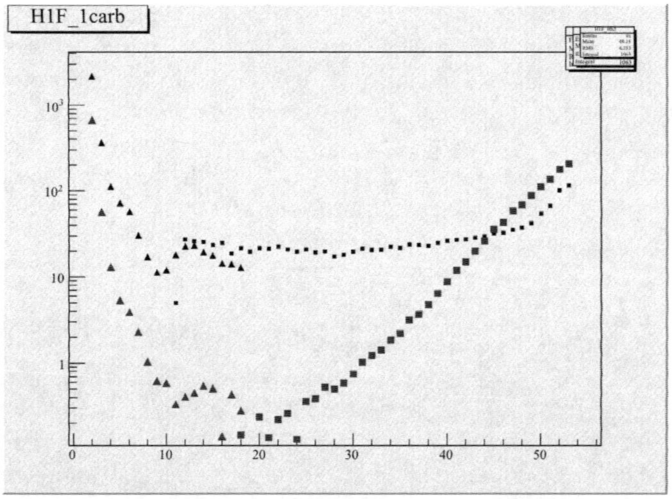

Figure 2. Preliminary element production cross sections (mbarn) for the reactions ^{136}Xe+p (blue) and ^{136}Xe + ^{12}C (black)

Journal of Physics: Conference Series **366** (2012) 012051 doi:10.1088/1742-6596/366/1/012051

3.2. $^{136}Xe+^{12}C$

At relativistic energies, the initial stages of the reactions of ^{136}Xe with proton and ^{12}C are described by two different approaches. The spallation process i.e. the reaction with protons is modeled by quasi-free nucleon-nucleon scattering at relativistic energies. The reactions with ^{12}C are more complex due to the higher number of nucleons involved in the interaction of the projectile- and target nucleus. Following the projectile-spectator picture, the initial abrasion stage of the fragmentation process proceeds via nucleon-nucleon scatterings in the zone of intersection of both nuclei. Expressed in geometrical terms, the size of the overlapping zone depends on the impact parameter. Due to the possible high number of nucleon-nucleon interactions, in the abrasion process the achievement of higher excitation energies of the prefragment becomes possible, compared to those obtained in single-proton-induced reactions. This is manifested in the measured increased production cross sections of light elements in the reaction ^{136}Xe + ^{12}C. An analysis of the coincident detection of a certain heavy residue in the forward ionization chamber and light charged particles in the ToF-wall, is performed also for the ^{136}Xe + ^{12}C reaction. The comparison of the distributions shows, that for a selected heavy residue, the mean value of the charge of particles detected in the ToF-wall is higher for the lH2-target than for the ^{12}C one. Considering a liquid hydrogen target, the production of a certain heavy reaction residue relies on the mass loss in the intranuclear- cascade- and the evaporation stage. In the case of impinging the ^{136}Xe on the ^{12}C target, the relevant mass loss occurs in the abrasion stage and the following evaporation. Concerning the possible amount of mass loss in the initial stages of the reaction, the processes of intranuclear cascade and abrasion are essentially different. The intranuclear cascade process in ^{136}Xe nucleus is triggered by one proton, while the abrasion process involves the nucleons from the overlapping zone of the projectile nucleus ^{136}Xe and the target nucleus ^{12}C. Consequently if compared to the intranuclear cascade, the abrasion process is capable to cause far bigger loss of mass, due to the larger number of possible nucleon-nucleon interactions. If ^{136}Xe is impinged on liquid hydrogen i.e. ^{136}Xe+p, the relevant mass loss for the production of a certain heavy residue is expected to occure in the evaporation stage of the reaction. If ^{136}Xe is impinged on ^{12}C, the production of the same heavy residue will be achieved by mass losses within the two processes of abrasion and evaporation. Thus, the contribution of evaporation into the required total mass loss will be lower in the case of the ^{136}Xe+^{12}C reaction than in the ^{136}Xe+p one. The mass loss achieved in evaporation is proportional to the excitation energy of the prefragment. The decrease of the mass to be lost in evaporation equates to a reduction of the excitation energy of the prefragment. A decreased prefragment excitation energy will reduce the probability for the emission of heavier charged residues, thus making the emission of light charged particles favorable. This effect is manifested in the different mean charges detected in the ToF-wall- the value $< Z >_{ToFWall}= 2.398$ obtained in coincidence with a heavy residue of charge $Z = 35$ from the ^{12}C-target data are lower if compared to the value $< Z >_{ToFWall}= 2.731$ obtained from data, collected with the liquid hydrogen target in coincidence with the same heavy residue of $Z = 35$. Experimental results of the measured element producion cross sections in the liquid hydrogen- and ^{12}C-target are shown on fig. 2.

4. Theory

The performed GEANT4 simulations combined the theoretically predicted residue production in the target with the particle-transport- and detection properties of the setup. The production of residues in the target was calculated by combining of the INCL4 [3] intranuclear-cascade simulation with the codes ABLA [5], SMM [4], GEMINI [6], and ISABEL [7] combined with SMM for simulating the de-excitation of the prefragment. The subsequent particle transport simulation uses the complete description of the setup including detector geometries and detection efficiencies, thus converting the calculated yields in the target into experimentally expected ones. The results of the comparative studies between experimental data and theoretical predictions on the coincident production of light- to intermediate charged particles and heavy residues are shown on figures 3 and 4. Calculations performed by combining of the INCL4 and ABLA

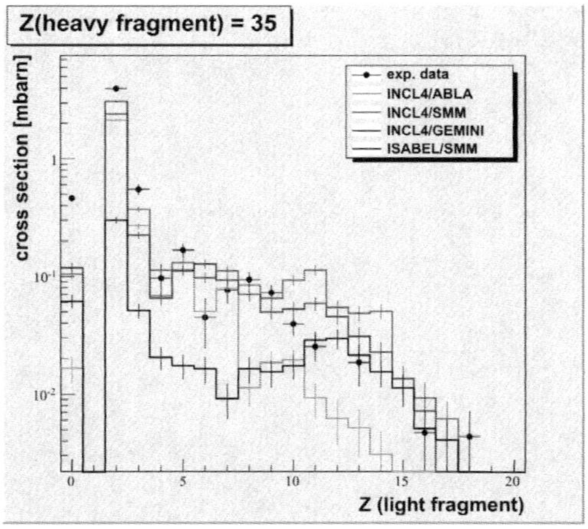

Figure 3. Light- to intermediate residues detected in coincidence with heavy residues of Z=35

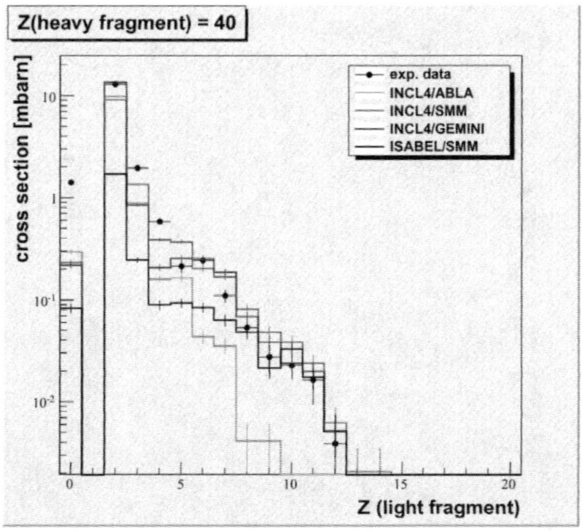

Figure 4. Light- to intermediate residues detected in coincidence with heavy residues of Z=40

codes show tendency to underestimate the production of charged particles. Applying the same coincidence conditions, an overestimated production of light charge particles was predicted by calculations combining INCL4 with SMM and GEMINI codes. The production cross sections of the heavy residues are well reproduced by combining ABLA-, SMM- and GEMINI codes with the INCL4. Significant deviation from the experimental elements cross section data were found in simulations performed using a combination of ISABEL and SMM codes.

5. Concluding remarks

Preliminary data on production cross sections and outgoing particle coincidences for ^{136}Xe+p and ^{136}Xe+^{12}C reactions are presented. A satisfactory description of the experimental data is obtained by means of a GEANT4 simulation of the setup, at which the residue production in proton-induced reactions is calculated by the INCL4 intranuclear cascade simulation, run together with the ABLA, SMM, GEMINI codes. A comparative study of the abrasion and intranuclear cascade mechanisms was performed by comparing the coincident production of heavy residue and light charged particles in the liquid hydrogen and ^{12}C-target. Although the particle removal channels of the nucleon-nucleon collisions occuring in the initial stages of the reactions are not directly observable, the observed differences in the charge distributions of evaporation residues in the ^{136}Xe+^{12}C- and ^{136}Xe+p-reactions can be explained as a result of an increased mass removal in the abrasion stage of the ^{136}Xe + ^{12}C reaction.

References

[1] Le Gentil E *et al.* 2008 *Phys. Rev. Lett.* **100** 022701
[2] Napolitani P *et al.* 2007 *Phys. Rev.* C **76** 064609
[3] Boudard A *et al.* 2002 *Phys. Rev.* C **66** 0044615
[4] Botvina A, Iljinov A S and Mishustin I N 1990 *Nucl. Phys.* A **507** 649
[5] Junghans A R *et al.* 1998 *Nucl. Phys.* A **629** 635
[6] Charity R J *et al.* 1988 *Nucl. Phys.* A **483** 371
[7] Yariv Y and Fraenkel Z 1979 *Phys. Rev.* C **20** 2227

Journal of Physics: Conference Series **366** (2012) 012052 doi:10.1088/1742-6596/366/1/012052

The ELI-Nuclear Physics Project

N V Zamfir

'Horia Hulubei' National Institute of Physics and Nuclear Engineering (IFIN-HH), Bucharest-Magurele, Romania

Abstract. The Extreme Light Infrastructure - Nuclear Physics facility (ELI-NP) to be constructed in Bucharest-Magurele will be one of the three pillars of ELI and will consist of two components: a very high intensity laser beam and a very intense (1013 γ/s), brilliant γ beam, 0.1% bandwidth, with Eγ = 19 MeV. ELI-NP will allow either combined experiments between the high-power laser and the γ beam or stand-alone experiments. This facility will create a new European laboratory with a broad range of science covering frontier fundamental physics, new nuclear physics and astrophysics as well as applications. A review of the Project and the Status of its implementation will be presented.

Keywords: Photonuclear Reaction, Laser-matter interaction

1. The ELI European Project

Initiated in 2005, the Extreme-Light-Infrastructure (ELI) will be an European research infrastructure that will use new and emerging laser technologies to generate the most intense light pulses in the world. Fully open to external users, it will be dedicated to the fundamental study and societal applications of laser-matter interaction in a new and unsurpassed regime of laser intensity: the ultra-relativistic regime. Identified by the European Strategy Forum on Research Infrastructures (ESFRI) as one of the top priority projects of research infrastructure for Europe, ELI has been brought to legal, organisational, financial and scientific maturity thanks to a 36-month Preparatory Phase (ELI-PP) FP7 project launched in November 2007 which involved nearly 40 research and academic institutions from 13 EU Member States. A major outcome of this work consisted in defining the scientific scope of ELI. The final report of the Scientific Advisory Committee, issued in May 2009, identified four main branches of research and applications:

- High Energy Beam Science: development and usage of dedicated beam lines with ultra short pulses of high energy radiation and particles reaching almost the speed of light

Journal of Physics: Conference Series **366** (2012) 012052 doi:10.1088/1742-6596/366/1/012052

- Attosecond Laser Science: temporal investigation of electron dynamics in atoms, molecules, plasmas and solids at attosecond scale (10-18 sec., i.e. a billionth of billionth of a second)
- Photonuclear Physics: nuclear physics methods to study laser-target interactions, new nuclear spectroscopy, new photonuclear physics, etc.
- Ultra High Field Science: investigation of laser-matter interaction in an energy range where relativistic laws could stop to be valid

In October 2009, the Steering Committee of the ELI-PP Consortium gave the mandate to the Czech Republic, Hungary and Romania to jointly implement ELI through the construction of three specialised research facilities dedicated to the first three scientific fields listed above: Beam Science near Prague, Attosecond Laser Science in Szeged and Photonuclear Physics in Bucharest-Magurele. The fourth "pillar" will be implemented later on through the upgrade of one of the three initial facilities or the construction of a fourth facility. ELI is the only facility on the ESFRI Roadmap proposed for construction in new EU member States (at the exception of some pan-European virtual facility nodes).
The three host countries intend to use Structural Funds for the project's implementation between 2011 - 2016.

The ELI facilities will form a unique infrastructure placed under the governance of a European Research Infrastructure Consortium (ERIC), a new legal structure specifically designed for infrastructures of the kind of ELI. This entity is expected to benefit from contributions from a large number of ELI-Preparatory Phase Partners. The location of the first three ELI facilities in new Member States represents a significant development in favour of a better balance in the distribution of Research Infrastructures in Europe. It should be noted that each of the individual pillars will have its own specific research, technology and construction characteristics and will not depend on the others to be successfully delivered.
With the likely involvement of major countries such as France, Germany, or the United Kingdom, apart from the three host countries, the ELI-ERIC will be a genuine international facility, which represents an unprecedented development in the field of laser science. The ELI-ERIC will be the legal entity coordinating the operation of the infrastructure, setting major strategies (research, IPR, training, access policy) and selecting users. Access will be delivered in collaboration with the local entities owning the facilities.

2. The ELI Nuclear Physics (ELI-NP) Project

The Romanian pillar, ELI-Nuclear Physics will be dedicated to nuclear physics studies of the laser-matter interaction and to photonuclear physics. According to the 'White Book for ELI-NP' (available on project web-site http://www.eli-np.ro), elaborated by more than 100 scientists from 30 research institutions from Europe, United States and Japan, the ELI Nuclear Physics facility (ELI-NP) will consist of two components:

- A very high intensity laser, where two 10 PW Apollon-type lasers are coherently added to intensities of 10^{23}-10^{24} W/cm2 or electrical fields of 10^{15} V/m
- A very intense (10^{13} γ/s), brilliant γ beam, 0.1% bandwidth, with $E\gamma$ up to 20 MeV, which is obtained by incoherent Compton back scattering of a laser light off a very brilliant, intense, electron beam produced by a warm LINAC (Ee up to 720 MeV).

ELI-NP will allow either combined experiments between the high-power laser and the γ beam or stand-alone experiments. ELI - NP will open up new research areas for laser, nuclear, astrophysics and fundamental research. The ultrahigh power and intensity of pulses will produce electric fields so strong that they may alter and sense the texture of the vacuum itself. The extremely performing gamma beam will allow the photonuclear reactions to become a powerful tool in nuclear structure studies. A very broad range of applications will be also studied, such as: industrial tomography and gamma radiography, surface characterization using positrons sources, medical imaging, radiation and proton cancer therapy. A very special class of applications will be based on photonuclear reactions produced by the very performing gamma-beam, namely the nuclear materials and radioactive waste management and new pharmaceutical radioisotopes.

A multitude of the experiments is already proposed and the experimental area is designed in a modular way to accommodate the planned experiments with a maximum of flexibility.

3. Conclusion

A new major infrastructure in nuclear physics is planned to be built in Bucharest-Magurele, Romania, namely ELI-NP, part of a larger European project. The ground breaking for ELI-NP should start as early as 2011 and the operation will start in 2016-2017. A very broad range of physics cases are targeted within ELI-NP, covering many facets of photonuclear reactions, but also astrophysics and fundamental physics. Moreover, various perspectives for applications ranging from medicine to material science, life science and radioactive materials management are envisaged to be studied.

4. Acknowledgement

The contribution of more than 100 people to the development of this project and to ELI-NP White Book is acknowledged.

Comparison of models of neutron and residue production in targets heavy elements irradiated by relativistic proton and deuteron beams

P Zhivkov[1,2], A Polanski[2,3], Ch Stoyanov[1], O Yordanov[1], N Gutev[1]
[1] Institute of Nuclear Research and Nuclear Energy, BAS, Sofia, Bulgaria
[2] JINR, LIT, Dubna, Russia
[3] National Centre for Nuclear Research Otwock-Swierk, Poland

E-mail: peter_jivkov@inrne.bas.bg, a.polanski@ncbj.gov.pl

Abstract. The improvement of both the precision of the nuclear models and the Monte Carlo transport requires a comparison of the simulated productions of residues, the spallation of neutrons and their transport, obtained in different scenarios. The calculations of the residue production and corresponding neutron double differential cross sections are performed using the different reaction models in the MCNPX 27d transport code. The aim of this study was to validate the calculation of nuclear spallation by using the latest version of MCNPX code and comparing the results of calculations with experimental data from GSI Darmstadt. This work focuses in particular on the study of generation of residue in uranium and lead during interactions of protons and deuterons. The cross sections of production of light ions and isotopes as a function of mass and atomic number during the impact of protons and deuterons of energy of 1 GeV/nuc. with the uranium and lead was also calculated. Finally the results of calculations obtained by different methods were compared with known experimental data.

1. Introduction

In a traditional model of an atomic nuclei spallation, production of particles is treated in three stages. In the first stage some nucleons from the target and projectile nuclei are knocked out, together with production of mesons. In the third stage, residual nuclei (generally, in an excited state) split into two remnants in the fission channel or evaporated protons, neutrons, and (or) light nuclei, including helium isotopes. The second stage intermediate between the first and the third called pre-equilibrium decay model is associated with the emission of nucleons. However, experimental data indicate that at the high energies competing process, multifragmentation comes into play, in which excited remnants break up into intermediate mass fragments. There are two main approaches to theoretical description of fast stage of spallation: statistical and dynamical. A statistical approach is based on intra nuclear cascade model. In dynamical models (Quantum Molecular Dynamic), multifragmentation is formed at the end of the first stage of nuclear collision via dynamical forces between nucleons during the evolution of the total system of interacting projectile and target. In this case, the intermediate mass fragments pass through states of thermal equilibrium. The excited intermediate mass fragments achieve a thermal equilibrium state and then expand, eventually reaching the freeze-out volume. At this point, it disintegrates into neutrons, protons, light charged particles (d, t, He) remnant nuclei. The possible scenarios behind fission and multifragmentation are indeed strongly different, because the first presupposes the slow decay of a compound nucleus, while the second one the passage through a fast break-up phase.

2. Models of calculations of multifragmentations

Basing on the analysis of experimental data and computational models we assumed that the most promising models to describe the phenomenon of fragmentation of nuclei are Quantum Molecular Dynamic QMD [1],[2], Ultra relativistic QMD (UrQMD) [3], Intra-nuclear Cascade Model ICM [4], and Quark-Gluon String Model QGSM [5], [6] in conjunction with General Evaporation Model GEM2 [7] ,Nuclear Lattice Model NLM [4], Statistical Multifragmentation Model SMM [8] and the abrasion-ablation model ABLA [9]. The QGSM and UrQMD models allow the calculation of both the production of particles (nucleons, mesons) and isotopes in heavy target for interaction of protons and nuclei of the energy from 1 GeV per nucleon (NUCLOTRON - Dubna) to the energy about 200 GeV per nucleon (RIC-BNL) with heavy nuclei (Au, Pb, U). Statistical cascade models available in MCNPX [10] were used for the calculation of the first phase namely: Bertini, ISABEL, CEM03, INCL4, LAQGSM. For the calculation of the second and third, longer phase of reaction Dresner, ABLA, GEM2 models were applied. There is one more approach to describing the process of multifragmentation: percolation theory [4] and statistical multifragmentation model [8]. Percolation models treat the nucleus as a lattice with nucleons located at nodes of the lattice. For reasons of computational convenience, the simple cubic lattice has been most frequently used in multifragmentation simulations , but several studies have found that the face-centered-cubic (FCC) lattice more accurately reproduces the experimental distributions of fragment masses and their energy spectra. That is, lattices were employed more as computational techniques, rather than as formal nuclear models. A new version of the modifed Dubna Intra Nuclear Cascade Model [4] for intermediate- and high-energy nucleus-nucleus collisions including multifragmentation channels has been developed. Multifragmentation is calculated in the framework of percolation theory with usage of a site-bond percolation model. Results of calculations mass distribution of residues in the reaction 208 Pb + p at 1 GeV and 208 Pb + d at 2 GeV are presented in [4]. We can find good agreement between results of calculations and experimental data. Statistical Multifragmentation Model SMM [8] uses the chemical potentials in order to determine the parameters resulting fragments disintegrating nucleus. The Los Alamos version of the Quark-Gluon String Model (LAQGSM) connected with SMM is a next generation of the Quark-Gluon String Model (QGSM) by Amelin et al.[5] and is intended to describe both particle- and nucleus-induced reactions at energies up to about 1 TeV/nucleon. The core of the QGSM is built on a time-dependent version of the intranuclear cascade model developed at Dubna, often referred in the literature simply as the Dubna intranuclear Cascade Model. The DCM models interactions of fast cascade particles ("participants") with nucleon spectators of both the target and projectile nuclei and include interactions of two participants (cascade particles) as well. It uses experimental cross sections (or those calculated by the Quark-Gluon String Model for energies above 3.5 GeV/nucleon) for these elementary interactions to simulate angular and energy distributions of cascade particles, also considering the Pauli exclusion principle. When the cascade stage of a reaction is completed, QGSM uses the coalescence model described in Toneev and Gudima to "create" high-energy d, t, 3He, and 4He by final state interactions among emitted cascade nucleons, already outside of the colliding nuclei. After calculating the coalescence stage of a reaction, QGSM moves to the description of the last slow stages of the interaction, namely to pre equilibrium decay and evaporation, with a possible competition of fission using different models of multi- fragmentation, GEM2 or SMM.

3. Results of calculations.

The production of light and intermediate-mass nuclides formed in the reaction 1H+238U at 1 GeV was measured at the Fragment Separator (FRS) at GSI, Darmstadt. The experiment was performed in inverse kinematics, shooting a 1 A GeV 238 U beam on a thin liquid-hydrogen target. Experimental technique, results and comparison with other data were presented in paper [11], [12]. The comparison of results of calculation of charge and mass distributions using INCL4+ABLA model CEM+GEM2,

LAQGSM+SMM with experimental data presented on Fig.1. and Fig.2. In the de-excitation stage the compound nucleus at each step of its evolution has three possible decay channels: evaporation, fission and intermediate mass fragments production IMF. In evaporation process we produce protons, neutrons, deuterons, tritons, He-3, He-4 and residual nuclei. The main fission region is from Z=20 to Z=65. The IMF region is for Z between 5 and 20. For Z greater than 65 we have evaporated residues. As shown in Fig.1 the difference between computational models and experimental data occur mainly in the IMF region and evaporated residues. It is associated with the use of different models of multi-fragmentation of atomic nuclei. As can be seen from the figure the best compatibility calculations with experimental data are obtained for a combination of QGSM+SMM. As expected, fragmentation generally produces nuclei on the neutron-deficient side of the beta-stability valley, while in fission processes more neutron-rich fragments are produced. Neutron-rich nucleus decay and can produce delayed neutrons. The analysis shows that the main role in the production of neutrons, plays the evaporation and fission of atomic nuclei.

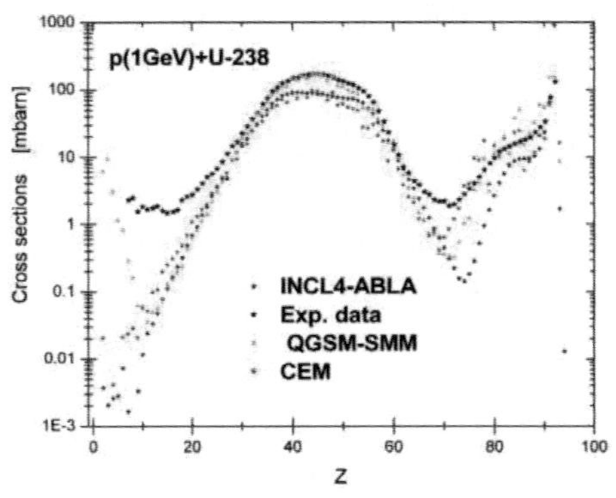

Figure 1. Charge distribution of cross sections for the nuclei produced in 238U (1 A GeV)+ p. Experimental data (stars).

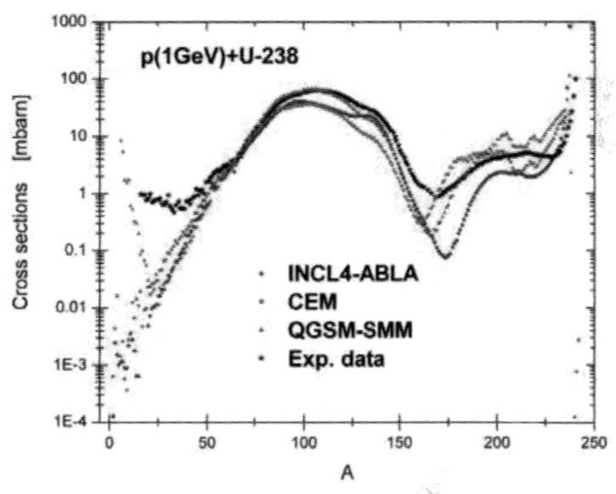

Figure 2. Mass distribution of cross sections for the nuclei produced in 238U (1 A GeV)+ p. Experimental data (stars).

The experimental numerical values of the isotopic production cross-sections for nuclides with Z between 7 and 37 is presented in the paper [11] and for Z between 74 and 92 is presented in the paper [12] from the spallation-evaporation reaction of 238 U by 1 GeV protons. The measurement was performed in inverse kinematics. The numbers give the nuclear charge Z, the mass number A, the measured cross-section as well as the systematic and the total relative uncertainty in percent. The results of calculations for selected isotopes are shown in Figure 3 and Figure 4. As is seen from the above plots the most adequate model to describe the experimental data is INCL4 + ABLA. The disadvantage of this model is that it does not describe the process of crushing nuclei for ions with an energy greater than 1 GeV per nucleon.

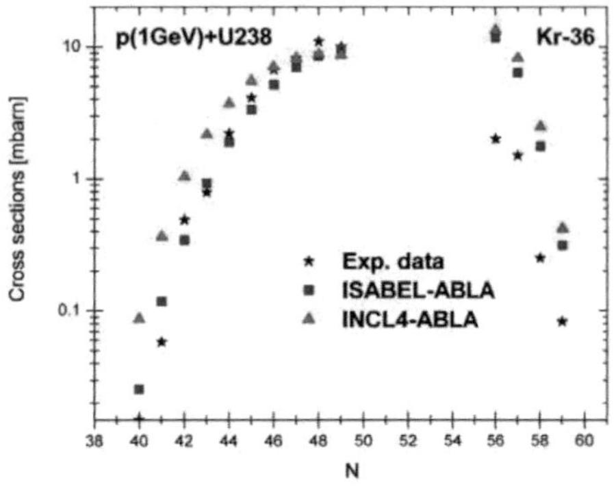

Figure 3. Isotopic cross sections for the products Z=36 produced in the reaction 238U on proton at 1 GeV per nucleon. Experimental data (stars)

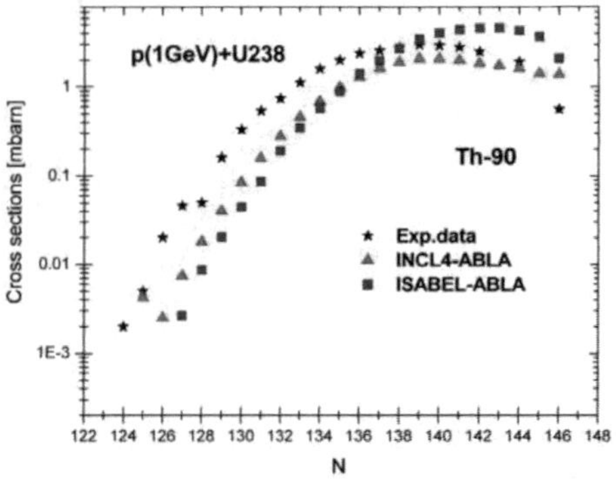

Figure 4. Isotopic cross sections for the products Z=90 produced in the reaction 238U on proton at 1 GeV per nucleon. Experimental data (stars)

Figures 5 and 6 show a comparison of calculation results with experimental data obtained at GSI where the interaction of uranium with deuterium was studied. The experimental data are taken from [13], [14]. It follows from the presented figures; INCL4-Dresner model best describes the experimental data.

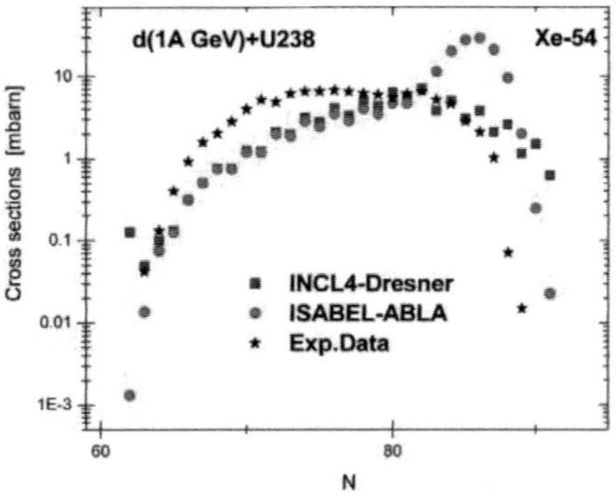

Figure 5. Isotopic cross sections for the products Z=54 produced in the reaction 238U on deuteron at 1 GeV per nucleon. Experimental data (stars)

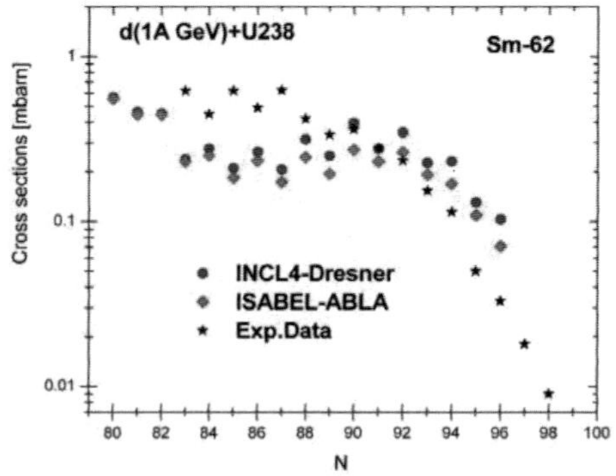

Figure 6. Isotopic cross sections for the products Z=62 produced in the reaction 238U on deuteron at 1 GeV per nucleon. Experimental data (stars)

In turn, Fig. 7 and Fig.8 depict the results of calculations and experimental data on cross sections describing isotopes production at the proton and deuteron impact of lead. It follows from the presented figures; ISABEl+ABLA model best describes the experimental data . From the presented data results that both cross sections for interaction of the proton and deuteron with lead have similar values. The process of fission of atomic nuclei plays a bigger role in the case of proton interactions with uranium than in the case of proton interactions with lead. In spallation target made of uranium we have ten times more fission products than in the case of lead. However, the production of neutrons is the largest in the uranium target in comparison with the other targets (Pb,W).

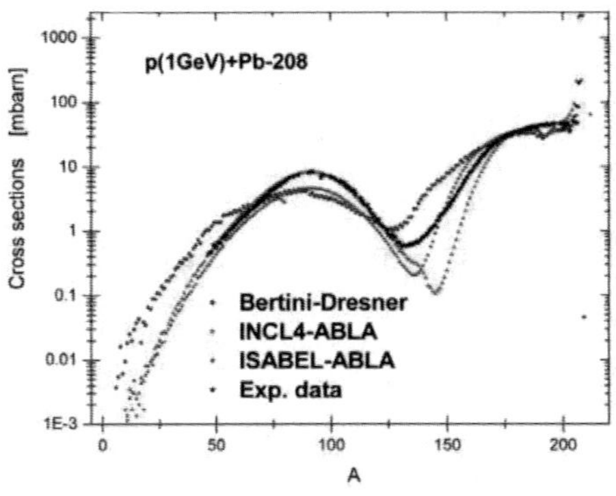

Figure 7. Mass distribution of cross sections for the nuclei produced in 208 Pb (1 A GeV)+ p. Experimental data (stars).

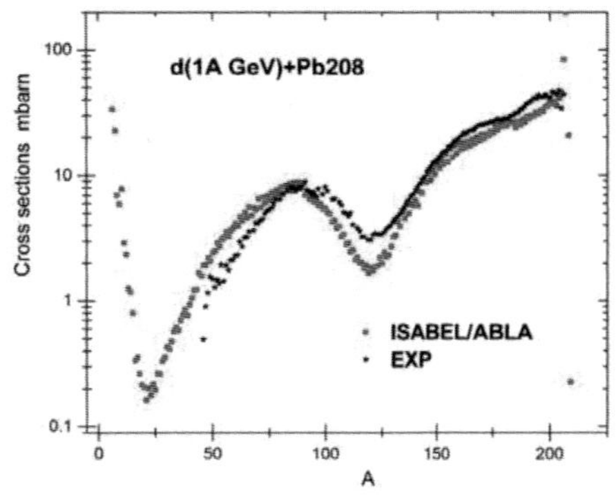

Figure 8. Mass distribution of cross sections for the nuclei produced in 208 Pb (1 A GeV)+ d. Experimental data (stars).

Summary and conclusion

We have analyzed the proton-induced and deuteron-induced spallation reactions with the INCL4 model merging the ABLA, ISABEL merging ABLA and LAQGSM model merging SMM. We compared these models with other models present in MCNPX code. The cross sections for products of isotopes in proton and deuteron-induced reactions on heavy targets can be reproduced quite well by statistical and dynamical models, especially by QMD+GEM2 model for low energy of ions less than 100 MeV/nuc , ISABEL +ABLA for energy from 100 MeV/nuc up to 1 GeV/nuc and by LAQGSM + SMM model for high energy of ions greater than 1 GeV/nuc. The double differential cross sections of emitted protons and neutrons in intermediate energy ion-induced spallation reactions are reproduced well by the intra-nuclear cascade models plus statistical evaporation-fission models, we find that they are not so sensitive to the merged statistical models. For the future, with more precise experimental data becoming available, we expect that the models based on quantum molecular dynamics (QMD and UrQMD) are promising to describe the production of particles and isotopes for a wide interval of energy protons and ions from 10 MeV/nuc to 200 GeV/nuc per nucleon on heavy targets.

References

[1] Polański A, Petrochenkov S, Uzhinsky V 2005 *Rad Prot Dos* vol **116** No (1-4) *Development of a quantum molecular dynamic model (QMD) to describe fission and fragment production* pp 582-585

[2] Polański A 2009 *High Energy Physics and Nuclear Safety, NATO Science for Peace and Security Series B: Physics and Biophysics Development of Methods for Simulation of Electronuclear Processes* pp 319 - 332

[3] Bass S, Belkacem M et al 1998 *Part Nucl Phys* **41** *Microscopic Models for Ultrarelativistic Heavy Ion Collisions* pp 225 - 370

[4] Musulmanbekov G, Al-Haidary A 2003 *Phys of Atomic Nuclei vol* **66** No 9 *Fragmentation of Nuclei at Intermediate and High Energies in Modied Cascade Model* pp 1671-1679

[5] Amelin S, Gudima K, Toneev V 1990 *Sov J Nucl Phys* **52** *Further development of the model of quark-gluon strings for the description of high-energy collisions with a target nucleus* pp 172-178

Journal of Physics: Conference Series **366** (2012) 012053 doi:10.1088/1742-6596/366/1/012053

[6] Gudima K, Mashnik S, Sierk A 2001 *User Manual for the Code LAQGSM, Los Alamos National* Report LA-UR-01-6804
[7] Furihata S 2000 *Nucl Instrum Methods Phys Res* B **171**, pp 251–258
[8] Bondorf P, Botvina S, Iljinov S, Mishustin N, Sneppen K 1995 *Phys Rep* pp 257 - 133
[9]Gaimard J, Schmidt K-H 1991 *Nucl Phys* **A531** *Reexamination of the abrasion-ablation model for the description of the nuclear fragmentation reaction* pp 709-745
[10] MCNPX *http://mcnpx.lanl.gov/*
[11] Ricciardi M, Armbruster P et al P 2006 *Phys Rev* C **73** 014607 *Light Nuclides produced in the Proton- induces Spallation of 238U at 1 GeV*
[12] Benlliure J et al 2004 *Phys Rev Lett* **93** 212701 *Measurement of a Complete Set of Nuclides Cross Section, and Kinetic Energies in Spallation of 238U 1A GeV with Protons*
[13] Pereira J, Benlliure J, et al 2007 *Phys Rev* C vol 75 *Isotopic production cross sections and recoil velocities of spallation-fission fragments in the reaction 238U(1AGeV) + d*
[14] Casarejos E, Benlliure J et al 2006 *Phys Rev* C **74** 044612 *Isotopic production cross sections of spallation-evaporation residues from reactions of 238U (1A GeV) with deuterium*
15] Enqvist T et al 2001 April *Nucl Phys A* vol **686**, Issues 1-4 *Isotopic yields and kinetic energies of primary residues in 1 A GeV 208Pb+p reactions* pp 481-524
[16] Enqvist T et al 2002 *Nucl Phys* A vol 703, Issues 1-2 *Primary-residue production cross sections and kinetic energies in 1AGeV 208Pb on deuteron reactions* pp 435-465

XIX International School on Nuclear Physics, Neutron Physics and Applications IOP Publishing
Journal of Physics: Conference Series **366** (2012) 012054 doi:10.1088/1742-6596/366/1/012054

Photon scattering off ^{94}Zr and ^{96}Zr

M. Zweidinger[1], J. Beller[1], J. Isaak[1], N. Pietralla[1], V. Yu. Ponomarev[1], C. Romig[1], D. Savran[2,3], M. Scheck[1], and K. Sonnabend[4]

[1] Institut für Kernphysik, Technische Universität Darmstadt, Germany
[2] ExtreMe Matter Institute EMMI, Darmstadt, Germany
[3] Frankfurt Institute for Advanced Studies, Germany
[4] Institut für angewandte Physik, Goethe-Universität Frankfurt, Germany

E-mail: zweidinger@ikp.tu-darmstadt.de

Abstract. Photon-scattering experiments off the nuclei ^{94}Zr and ^{96}Zr were performed at the 10-MeV bremsstrahlung facility DHIPS at the super-conducting electron linear accelerator S-DALINAC at Darmstadt. The dipole excitation-strength distributions were determined in an energy range between 2 and 8 MeV, including the 1^- quadrupole-octupole coupled Two-Phonon-State and parts of the Pygmy Dipole Resonance. In total 74 transitions in ^{96}Zr were observed and for most of the excited states the spin quantum number could be firmly assigned to $J = 1$. Calculations within the Quasiparticle Phonon Model show a good agreement.

1. Introduction

The Nuclear Resonance Fluorescence (NRF) method is sensitive to dipole and quadrupole excitations [1]. For that reason, the NRF is an excellent tool to investigate the so-called "*Pygmy Dipole Resonance*" (PDR) and the $\left[2_1^+ \otimes 3_1^-\right]_{1^-}$ state [2]. The latter originates from a coupling of the low-energy 2^+ quadrupole phonon and the 3^- octupole phonon which leads to a quintuplet of states with spin and parity quantum numbers ranging from 1^- to 5^-. Furthermore, magnetic dipole excitations [3] such as the scissors mode [4] and spin flip-excitations can be observed in NRF.

The stable Zr isotopes (Z = 40, N = 50 - 56) exhibit a pronounced proton subshell closure towards the empty $\pi(1g_{9/2})$ subshell with the Fermi level just above the $\pi(2p_{1/2})$ orbital. For this reason, it is very interesting to investigate the PDR in these nuclei, because, until now, the PDR is mainly investigated in nuclei at or near well-developed shell closures (see e.g. [5, 6, 7]). In a geometrical picture, the PDR is described by an oscillation of a neutron skin against an isospin saturated N = Z core. The PDR is also described in an hydrodynamic model [8]. The centroid can be calculated by the expression $E_{\mathrm{PDR}} = 31 \cdot A^{-1/3}$ MeV.

Experiments performed at the Kernfysisch Versneller Instituut (KVI) in Groningen (NL), using the isoscalar probe in an α-scattering experiment suggest a fragmentation of the PDR in an isoscalar and an isovector part. While the first one is observed in $(\alpha, \alpha'\gamma)$ reactions, the isovector part becomes dominant in (γ, γ') experiments [9].

Indications of the PDR appear in many nuclei in the excitation energy range between 6 and 10 MeV. Dipole excitation cross sections in this energy region may have a large impact on the neutron-capture cross sections and, consequently, on the astrophysical reaction rates, that are

Journal of Physics: Conference Series **366** (2012) 012054 doi:10.1088/1742-6596/366/1/012054

relevant for the r-process [10].
Below the energy region where the PDR is observed, multi-phonon excitations (MPE) can be found. The best studied examples are given by the coupling of the low-lying quadrupole excitations in spherical nuclei forming the two-quadrupole phonon triplet of states with spin and parity quantum numbers 0^+, 2^+ and 4^+ [11]. MPEs may even result of the coupling of two phonons built on different nucleon configurations. In the case of ^{96}Zr the phonon of the octupole vibration built on the ground state might be able to couple to the phonons of the quadrupole excitations built on top of the ground state or on top of the low-lying 0^+ intruder state. The properties of MPEs depend on correlations induced by the nuclear forces between several nucleons and the fermionic nature of the latter. Therefore, multi-phonon states are very sensitive to the underlying shell structure. Nuclear multi-phonon structures represent fascinating objects of research at the heart of nuclear structure physics.
Beside the mentioned electric dipole transitions, NRF offers the possibility to observe magnetic dipole transitions. The scissors mode of deformed nuclei [12] and spin-flip excitations belong to this kind of M1 excited states. The latter describe the excitation of at least one nucleon from a subshell $j\pm1/2$ to its spin-orbit partner in a $j\mp1/2$ shell with a change of the spin orientation.

2. Experiments

The photon scattering experiments have been performed using the Darmstadt High Intensity Photon Setup (DHIPS) [13] at the super-conducting electron linear accelerator S-DALINAC at the Technische Universität Darmstadt, Germany.
The injector of the S-DALINAC can accelerate the electrons up to 10 MeV with beam currents up to 40 μA. These electrons are completely stopped in a thick copper radiator and converted into bremsstrahlung. The maximum energy of the continuous-energy bremsstrahlung corresponds to the energy of the electron beam. The photons are collimated with a copper collimator with a diameter of 2.5 cm and 3 cm, respectively, at two target positions. The whole setup is passively shielded against background stemming from the radiator and natural sources by Lead.
In order to detect the scattered photons at the first target position, three High-Purity Germanium detectors (HPGe) equipped with active Anti-Compton shields are available. In the case of the experiments on ^{96}Zr, one detector was placed at 90° and two detectors at 130°. Approximately 4 g of the very expensive target material (ZrO_2 enriched to 96 % in ^{96}Zr) was loosely packed in an Aluminum container, to which the calibration standard ^{11}B was mounted. In the measurement on ^{94}Zr, one detector was placed at 90°, a second at 95° and the third one at 130°. The target was composed of 300 mg of metallic Zr enriched to 91.2 % in the isotope ^{94}Zr and 800 mg of ZrO_2 enriched to 92.6 %. In addition, about 180 mg of the calibration standard ^{11}B was added to the target.

3. Results

The dipole-strength distributions resulting from the measurements are presented in Figure 1. For ^{96}Zr calculations within the Quasiparticle Phonon Model [14] were performed. These are shown in the lower panel of Figure 1.
The comparison of the dipole-strength distributions of ^{94}Zr and ^{96}Zr up to an energy of 5 MeV exhibits that in ^{96}Zr the transitions are stronger and the level density is lower. In the energy range of 4.7 and 5.2 MeV, the two isotopes exhibit a similar structure. There appears a group of states in the respective isotope with almost identical transition strengths. Furthermore, the measurement on the isotope ^{96}Zr with an end-point energy of 7.7 MeV shows an accumulation of dipole strength starting from an energy of 5.5 MeV. This is characteristic for the E1 Pygmy Dipole Resonance. However, in the experiments presented here, the parity quantum numbers cannot be determined. Therefore, it is not for sure that all shown transitions are of E1 character.

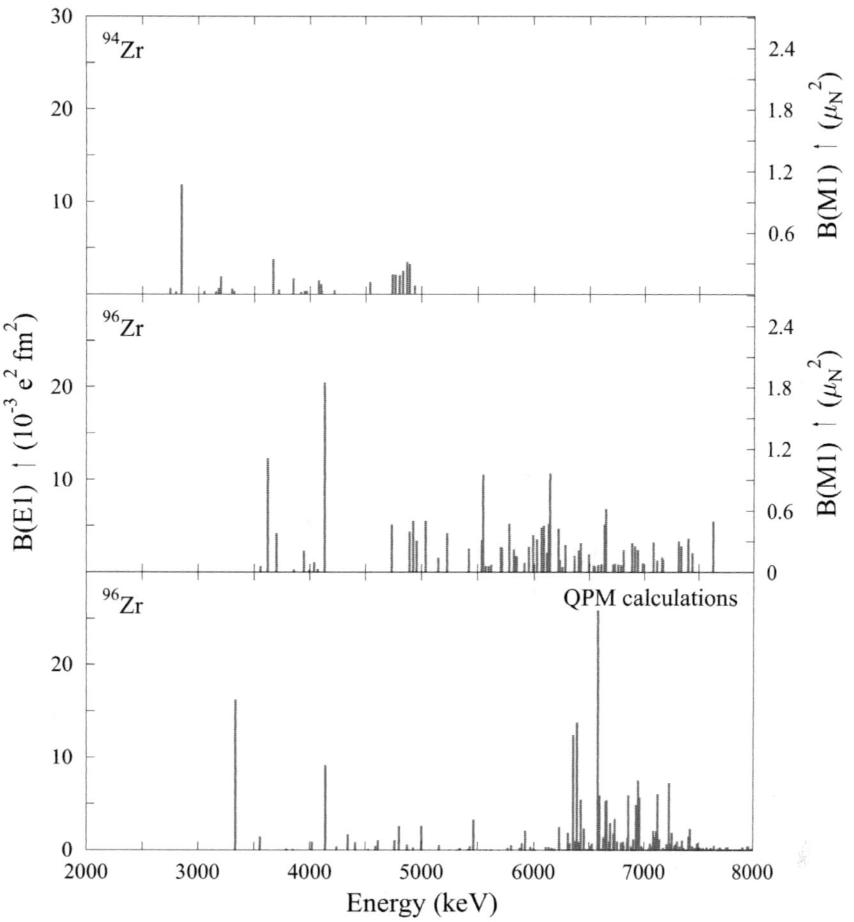

Figure 1. Preliminary dipole excitation strength distributions of ^{94}Zr and ^{96}Zr: The upper panel shows the observations on ^{94}Zr. The panel in the middle presents the measured dipole strength distribution of ^{96}Zr, while the lower panel shows the one calculated within the QPM (see text). The data analysis is still on-going.

The E1 strength distribution calculated within the QPM shows a resonance-like structure above an energy of about 6 MeV. To compare the calculations with the data, the centroid energies for the resonance-like part of the distributions were estimated as strength-weighted averages from an energy of 5.4 MeV to 7.63 MeV. The resulting value for the QPM calculation is $\bar{E}_{theo} = 6.59$ MeV while the experimental one is $\bar{E}_{exp} = 6.33$ MeV. The hydrodynamical model predicts a value of about 6.77 MeV. Furthermore, the latter predicts the full transition strength to 0.141 e^{2}fm^2 while the QPM calculates the summed E1 strength to 0.186 e^{2}fm^2. The preliminary experimental value is 0.131(17) e^{2}fm^2. Infact, the given experimental value can be considered as a lower limit for the total strength. Because of the limited sensitivity there may be unresolved strength hidden in the non-resonant background. Furthermore, in the summation M1 and E2 excited states could have been included.

The low-energy part of the strength distributions in both isotopes is dominated by strong dipole excitations at 2846 keV in ^{94}Zr and 4132 keV in ^{96}Zr, respectively. Both states are strong candidates for the 1^- state of the $2_1^+ \otimes 3_1^-$ quintuplet. The parity of the 4132 keV state in ^{96}Zr was determined to be negative [18] and we tentatively assign a negative parity for the 2846 keV

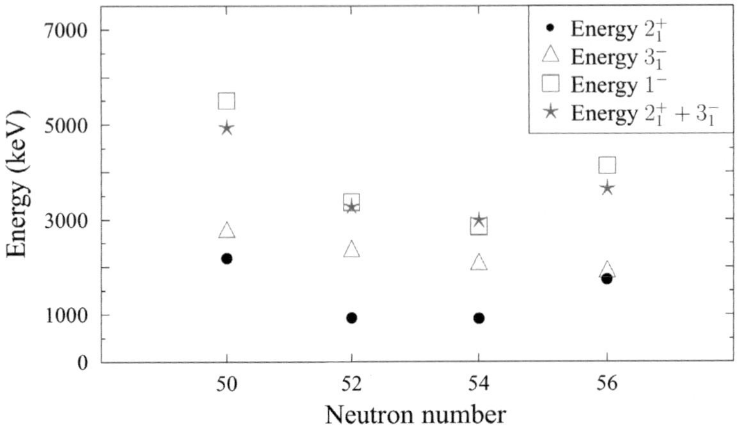

Figure 2. Systematics of the values of the sum energy of the 2_1^+ and the 3_1^- state and the energy of the prominent low-energy 1^- states in the even-even Zirconium isotopes: The excitation energies of the doubly-closed shell isotopes ^{90}Zr and ^{96}Zr is higher than in the two other isotopes. Data for ^{90}Zr are taken from [2] and those for ^{92}Zr from [15]. The energy of the 3_1^- state for ^{94}Zr is taken from [16] and for ^{96}Zr from [17].

state in ^{94}Zr from systematics, because it appears in the region of the sum energy of the 2_1^+ and the 3_1^- state. In Figure 2 the systematics of the deviation of the values of the sum energy of the 2_1^+ and the 3_1^- state and the energy of the observed 1_1^- states in the even-even Zirconium isotopes is shown.
The two isotopes ^{90}Zr and ^{96}Zr have high excitation energies of the 2_1^+ state. The latter indicates, once more, the doubly-magic character of ^{96}Zr. For that reason, the excitation energy of the 1^- states is high as well. The difference between the sum energy of the 2_1^+ and the 3_1^- state and the energy of the prominent low-energy 1^- state in these two isotopes might be explained by blocking effects.
Figure 3 shows the comparison of the quadrupole-octupole coupled E1 excitation strength in nuclei in the Z = 38 - 50 region. Especially the transition from the 1^- to the ground state in ^{96}Zr is much stronger than the common trend in this mass region. One reason might be the large octupole collectivity of this isotope. This octupole collectivity is expected to be dominated by the available low-energy $\Delta L = 3, \Delta J = 3$: $\pi\left(p_{3/2} \rightarrow g_{9/2}\right)$ and $\nu\left(d_{5/2} \rightarrow h_{11/2}\right)$ particle-hole excitations [19]. Hence, a strongly excited $[2^+ \otimes 3^-]_{1^-}$-state is expected in ^{96}Zr. The excitation should get weaker with decreasing neutron number. These expections can be confirmed as shown in Figure 3.

Acknowledgments
We thank C. Fransen and G. Tagliente for the loan of the ^{94}Zr and ^{96}Zr targets, respectively. Furthermore, we thank our groupmembers for the help on the experiment. We are grateful to the crew at the S-DALINAC for help on the operation of the electron gun and the injector linac. Support of the DFG under grant SFB 634 is gratefully acknowledged. This presentation was supported by the DAAD (Deutscher Akademischer Austausch Dienst).

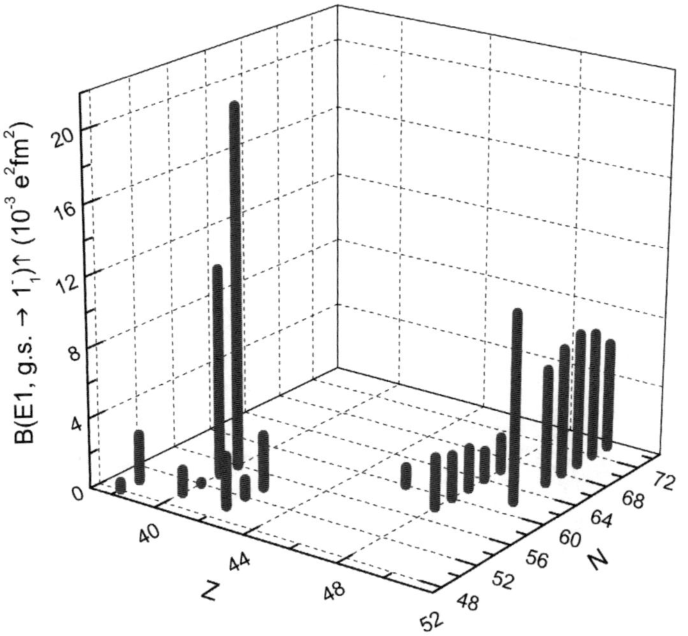

Figure 3. Systematics of the B(E1, g.s. $\rightarrow 1_1^-$)-transition strengths in the Z = 38 - 50 mass region: The isotopes ^{94}Zr and ^{96}Zr have the strongest transitions. This fact might be caused by their large octupole collectivity. Data for ^{86,88}Sr, ^{90}Zr, ^{92,94}Mo, ^{110,112,114}Cd and ^{116,118,120,122,124}Sn have been taken from [2] and for ^{92}Zr from [15]. Furthermore the transitions strengths for ^{96}Mo, ^{106}Pd, ^{106}Cd, ^{108}Cd and ^{112}Sn have been taken from [20] - [24].

[1] Kneissl U, Pietralla N and Zilges A 2006 *J. Phys. G: Nucl. Part. Phys.* **32** R217
[2] Andrejtscheff W et al. 2001 *Phys. Lett.* B **506** 239
[3] Heyde K, von Neumann Cosel P and Richter A 2010 *Rev. Mod. Phys.* **82** 2365
[4] Bohle D, Richter A et al. 1984 *Phys. Lett.* B **137** 27
[5] Volz S et al. 2006 *Nucl. Phys.* A **779** 1
[6] Zilges A et al. 2002 *Phys. Lett.* B **542** 43
[7] Rusev G et al. 2009 *Phys. Rev.* C **79** 061302(R)
[8] Bastrukov S I et al. 2008 *Phys. Lett.* B **664** 258
[9] Savran D et al. 2007 *Nucl. Phys.* A **788** 165
[10] Litvinova E et al. 2009 *Nucl. Phys.* A **823** 159
[11] Casten R F 1990 *Nuclear Structure from a Simple Perspective* (Oxford, Oxford University Press); and references therein
[12] Heyde K, von Neumann-Cosel P and Richter A 2010 *Rev. Mod. Phys.* **82** 2365
[13] Sonnabend K et al. 2011 *Nucl. Instr. and Meth.* A **640** 6
[14] Soloviev V G 1992 *Theory of Atomic Nuclei: Quasiparticles and Phonons* (Bristol: Institute of Physics Publishing)
[15] Werner V et al. 2002 *Phys. Lett.* B **550** 140-146
[16] Fransen C et al. 2005 *Phys. Rev.* C **71** 054304
[17] Kumbartzki G et al. 2003 *Phys. Lett.* B **562** 193-200
[18] Pietralla N et al. *Nuclear Structure and Related Topics*, Proc. of the International Conference, June 30-July 4 2009, Dubna, Volume II
[19] Nazarewicz W and Butler P A 1996 *Rev. Mod. Phys.* **68** 2
[20] Lesher S R et al. 2007 *Phys. Rev.* C **75** 034318
[21] Peters E E 2011 *priv. comm.*
[22] Linnemann A 2005 *Ph.D. Thesis* University of Cologne
[23] Gade A et al. 2003 *Phys. Rev.* C **67** 034304
[24] Pysmenetska I et al. 2006 *Phys. Rev.* C **73** 017302

AUTHOR INDEX

AUTHOR INDEX

AUTHOR INDEX